Ashutosh Gupta   Thomas A. Henzinger (Eds.)

Computational Methods in Systems Biology

11th International Conference, CMSB 2013
Klosterneuburg, Austria, September 22-24, 2013
Proceedings

Springer

Volume Editors

Ashutosh Gupta
Thomas A. Henzinger
IST Austria
3400 Klosterneuburg, Austria
E-mail: {agupta,tah}@ist.ac.at

ISSN 0302-9743　　　　　　　　　　e-ISSN 1611-3349
ISBN 978-3-642-40707-9　　　　　　e-ISBN 978-3-642-40708-6
DOI 10.1007/978-3-642-40708-6
Springer Heidelberg New York Dordrecht London

Library of Congress Control Number: 2013946852

CR Subject Classification (1998): F.1.1-2, I.6.3-5, J.3, I.2.3, D.2.2, D.2.4, F.4.3, I.1.3

LNCS Sublibrary: SL 8 – Bioinformatics

Typesetting: Camera-ready by author, data conversion by Scientific Publishing Services, Chennai, India

Printed on acid-free paper

Springer is part of Springer Science+Business Media (www.springer.com)

Preface

This volume contains the papers presented at CMSB 2013. The 11th International Conference on Computational Methods in Systems Biology was held during September 22–24, 2013, at IST Austria in Klosterneuburg.

The conference is an annual event that brings together computer scientists, biologists, mathematicians, engineers, and physicists from all over the world who share an interest in the computational modeling and analysis of biological systems, pathways, and networks. It covers computational models for all levels, from molecular and cellular, to organs and entire organisms.

There were 27 regular and 19 poster submissions. Each regular submission was reviewed by at least three, and on average 3.96, Program Committee members. Each poster submission was reviewed by at least two, and on average 2.95, Program Committee members. The committee decided to accept 15 regular papers, four regular submissions as posters, and all submitted posters. The program also included five invited talks, by Jürg Bähler, Flavio H. Fenton, John Lygeros, Nassos Typas, and Verena Wolf.

We thank the Program Committee for their hard work in reviewing submissions. We especially thank Calin Guet and Monika Heiner for their advice on the academic program and organization of the conference. We thank Marie Trappl for her help with the organization of the meeting, and Sebastian Nozzi and Moritz Schepp for their assistance with the website. We acknowledge support by the EasyChair conference system, see http://www.easychair.org, during the reviewing process and the production of these proceedings.

We thank Tommaso Mazza and the IEEE Computer Society Technical Committee on Simulation for supporting the best student paper award. We thank IST Austria for providing support for the conference and the travel of student participants. We thank the European Research Council for providing support for the meeting through the ERC Advanced Grant QUAREM (Quantitative Reactive Modeling). We thank the *ACM Transactions on Modeling and Simulation* for inviting the best papers of the conference to a special issue of the journal.

July 2013

Ashutosh Gupta
Thomas A. Henzinger

Organization

Program Committee

Luca Cardelli	Microsoft Research, Cambridge, UK
Vincent Danos	University of Edinburgh, UK
Hidde De Jong	INRIA, Grenoble, France
Finn Drablos	NTNU, Trondheim, Norway
François Fages	NRIA, Rocquencourt, France
Jérôme Feret	INRIA, Paris, France
Jasmin Fisher	Microsoft Research, Cambridge, UK
Walter Fontana	Harvard Medical School, Boston, USA
Radu Grosu	Technical University, Vienna, Austria
Calin Guet	IST Austria
Ashutosh Gupta	IST Austria
Monika Heiner	Brandenburg University at Cottbus, Germany
Thomas A. Henzinger (Chair)	IST Austria
Jane Hillston	University of Edinburgh, UK
William Hlavacek	Los Alamos National Laboratory, USA
Eric Klavins	University of Washington, Seattle, USA
Heinz Koeppl	ETH Zurich, Switzerland
Marta Kwiatkowska	University of Oxford, UK
Oded Maler	CNRS, Grenoble, France
Wolfgang Marwan	University of Magdeburg, Germany
Tommaso Mazza	IRCCS Casa Sollievo della Sofferenza - Mendel, Rome, Italy
Ilya Nemenman	Emory University, Atlanta, USA
Alberto Policriti	University of Udine, Italy
Corrado Priami	Microsoft Research and University of Trento, Italy
Saurabh Srivastava	University of California, Berkeley, USA
Joerg Stelling	ETH Zurich, Switzerland
Carolyn Talcott	SRI International, Menlo Park, USA
P.S. Thiagarajan	National University of Singapore, Singapore
Adelinde Uhrmacher	University of Rostock, Germany
Jose Vilar	University of the Basque Country, Bilbao, Spain
Verena Wolf	Saarland University, Saarbrücken, Germany

Steering Committee

Finn Drablos	NTNU, Trondheim, Norway
François Fages	NRIA, Rocquencourt, France
Monika Heiner	Brandenburg University at Cottbus, Germany
Tommaso Mazza	IRCCS Casa Sollievo della Sofferenza - Mendel, Rome, Italy
Satoru Miyano	University of Tokyo, Japan
Gordon Plotkin	University of Edinburgh, UK
Corrado Priami	Microsoft Research and University of Trento, Italy
Carolyn Talcott	SRI International, Menlo Park, USA
Adelinde Uhrmacher	University of Rostock, Germany

Organizing Committee

Calin Guet	IST Austria
Ashutosh Gupta	IST Austria
Thomas A. Henzinger	IST Austria
Marie Trappl	IST Austria

Additional Reviewers

Andreychenko, Alexander	Leye, Stefan
Batt, Gregory	Liu, Bing
Bittig, Arne T.	Madsen, Curtis
Blätke, Mary Ann	Murthy, Abhishek
Bortolussi, Luca	Palaniappan, Sucheendra Kumar
Castellana, Stefano	Piterman, Nir
Fanchon, Eric	Rohr, Christian
Galpin, Vashti	Rybacki, Stefan
Georgoulas, Anastasis	Schwarick, Martin
Guet, Calin	Sezgin, Ali
Gyori, Benjamin	Shepherd, The
Hahn, Ernst Moritz	Soliman, Sylvain
Harmer, Russ	Sorine, Michel
Islam, Md. Ariful	Spieler, David
Krivine, Jean	Stephanou, Angelique
Krüger, Thilo	Thomas, Randy
Kugler, Hillel	Yordanov, Boyan

Invited Talks
(Abstracts)

Genome Regulation in Fission Yeast

Jurg Bahler

University College London, UK

Abstract. Data on absolute molecule numbers can empower the modelling, understanding, and comparison of cellular functions and biological systems. We quantified transcripts and proteins in fission yeast during cell proliferation and quiescence. This data set provides the first comprehensive reference for all RNA and most protein concentrations in a eukaryote under two distinct physiological conditions. The integrated data supports quantitative biology and affords unique insights into cell regulation. Although mRNAs are typically expressed in a narrow range above 1 copy per cell, most long non-coding RNAs are tightly repressed below 1 copy/cell. Proteins greatly exceed mRNAs in both abundance and dynamic range, and their concentrations are regulated to functional demands. During the transition to quiescence, the proteome is substantially remodelled, but, in stark contrast to mRNAs, proteins do not uniformly decrease but scale with cell volume.

Complexity, Pattern Formation and Chaos in the Heart

Flavio H. Fenton

Cornell University, Ithaca, US

Abstract. The heart is an electro-mechanical system in which, under normal conditions, electrical waves propagate in a coordinated manner to initiate an efficient contraction. In pathologic states, single and multiple rapidly rotating spiral and scroll waves of electrical activity can appear and generate complex spatiotemporal patterns of activation that inhibit contraction and can be lethal if untreated. Despite much study, many questions remain regarding the mechanisms that initiate, perpetuate, and terminate reentrant waves in cardiac tissue.

In this talk, we will show how a combined experimental and computational approach is used to better understand the dynamics of cardiac arrhythmias. From a computational point of view we will discuss from the numerical models derived to represent the dynamics of single cells to the coupling of millions of cells to represent the three-dimensional structure of a working heart. Some of the major difficulties of computer simulations for these kinds of systems include: i) Different orders of magnitude in time scales, from milliseconds to seconds; ii) millions of degrees of freedom over millions of integration steps within irregular domains; and iii) the need for near-real-time simulations. Advances in these areas will be discussed as well as the use of GPUs for large scale simulations. Finally we will show how computer simulations guide the development of new low energy defibrillation methods that are being tested experimentally that require only 10 percent the energy of current standard methods.

On the Use of the Moment Equations for Parameter Inference, Control and Experimental Design in Stochastic Biochemical Reaction Networks

Jakob Ruess and John Lygeros

Automatic Control Laboratory,
CH-8092 Zurich, ETH Zurich,
Switzerland

Abstract. Variability is present at all levels of biological systems. At the molecular level Brownian motion of the molecules leads to randomness of the biochemical reactions inside the cells. On a higher level, the molecular noise and other stochastic effects can lead to fundamentally different behavior of the cells in a population. As a consequence, average dynamics of a cell population are often not adequate to understand or control the dynamics of the population as a whole. We discuss how stochastic models of biochemical reaction networks which enable one to study heterogeneous cell populations can be identified from data and how experiments which make this identification as easy as possible can be designed.

From High-throughput Approaches to Molecular Mechanism

Nassos Typas

European Molecular Biology Laboratory,
Heidelberg, Germany

Abstract. A combination of new technologies, resources and methodologies has enabled researchers to move traditional reverse genetics approaches to a genome-wide level. High-throughput gene-gene, gene-drug and drug-drug interaction maps, pioneered mostly in yeast, have provided a plethora of mechanistic insights in gene function, pathway architecture and drug mode of action. Starting with *E. coli*, we have implemented analogous high-throughput approaches in a number of bacteria and used them to study different aspects of their biology. Here I will illustrate how these system-approaches can be used to assign function to uncharacterized genes, discover new layers of regulation for known biological processes-pathways, map higher-order interconnections in the genetic network, and identify the molecular mechanism behind drug mode-of-action and drug-drug synergy.

Numerical Approximation of Rare Event Probabilities in Biochemically Reacting Systems

Linar Mikeev, Werner Sandmann, and Verena Wolf

Saarland University, Department of Computer Science,
Campus E1 3, 66123 Saarbrücken, Germany

Abstract. In stochastic biochemically reacting systems, certain rare events can cause serious consequences, which makes their probabilities important to analyze. We solve the chemical master equation using a four-stage fourth order Runge-Kutta integration scheme in combination with a guided state space exploration and a dynamical state space truncation in order to approximate the unknown probabilities of rare but important events numerically. The guided state space exploration biases the system parameters such that the rare event of interest becomes less rare. For each numerical integration step, the portion of the state space to be truncated is then dynamically obtained using information from the biased model and the numerical integration of the unbiased model is conducted only on the remaining significant part of the state space. The efficiency and the accuracy of our method are studied through a benchmark model that recently received considerable attention in the literature.

Table of Contents

Regular Papers

Posters

On the Use of the Moment Equations for Parameter Inference, Control and Experimental Design in Stochastic Biochemical Reaction Networks

Jakob Ruess and John Lygeros

Automatic Control Laboratory, CH-8092 Zurich, ETH Zurich, Switzerland

Abstract. Variability is present at all levels of biological systems. At the molecular level Brownian motion of the molecules leads to randomness of the biochemical reactions inside the cells. On a higher level, the molecular noise and other stochastic effects can lead to fundamentally different behavior of the cells in a population. As a consequence, average dynamics of a cell population are often not adequate to understand or control the dynamics of the population as a whole. We discuss how stochastic models of biochemical reaction networks which enable one to study heterogeneous cell populations can be identified from data and how experiments which make this identification as easy as possible can be designed.

The most widely used model class to capture the molecular noise of biochemical reaction networks are continuous-time Markov chains (CTMC), where the state corresponds to the number of molecules of the different chemical species and random transitions occur when molecules react. It is easy to draw sample paths from such models using Gillespie's stochastic simulation algorithm [1]. Computing the time evolution of the probability distribution of the CTMC, however, is usually very difficult since it requires one to either solve the chemical master equation (CME) or to simulate a very large number of sample paths and compute statistics from the simulations.

Identifying the parameters of such a model from measurements of the dynamics of a cell population is even more difficult. This is because methods for parameter inference are typically built on iterative schemes where the dynamics have to be investigated for many different sampled parameter values in order to find the best agreement of model predictions and measurements (see Figure 1). Clearly, simulating many sample paths of a CTMC at each iteration of a search on a potentially high-dimensional parameter space is computationally challenging. More efficient would be to compute the time evolution of the probability distribution by solving the CME. Unfortunately, solving the CME or approximating its solution is only possibly in relatively small and simple systems [2,3]. This (arguably) still poses the main limitation in building stochastic models from single cell measurements; in contrast to non-identifiable parameters

A. Gupta and T.A. Henzinger (Eds.): CMSB 2013, LNBI 8130, pp. 1–4, 2013.

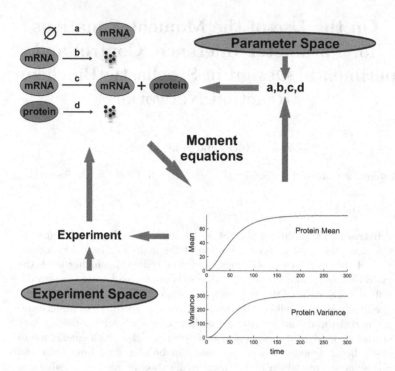

Fig. 1. Schematic illustration of parameter identification and experimental design for a simple model of gene expression. Parameter identification requires searching the parameter space for the parameters that lead to the best agreement of model predictions with the experimental measurements. Experimental design requires searching the space of possible experiments for the most informative experiment. Both these tasks require iteratively computing the model dynamics. Use of the moment equations allows to perform this fast enough to make parameter inference and experimental design computationally feasible.

and not informative enough measurements which are usually the major limitations in building ordinary differential equation models from averaged population measurements.

Many recent experimental results have shown that molecular noise alone cannot adequately explain the amount of variability which is present in biological systems. This is often a consequence of differences between the cells such as size or expression capacity, but can also be due to different local growth conditions which the cells in a population encounter. It has therefore been suggested that a CTMC model should be allowed to have (at least some) parameters which vary between the cells according to some probability distribution and to identify parameters of these distributions (e.g. moments) along with the original parameters of the CTMC from the measurements. While such models can be expected to represent the true experimental situation better, they pose several additional challenges. On the one hand, if the variable parameters of the CTMC come from a continuous distribution, the time evolution of the probability distribution of the population is not governed by a CME anymore. On the other hand, such models will have more free parameters and one has to make sure that the measured data is informative enough to identify the additional parameters. That is, one has to make sure that the data allows one to distinguish between molecular noise and variability stemming from randomness of the parameters.

All of the above difficulties can be addressed (at least in some cases) if one refrains from trying to compute the time evolution of the entire probability distribution of the process and focuses on some lower-order moments instead. In [4] we showed how ordinary differential equations describing the time evolution of the moments of a CTMC with variable parameters can be derived. The resulting equations can usually not be solved exactly but reasonable approximations can often be obtained by using approximation techniques such as moment closure [5] or the technique we presented in [6]. These approximations can be computed very fast for systems of at least moderate size and can therefore be used efficiently in iterative parameter identification schemes. First results in [4] show that this approach is indeed applicable for real biological systems and can potentially even be used in cases where the measured population distributions are bimodal and cannot be adequately described by lower-order moments only.

A way to address the question of whether the data is informative enough to identify the model parameters and to design informative experiments is through the computation of the Fisher information [7]. The Fisher information gives lower bounds for the variances of any unbiased estimators of the model parameters and thus tells us to which accuracy the model parameters can at best be estimated in a given experimental setting. Computing the Fisher information for CTMC models, however, is usually very difficult and experimental design schemes which aim at searching the space of possible experiments for the most informative experiment (see Figure 1) suffer from similar computational problems as parameter inference schemes: if the information cannot be evaluated in a fast and efficient way, searching a potentially large space of possible experiments is computationally prohibitive. In [8] we showed that the moment equations can be used to approximate the Fisher information and can therefore serve as the basis of experimental design. Hence, also the question of how informative enough experiments can be obtained can be addressed.

Finally, also control of cell populations can be addressed in the setting of the moment equations. Model-based feedback control schemes require online computation of control actions and one has to be able to predict the response of the cell population to the control actions very fast if such schemes are to be used. Contrary to parameter estimation and experimental design, where the goal is to gain a mechanistic understanding of the underlying biochemical process, in the context of control accurate models are not always necessary, since feedback tends to correct modeling errors. Still, even for simplified models, the CME is usually too difficult to solve for practical use. One has to note, however, that control schemes based on the moment equations can at best allow one to control the moments up to the order where the moment equations have been truncated, since higher order moments are (in contrast to the CME) not included anymore.

In practice, however, this is hardly the real limitation, since the control inputs which have so far been implemented in real cells are anyways too limited to allow joint control of more than one moment. The best that has been achieved so far in practice is control of population averages neglecting all higher order moments [9,10].

Concluding, we can say that the fundamental (computational) difficulties in parameter inference, control and experimental design in stochastic biochemical reaction networks can all be addressed very well and efficiently with the moment equations. This is of course all based on being able to accurately compute the solution of the moment equations. While this is often possible, there are also many situations where the currently existing approximation techniques for the moment equations fail. In such situations not much is possible at the moment and good alternatives remain to be developed.

References

1. Gillespie, D.: A general method for numerically simulating the stochastic time evolution of coupled chemical reactions. J. Comput. Phys. 22, 403–434 (1976)
2. Munsky, B., Khammash, M.: The finite state projection algorithm for the solution of the chemical master equation. J. Chem. Phys. 124, 044104 (2006)
3. Munsky, B., Trinh, B., Khammash, M.: Listening to the noise: random fluctuations reveal gene network parameters. Mol. Syst. Biol. 5, 318 (2009)
4. Zechner, C., Ruess, J., Krenn, P., Pelet, S., Peter, M., Lygeros, J., Koeppl, H.: Moment-based inference predicts bimodality in transient gene expression. Proc. Natl. Acad. Sci. USA 109, 8340–8345 (2012)
5. Singh, A., Hespanha, J.: Approximate moment dynamics for chemically reacting systems. IEEE Trans. Automat. Contr. 56, 414–418 (2011)
6. Ruess, J., Milias-Argeitis, A., Summers, S., Lygeros, J.: Moment estimation for chemically reacting systems by extended Kalman filtering. J. Chem. Phys. 135, 165102 (2011)
7. Komorowski, M., Costa, M., Rand, D., Stumpf, M.: Sensitivity, robustness, and identifiability in stochastic chemical kinetics models. Proc. Natl. Acad. Sci. USA 108, 8645–8650 (2011)
8. Ruess, J., Milias-Argeitis, A., Lygeros, J.: Designing experiments to understand the variability in biochemical reaction networks. ArXiv e-prints. 1304.1455. q-bio.QM (2013)
9. Milias-Argeitis, A., Summers, S., Stewart-Ornstein, J., Zuleta, I., Pincus, D., El-Samad, H., Khammash, M., Lygeros, J.: In silico feedback for in vivo regulation of a gene expression circuit. Nat. Biotechnol. 29, 1114–1116 (2011)
10. Uhlendorf, J., Miermont, A., Delaveau, T., Charvin, G., Fages, F., Bottani, S., Batt, G., Hersen, P.: Long-term model predictive control of gene expression at the population and single-cell levels. Proc. Nat. Acad. Sci. USA 109, 14271–14276 (2012)

Numerical Approximation of Rare Event Probabilities in Biochemically Reacting Systems

Linar Mikeev, Werner Sandmann, and Verena Wolf

Saarland University, Department of Computer Science,
Campus E1 3, 66123 Saarbrücken, Germany
{mikeev,sandmann,wolf}@cs.uni-saarland.de

Abstract. In stochastic biochemically reacting systems, certain rare events can cause serious consequences, which makes their probabilities important to analyze. We solve the chemical master equation using a four-stage fourth order Runge-Kutta integration scheme in combination with a guided state space exploration and a dynamical state space truncation in order to approximate the unknown probabilities of rare but important events numerically. The guided state space exploration biases the system parameters such that the rare event of interest becomes less rare. For each numerical integration step, the portion of the state space to be truncated is then dynamically obtained using information from the biased model and the numerical integration of the unbiased model is conducted only on the remaining significant part of the state space. The efficiency and the accuracy of our method are studied through a benchmark model that recently received considerable attention in the literature.

Keywords: Biochemically Reacting Systems, Stochastic Chemical Kinetics, Rare Events, Chemical Master Equation, Numerical Approximation.

1 Introduction

Stochastic modeling of biochemically reacting systems has a long tradition [10, 20, 24, 26, 33] and is today well established since the inherent randomness in biochemical reactions has become more and more evident [2, 6, 13, 22, 23, 34, 36]. The most common stochastic approach is to model biochemically reacting systems by multi-dimensional continuous-time Markov chains (CTMCs), where the system state is represented by a vector of the numbers of each molecular species. Then the transient, time dependent state probabilities are given by the chemical master equation (CME) [14–16, 37], well known as the system of Kolmogorov forward differential equations in the general theory and applications of Markov processes [1, 5, 12, 18]. A physical justification [14, 15], rigorously derived in [16], provides evidence that these stochastic chemical kinetics are in accordance with the theory of thermodynamics.

In many cases, rare events, that is events that occur with a very small probability, are particularly important, for instance because they describe a system

A. Gupta and T.A. Henzinger (Eds.): CMSB 2013, LNBI 8130, pp. 5–18, 2013.

behavior of high practical relevance or because they may have serious consequences. Examples include population sizes exceeding an exceptionally high level or falling below an exceptionally low level during some fixed time period, extinction of molecular species, outbreak of infectious diseases, apoptosis (cell death), or rare but important transitions between different long-lived stable regions in metastable systems, amongst many others. Determining the probabilities of such rare but important events is highly desirable.

Explicit closed-form solutions of the CME are usually not available such that it has to be solved numerically. However, the size of the multi-dimensional state space of the underlying CTMC typically increases exponentially with the number of molecular species, hence with the model dimensionality. This effect is known as state space explosion and often causes models to be numerically intractable due to the prohibitively large, often even infinite state space.

Therefore, the most widespread approach to analyzing stochastic chemical kinetics is stochastic simulation [14, 15], which means to mimic the time evolution of a biochemically reacting system by repeatedly generating trajectories (sample paths) of the underlying CTMC with the help of computer-generated random numbers. Mathematically, this constitutes a statistical estimation procedure for system properties such as expectations, moments and cumulants of molecular population sizes, or probabilities of certain events of interests. Proper statistical output analysis yields point estimators and confidence intervals [4, 19, 32].

Stochastic simulation does not suffer from state space explosion because the state space need not be explicitly enumerated, but stochastic simulation tends to be computationally expensive and can only provide estimates whose reliability and accuracy in terms of relative errors or confidence interval half widths depend on the variance of the corresponding simulation estimator. In particular, estimating rare event probabilities by 'standard' simulation is inefficient, because rare events are simulated too infrequently. The variance and the relative error of the corresponding standard estimators are much too large to obtain statistically reliable estimates in reasonable time. Variance reduction and specific rare event simulation techniques are required [4, 7, 28].

Despite recent progress in the application of such techniques to biochemically reacting systems [31, 21, 17, 9, 27], as outlined above a clear disadvantage of stochastic simulation compared to numerical analysis, provided that such an analysis would be possible, is the inherent statistical uncertainty of simulation results, that is estimates of the probabilities of interest. Thus, we argue that if a problem may be tackled both by stochastic simulation and by numerical analysis, the latter should be preferred.

In this paper, we consider a numerical solution approach that overcomes the state space explosion by using a dynamical state space truncation. The underlying principle is a guided state space exploration where paths that contribute significantly to the rare event probability are not truncated. We use parameter biasing strategies similarly as in rare event simulation to identify the significant parts of the state space and 'guide' the exploration of the state space in such a way that an accurate approximation of the rare event probability is obtained.

Our method approximates the solution of the CME by truncating large, possibly infinite state spaces dynamically in an iterative fashion. At a particular time instant t, we consider an approximation of the transient distribution and temporarily neglect states with a probability smaller than a threshold δ, that is, their probability at time t is set to zero. The CME is then solved for an (adaptively chosen) time step h during which the truncated state space is adapted to the distribution at time $t + h$. More precisely, certain states that do not belong to the truncated space at time t are added at time $t + h$, when in the meantime they receive a significant amount of probability which exceeds δ. Other states whose probability drops below δ are temporarily neglected. The smaller the significance threshold δ is chosen the more accurate the approximation becomes. A similar approach was previously applied in [25] to the computation of certain transient rare event probabilities in queueing networks.

In the next section, stochastic chemical kinetics are briefly recapitulated. Section 3 describes the dynamical state space truncation and the guided state space exploration as well as the choice of the parameter biasing for the 'guiding' system. Numerical results are presented in Section 4. Section 5 concludes the paper and outlines further research directions.

2 Stochastic Chemical Kinetics

Consider a well-stirred mixture of $d \in \mathbb{N}$ molecular species $S_1, \ldots, S_d$ interacting through $M \in \mathbb{N}$ chemical reaction channels $R_1, \ldots, R_M$ in a thermally equilibrated system of fixed volume. Each reaction channel $R_m, m = 1, \ldots, M$, is defined by a corresponding stoichiometric equation

$$R_m : \; s_{m_1} S_{m_1} + \cdots + s_{m_r} S_{m_r} \xrightarrow{c_m} s_{m_{r+1}} S_{m_{r+1}} + \cdots + s_{m_\ell} S_{m_\ell} \tag{1}$$

with an associated stochastic rate constant c_m, reactants $S_{m_1}, \ldots, S_{m_r}$, products $S_{m_{r+1}}, \ldots, S_{m_\ell}$, and corresponding stoichiometric coefficients $s_{m_1}, \ldots, s_{m_\ell} \in \mathbb{N}$, where $m_1, \ldots, m_\ell$ indexes those species involved in the reaction. Mathematically, the stoichiometry is described by the state change vector $v_m = (v_{m1}, \ldots, v_{md})$, where v_{mk} is the change of molecules of species S_k due to R_m. At any time $t \geq 0$ a discrete random variable $X_k(t)$ describes the number of molecules of species S_k and the system state is given by the random vector $X(t) = (X_1(t), \ldots, X_d(t))$.

The system changes its state due to one of the possible reactions and for each reaction channel R_m the reaction rate is given by a state dependent propensity function α_m, where $\alpha_m(x)dt$ is the conditional probability that a reaction of type R_m occurs in the time interval $[t, t + dt)$, given that the system is in state x at time t. That is

$$\alpha_m(x)dt = P\left(R_m \text{ occurs in } [t, t + dt) \mid X(t) = x\right). \tag{2}$$

The propensity function is given by c_m times the number of possible combinations of the required reactants and thus computes as

$$\alpha_m(x) = c_m \prod_{j=1}^{m_r} \binom{x_{m_j}}{s_{m_j}}, \tag{3}$$

where x_{m_j} is the number of molecules of species S_{m_j} present in state x, and s_{m_j} is the stoichiometric coefficient of S_{m_j} according to (1). Because at any time the system's future evolution only depends on the current state, $(X(t))_{t\geq 0}$ is a time-homogeneous continuous-time Markov chain (CTMC) with d-dimensional state space $\mathcal{X} \subseteq \mathbb{N}^d$.

The conditional transient (time dependent) probability that the system is in state $x \in \mathcal{X}$ at time t, given that the system starts in an initial state $x_0 \in \mathcal{X}$ at time t_0, is denoted by

$$p^{(t)}(x) := p^{(t)}(x|x_0, t_0) = P\left(X(t) = x \mid X(t_0) = x_0\right) \qquad (4)$$

and the system dynamics in terms of the state probabilities' time derivatives are described by the chemical master equation (CME)

$$\frac{\partial p^{(t)}(x)}{\partial t} = \sum_{m=1}^{M} \left(\alpha_m(x - v_m)p^{(t)}(x - v_m) - \alpha_m(x)p^{(t)}(x) \right) =: \mathcal{M}(p^{(t)})(x), \quad (5)$$

which is also well known as the system of Kolmogorov forward differential equations for Markov processes. Note that (5) is the most common way to write the CME, namely as a partial differential equation (PDE), where t as well as $x_1, \ldots, x_d$ are variables. However, for any fixed state $x = (x_1, \ldots, x_d)$ the only free parameter is the time parameter t such that (5) with fixed x is an ordinary differential equation (ODE) with variable t. In particular, when solving for the transient state probabilities numerical ODE solvers can be applied.

3   Numerical Computation of Rare Event Probabilities

In this section we first focus on the solution of (5) using numerical integration methods where the state probabilities $p^{(t)}(x)$ are approximated up to a certain absolute error $\epsilon > 0$. This technique is inefficient if we aim at a very small error ϵ that is several orders of magnitude smaller than the probability $P(A)$ of a rare event A. Therefore, we extend the approximate numerical integration in such a way that the state probabilities $p^{(t)}(x)$ contributing to $P(A)$ are very accurate while for states x that do not significantly contribute we obtain high relative errors. In this way we obtain an accurate approximation of $P(A)$ while the probability of other events may be very inaccurate.

As a benchmark example throughout the paper we consider the enzymatic futile cycle

$$S_1 + S_2 \xrightarrow{c_1} S_3,$$
$$S_3 \xrightarrow{c_2} S_1 + S_2,$$
$$S_3 \xrightarrow{c_3} S_1 + S_5,$$
$$S_4 + S_5 \xrightarrow{c_4} S_6,$$
$$S_6 \xrightarrow{c_5} S_4 + S_5,$$
$$S_6 \xrightarrow{c_6} S_4 + S_2,$$

with stochastic rate constants $c_1 = c_2 = c_4 = c_5 = 1, c_3 = c_6 = 0.1$ and initial state $x_0 = (1, 50, 0, 1, 50, 0)$, described by [30] and considered in the context of weighted stochastic simulation algorithms by [21, 17, 9]. The goal is to approximate the probability that before time $t = 100$ the number of molecules of species S_5 drops to ℓ for some $\ell \in \{5, 15, 25\}$.

3.1   Dynamical State Space Truncation

The system of linear differential equations in (5) is typically large or even infinite such that its solution with standard numerical integration methods becomes computationally infeasible. For most systems, however, we can exploit that only a tractable number of states have 'significant' probability, that is, only relatively few states have a probability that is greater than a small threshold.

The main idea of our dynamical state truncation for numerical integration methods is to integrate only those differential equations in Equation (5) that correspond to significant states. All other state probabilities are (temporarily) set to zero. This reduces the computational effort significantly since in each iteration step only a comparatively small subset of states is considered. Based on the fixed probability threshold $\delta > 0$, we dynamically decide which states to drop or add, respectively. Due to the regular structure of the CTMC the approximation error of the algorithm remains small since probability mass is usually concentrated at certain parts of the state space. The farther away a state is from a 'significant set' the smaller is its probability. Thus, in most cases the total error of the approximation remains small. Since in each iteration step probability mass may be 'lost' the approximation error at step i is the sum of all probability mass lost (provided that the numerical integration could be performed without any errors), that is,

$$\epsilon := 1 - \sum_{x \in S} \hat{p}^{(t)}(x) \tag{6}$$

where $\hat{p}^{(t)}$ is the approximation at time t.

The standard explicit four-stage fourth-order Runge-Kutta method (cf., e.g., [3, 8, 11]) applied to Eq. (5) yields the integration step

$$p^{(t+h)}(x) = p^{(t)}(x) + \frac{h}{6}\left(k^{(1)}(x) + 2k^{(2)}(x) + 2k^{(3)}(x) + k^{(4)}(x)\right), \tag{7}$$

where $h > 0$ is the time step of the method. For $i \in \{1, 2, 3, 4\}$ the values $k^{(i)}(x)$ are defined recursively as

$$
\begin{aligned}
k^{(1)}(x) &= \mathcal{M}(p^{(t)})(x), \\
k^{(2)}(x) &= k^{(1)}(x) + \tfrac{h}{2}\mathcal{M}(k^{(1)})(x), \\
k^{(3)}(x) &= k^{(1)}(x) + \tfrac{h}{2}\mathcal{M}(k^{(2)})(x), \\
k^{(4)}(x) &= k^{(1)}(x) + h\,\mathcal{M}(k^{(3)})(x).
\end{aligned}
\tag{8}
$$

Table 1. A single iteration step of the fast RK4 algorithm, which approximates the solution of the CME

```
1   choose step size h;
2   for i = 1, 2, 3, 4 do //traverse Sig four times
3      //decide which fields from state data structure
4      //are needed for k_i
5      switch i
6         case i = 1: coeff := 1; field := p;
7         case i ∈ {2, 3}: coeff := h/2; field := k_{i-1};
8         case i = 4: coeff := h; field := k_{i-1};
9      for all x ∈ Sig do
10        x.k_i := x.k_i + x.k_1;
11        for j = 1, ..., m with α_j(x) > 0 do
12           x.k_i := x.k_i − coeff · x.field · α_j(x);
13           if x + v_j ∉ Sig then
14              Sig := Sig ∪ {x + v_j};
15           (x + v_j).k_i := (x + v_j).k_i + coeff · x.field · α_j(x);
16  for all x ∈ Sig do
17     x.p := x.p + h/6 ·(x.k_1 + 2 · x.k_2 + 2 · x.k_3 + x.k_4);
18     x.k_1 := 0; x.k_2 := 0; x.k_3 := 0; x.k_4 := 0;
19     if x.p < δ then
20        Sig := Sig \ {x};
```

In order to avoid the explicit construction of a matrix and in order to work with a dynamic set Sig of significant states that changes in each step, we use for a state x a data structure with the following components:

- a field $x.p$ for the current probability of state x,
- fields $x.k_1, \ldots, x.k_s$ for the stage terms $k^{(1)}(x), \ldots, k^{(s)}(x)$,
- for all m with $\alpha_m(x) > 0$ a pointer to the successor state $x + v_m$ as well as a field with the rate $\alpha_m(x)$.

We start at time $t = 0$ and initialize the set Sig as the set of all states that have initially a probability greater than δ, i.e. $Sig := \{x \mid p^{(0)}(x) > \delta\}$. Note that the probability of all states $x \notin Sig$ is approximated as zero. We perform a step of the iteration in Eq. (7) by traversing the set Sig five times. In the first four rounds we compute $x.k_1, \ldots, x.k_4$ and in the final round we accumulate the summands. While processing state x in round i, $i < 5$, for each reaction j, we transfer probability mass from state x to its successor $x + v_j$, by subtracting a term from $x.k_i$ and adding the same term to $(x + v_j).k_i$. Note that this exactly gives the result of applying the operator $\mathcal{M}(\cdot)$ to k_i. A single iteration step is given in pseudocode in Table 1. In line 20, we ensure that Sig does not contain states with a probability less than δ. As step size h in line 1 of the algorithm,

Table 2. Dynamical state space truncation results for the enzymatic futile cycle

δ	$|Sig|$	ϵ
0	378	-
1e-20	330	3e-15
1e-15	266	2e-11
1e-10	190	2e-6

we choose the smallest average sojourn time of all states in Sig, that is,

$$h = \min_{x \in Sig} \left(\sum_{j=1}^{m} \alpha_j(x) \right)^{-1}. \tag{9}$$

In lines 2-15 we compute the values $k^{(1)}(x), \ldots, k^{(4)}(x)$ for all $x \in Sig$. The fifth round starts in line 16 and in line 17 the approximation of the probability $p^{(t+h)}(x)$ is calculated. Note that the fields $x.k_1, \ldots, x.k_4$ are initialized with zero.

The performance of the algorithm can be further improved if we additionally check in line 13 whether it is worthwhile to add state $x + v_j$ to Sig, that is, we guarantee that $x + v_j$ will receive enough probability mass and that $x + v_j$ will not be removed in the same iteration due to the check in line 19. Thus, we add $x + v_j$ only if the inflow $coeff \cdot x.field \cdot \alpha_j(x)$ to $x + v_j$ is greater or equal than a certain threshold $\tilde{\delta} > 0$. Obviously, $x + v_j$ may receive more probability mass from other states and the total inflow may be greater than $\tilde{\delta}$. Thus, if a state is not a member of Sig and if for each incoming transition the inflow probability is less than $\tilde{\delta}$, then this state will not be added to Sig even if the total inflow is greater or equal than $\tilde{\delta}$. This small modification yields a significant speed-up since otherwise all states that are reachable within at most four transitions will always be added to Sig because of line 13, but many of the newly added states will be removed in the same iteration because of line 19. For our numerical results we simply chose $\tilde{\delta} = \delta$.

In Table 2 we give the results of the numerical approximation of the enzymatic futile cycle model with the approach outlined above. We list the approximation error ϵ and the size of the set Sig of significant states (averaged over all iteration steps) for different values of δ. The running time was less than one second. Note that this system is small enough to perform a full exploration of the state space ($\delta = 0$) which corresponds to a numerical integration of Eq. (5) where all equations are considered. For more complex models this is typically not possible and the choice of δ is critical since it affects the number of significant states and therefore the running time of the method.

For many practical applications, the accuracy of the approximation is sufficient for a moderately small choice of the truncation thresholds δ and $\tilde{\delta}$, respectively. If, however, the probabilities of rare events have to be calculated, then the truncation approach above is no longer appropriate. As it stands now, the main drawback of the truncation approach is that rare events of interest may be neglected, that is, the truncated state space may not include those paths that lead to a certain rare event because their probability is smaller than the corresponding truncation threshold. If smaller truncation thresholds are chosen then the paths that significantly contribute to the rare event probability may not be truncated, but the number of states that have to be considered may become too large to be manageable.

3.2   Guided State Space Exploration

In this section we propose an extension of the truncation approach presented in Section 3.1 that is inspired by ideas from importance sampling and recent weighted stochastic simulation algorithms for estimating rare event probabilities [31, 21, 17, 9, 27]. Assume that we are interested in the probability $P(A)$ of a rare event A. Besides the CTMC $(X(t))_{t\geq 0}$, we consider another CTMC $(Y(t))_{t\geq 0}$ with the same state space and the same reaction channels but with different propensity functions $\beta_1, \ldots, \beta_M$ instead of the true propensity functions $\alpha_1, \ldots, \alpha_M$. We choose these 'biased' propensity functions $\beta_1, \ldots, \beta_M$ such that the occurence of A is more likely than with $\alpha_1, \ldots, \alpha_M$. Then we use the CTMC $(Y(t))_{t\geq 0}$ as 'guide' with regard to the state space truncation, that is essentially the propensity functions $\beta_1, \ldots, \beta_M$ guide us through the state space exploration in such a way that paths to the rare event of interest are not truncated. Therefore, we refer to $\beta_1, \ldots, \beta_M$ as guidance functions.

The idea is to solve X and Y simultaneously using the dynamical state space truncation. Let $\hat{p}^{(t)}$ ($\hat{q}^{(t)}$) be the corresponding numerical approximation of the distribution of X (of Y) at time t, respectively. The algorithm for solving Y is exactly as in Section 3.1 whereas for X we slightly modify the dynamical state space truncation algorithm. The decision whether we remove a state x from the set Sig at time t depends only on $\hat{q}^{(t)}$ and not on $\hat{p}^{(t)}$. Thus, at all time instances t for both the solution of X and Y we use the same sets Sig. This ensures that we do not truncate the paths leading to the rare event A. Intuitively, Y shows the direction to the rare event. Therefore, we refer to this approach as guided state space exploration. If the guidance functions are chosen appropriately, then the vectors $\hat{q}^{(t)}$ are computed using those paths that contribute most to $P(A)$. Hence, the vector $\hat{p}^{(t)}$ may loose a lot of probability mass over time, that is $\sum_{x\in S} \hat{p}^{(t)}(x) \ll 1$. The probability mass that remains in $\hat{p}^{(t)}$ then contains those parts that contribute most to $P(A)$.

The guided state space exploration differs from the 'pure' truncation algorithm in the following aspects:

– Instead of a single field $x.p$ for the current probability of state x we use two fields $x.p$ and $x.q$. The former refers to the current probability of state x in X and the latter refers to the probability of x in Y.
– In each iteration, we compute two different values for each field $x.k_i$, one for the probability flow in X and one for the flow in Y. Obviously, for Y we replace α_j by β_j (see line 12 of Table 1) and $x.p$ by $x.q$.

– We execute the two for-loops in lines 2-18 of Table 1 twice in order to compute $x.p$ and $x.q$, respectively. Lines 19 and 20, however, are only executed once in each iteration where we check whether $x.q < \delta$ (instead of $x.p < \delta$).

Note that the rare event probability $P(A)$ is directly approximated by the probabilities $x.p$ and the values $x.q$ are only used to determine the set of states that are considered in each step of the numerical integration. Actually, it would even be possible to solve Y and X not simultaneously but one after another. During the solution of Y, we would then record the elements of Sig for each time interval and use this information for the subsequent solution of X during which we truncate the state space in the same way as for Y. The simultaneous solution, however, has the advantage that it is faster than two subsequent solutions.

Let $P_\delta(A)$ denote the computed approximation with the approach outlined above where δ is the chosen significance threshold. It is important to note that, if we ignore errors of the numerical integration method, it holds that

$$P_\delta(A) \leq P(A) \tag{10}$$

because some paths leading to A may be truncated. Moreover, as $\delta \to 0$ our approximation approaches $P(A)$, that is, $\lim_{\delta \to 0} P_\delta(A) = P(A)$. In particular, when we decrease δ the accuracy will improve. Thus, if we apply the guided state space exploration for decreasing values of δ and see that $P_\delta(A)$ converges, we can estimate the approximation error as $P_\delta(A) - P_{\tilde{\delta}}(A)$ where $\delta < \tilde{\delta}$.

3.3 Choice of the Guidance Functions

It remains to choose the guidance functions $\beta_1, \ldots, \beta_M$ such that the occurrence of A is more likely than with $\alpha_1, \ldots, \alpha_M$ and the CTMC $(Y(t))_{t \geq 0}$ properly guides us to the rare event. For this purpose we borrow ideas from recent weighted stochastic simulation algorithms (wSSAs) for estimating rare event probabilities based on the well known importance sampling technique for variance reduction [31, 21, 17, 9, 27]. Actually, with importance sampling the goal is to change the probability measure underlying a stochastic system such that certain target events of interest become more likely to occur in simulations. Hence, it is nearby that choices of $\beta_1, \ldots, \beta_M$ that are useful for importance sampling simulations are also useful for our guided state space exploration. As we shall see, our guided state space exploration is far less sensitive against the choice of $\beta_1, \ldots, \beta_M$ than importance sampling and weighted stochastic simulation approaches.

In the next section we present experimental results where our choice of the guidance functions is inspired by approaches to state-independent importance sampling as taken in weighted stochastic simulation algorithms (wSSAs) [21, 17, 9]. We choose guidance functions

$$\beta_j(x) := \gamma_j \alpha_j(x) \tag{11}$$

with positive constants $\gamma_j > 0$. Hence, the parameter biasing consists in assigning a constant factor to each reaction type and multiplying the reaction propensity function by this factor, independent of the specific state in which the reaction occurs. The factors are collected in the biasing parameter vector $\gamma = (\gamma_1, \ldots, \gamma_M)$. Note that $\gamma_j = 1$ for each reaction R_j whose propensity function is not changed.

In order to keep the choice of the guidance functions as simple as possible, we do not change all propensity functions $\alpha_1, \ldots, \alpha_M$ but only those for which an increase or decrease obviously increases the probability of the rare event of interest. This can be often seen just by inspection. For instance, if we are interested in a certain species reaching a high or low population level, then we can select the reactions where this species is involved as reactant or product, respectively. In many cases, also 'indirect' impacts of other species on a certain target species can be easily seen.

More specifically, for the probability that before time $t = 100$ the number of molecules of species S_5 drops to ℓ for some $\ell \in \{5, 15, 25\}$ in the enzymatic futile cycle with initial state $x_0 = (1, 50, 0, 1, 50, 0)$ we should suppress the creation of S_5 molecules. This can be accomplished by decreasing the propensity function of reaction R_3, which creates S_5 molecules, and increasing the propensity function of reaction R_6, which by creating S_4 molecules encourages the consumption of S_5 molecules via reaction R_4. A similar approach has been taken in [21, 17] by setting the parameter biasing vector to $\gamma = (1, 1, 0.5, 1, 1, 2.0)$.

We shall generalize this and study $\gamma = (1, 1, \gamma_3, 1, 1, 1/\gamma_3)$ for various choices of γ_3 with $0 < \gamma_3 \leq 1$. Hence, in essence the propensity function of reaction R_3 is decreased by the factor γ_j and the propensity function of reaction R_6 is increased correspondingly by the factor $1/\gamma_3$, while all other propensity functions remain unchanged. In addition we apply guidance functions proposed by [9], where the parameter biasing vector γ was obtained via the cross-entropy method [9, 29].

4   Numerical Results

In this section we provide numerical results from comprehensive studies of our guided state space exploration for the enzymatic futile cycle benchmark model. In particular, this model is small enough to obtain an (up to numerical errors) exact solution. That is, by setting $\delta = 0$ no truncation error is introduced and if we neglect errors due to the numerical integration method (which is reasonable for this model with our step size chosen as the smallest average sojourn time of all states), then we have a 'quasi-exact' solution. Hence, we are able to evaluate the accuracy of the approximations obtained by the guided state space exploration in terms of their relative errors.

In Table 3 we list the probabilities $P(A)$ for three different choices of ℓ as well as the size of the set Sig of significant states. For the parameter biasing vector we chose $\gamma = (1, 1, \gamma_3, 1, 1, 1/\gamma_3)$, as explained before. In Table 4 we list the relative error of the approximated rare event probability and the size of the set Sig of significant states for different values of γ_3, δ and ℓ. Note that a relative error of one corresponds to approximating the rare event probability as zero. This is actually the same as what happens in direct simulation when due to the small probability the rare event is not observed and the probability of the non-observed event is estimated as zero. We also considered the parameter biasing

Table 3. Exact solution results for the enzymatic futile cycle

ℓ	Pr	$|Sig|$
25	1.7382e-07	298
15	6.2866e-13	338
5	1.0015e-19	378

Table 4. Guided state space exploration results for the enzymatic futile cycle and parameter biasing vector $\gamma = (1, 1, \gamma_3, 1, 1, 1/\gamma_3)$ with varying γ_3

ℓ	δ	$\gamma_3 = 1$		$\gamma_3 = 0.8$		$\gamma_3 = 0.65$		$\gamma_3 = 0.5$		$\gamma_3 = 0.35$		$\gamma_3 = 0.2$	
		rel.err.	$|Sig|$	rel.err.	$|Sig|$	rel.err.	$|Sig|$	rel.err.	$|Sig|$	rel.err.	$|Sig|$	rel.err.	$|Sig|$
25	1e-20	1.06e-6	263	1.06e-6	239	1.06e-6	220	1.06e-6	198	1.06e-6	167	1.06e-6	114
	1e-15	1.17e-5	231	1.12e-6	211	1.06e-6	191	1.06e-6	167	1.06e-6	142	1.06e-6	86
	1e-10	9.55e-1	190	1.53e-2	173	2.46e-4	155	3.88e-6	135	1.14e-6	109	6.51e-5	48
15	1e-20	2.77e-5	303	1.82e-6	279	1.80e-6	260	1.80e-6	238	1.80e-6	207	1.80e-6	154
	1e-15	1	266	2.78e-3	251	8.26e-6	231	1.81e-6	207	1.80e-6	182	1.80e-6	126
	1e-10	1	190	1	191	6.18e-1	193	1.90e-3	175	3.42e-6	149	2.60e-6	88
5	1e-20	1	330	1.23e-2	319	6.44e-6	300	2.63e-6	278	2.63e-6	247	2.63e-6	194
	1e-15	1	266	1	270	3.60e-1	269	6.11e-5	247	2.63e-6	222	2.63e-6	166
	1e-10	1	190	1	191	1	194	1	205	4.93e-4	189	2.72e-6	128

vector $\gamma = (1.000, 1.003, 0.320, 1.003, 0.993, 3.008)$ proposed in [9] for $\ell = 25$. The results are given in Table 5. We observe a slightly better approximation compared to the one corresponding to $\gamma_3 = 0.35$ even though the number of significant states was less. For all parameters that we chose, the running time of our algorithm was less than one second.

5   Conclusion

We have presented an accurate and computationally efficient numerical method for approximating rare event probabilities in stochastic models of biochemically reacting systems. Rather than estimating such probabilities via stochastic simulation, we numerically integrate the chemical master equation. In order to render the numerical computations possible we have to truncate the state space such that rare event probabilities can be approximated efficiently.

For this purpose, our method equips the well established explicit four-stage fourth order Runge-Kutta (RK4) method with a dynamical state space truncation and a guided state space exploration. The latter borrows ideas from importance sampling and corresponding weighted stochastic simulation algorithms for biochemically reacting systems in that we bias the true propensity functions such that for the biased model the probability of the rare event of interest is

Table 5. Guided state space exploration results for the enzymatic futile cycle for $\ell = 25$ and the parameter biasing vector $\gamma = (1.000, 1.003, 0.320, 1.003, 0.993, 3.008)$

δ	rel.err.	$|Sig|$
1e-20	5.45e-7	162
1e-15	7.38e-7	138
1e-10	1.10e-6	103

increased. We use the biased model as guide to the rare event, which yields a guided state exploration that avoids truncating paths to the rare event.

Our method has the general advantages of numerical methods over stochastic simulation that it does not require the generation of Markov chain trajectories and has only a numerical error but no statistical error. Moreover, our experimental results show that it is not very sensitive to the specific parameter biasing, that is, our method performs well for a quite broad range of the biasing factors. This is a significant advantage over weighted stochastic simulation algorithms, which are known to require very specific biasing parameters in order to estimate rare event probabilities efficiently and with high statistical accuracy. Obtaining such biasing parameters for weighted stochastic simulation algorithms by hand is intricate and despite recent advances in automated parameter selection via the cross-entropy method still often the determination of suitable biasing parameters takes a substantial amount of computational time.

The accuracy of our method is controlled by the truncation threshold δ. Obviously, as the truncation threshold δ approaches zero the approximation becomes 'quasi-exact' in that truncation errors are avoided and there is only the error introduced by the numerical integration method, which is negligible for sufficiently small step sizes. For too large values of δ the accuracy of the approximation can be degraded by badly chosen biasing parameters, but even then the degradation is far less extreme than for weighted stochastic simulation algorithms. A sufficiently small truncation threshold yields accurate results.

A couple of further research topics arise. We are currently intensely studying more complex benchmark models. Besides, the use of more advanced numerical integration methods than the explicit RK4 method seems to be promising, in particular implicit integration schemes for stiff models. Furthermore, it is reasonable that an improved adaptive step size selection based on local error estimates allows for larger steps than our choice of the minimum average state sojourn time. This will substantially improve the computational efficiency for extremely large and stiff models. Our guided state space exploration does not even require to solve the chemical master equation by numerical integration. It can similarly combined with other approaches to the transient solution of continuous-time Markov chains, e.g. with the uniformization method [35].

Another research direction is targeted towards formulas for the approximation error of the guided state space exploration, including approaches to obtain a priori error bounds for given guidance functions and truncation thresholds, to determine the required truncation threshold for a prescribed maximum relative error, and to determine the relative error for results obtained with certain guidance functions and an a priori fixed truncation threshold.

Acknowledgments. This work been partially funded by the German Research Council (DFG) as part of the Cluster of Excellence on Multimodal Computing and Interaction at Saarland University and the Transregional Collaborative Research Center 'Automatic Verification and Analysis of Complex Systems' (SFB/TR 14 AVACS).

References

1. Anderson, W.J.: Continuous-time Markov Chains: An Applications-Oriented Approach. Springer (1991)
2. Arkin, A., Ross, J., McAdams, H.H.: Stochastic kinetic analysis of developmental pathway bifurcation in phage λ-infected escherichia coli cells. Genetics 149, 1633–1648 (1998)
3. Ascher, U.M., Petzold, L.R.: Computer Methods for Ordinary Differential Equations and Differential-Algebraic Equations. SIAM (1998)
4. Asmussen, S., Glynn, P.W.: Stochastic Simulation: Algorithms and Analysis. Springer (2007)
5. Bharucha-Reid, A.T.: Elements of the Theory of Markov Processes and Their Applications. McGraw-Hill (1960)
6. Blake, W.J., Kaern, M., Cantor, C.R., Collins, J.J.: Noise in eukaryotic gene expression. Nature 422, 633–637 (2003)
7. Bucklew, J.A.: Introduction to Rare Event Simulation. Springer (2004)
8. Butcher, J.C.: Numerical Methods for Ordinary Differential Equations, 2nd edn. John Wiley & Sons (2008)
9. Daigle Jr., B.J., Roh, M.K., Gillespie, D.T., Petzold, L.R.: Automated estimation of rare event probabilities in biochemical systems. Journal of Chemical Physics 134, 044110 (2011)
10. Delbrück, M.: Statistical fluctuations in autocatalytic reactions. Journal of Chemical Physics 8, 120–124 (1940)
11. Deuflhard, P., Bornemann, F.: Scientific Computing with Ordinary Differential Equations. Springer (2002)
12. Ethier, S.N., Kurtz, T.G.: Markov Processes: Characterization and Convergence, 2nd edn. John Wiley & Sons (2005)
13. Fedoroff, N., Fontana, W.: Small numbers of big molecules. Science 297, 1129–1131 (2002)
14. Gillespie, D.T.: A general method for numerically simulating the time evolution of coupled chemical reactions. Journal of Computational Physics 22, 403–434 (1976)
15. Gillespie, D.T.: Exact stochastic simulation of coupled chemical reactions. Journal of Physical Chemistry 71(25), 2340–2361 (1977)
16. Gillespie, D.T.: A rigorous derivation of the chemical master equation. Physica A 188, 404–425 (1992)
17. Gillespie, D.T., Roh, M., Petzold, L.R.: Refining the weighted stochastic simulation algorithm. Journal of Chemical Physics 130, 174103 (2009)
18. Karlin, S., Taylor, H.M.: A First Course in Stochastic Processes, 2nd edn. Academic Press (1975)

19. Kroese, D.P., Taimre, T., Botev, Z.I.: Handbook of Monte Carlo Methods. John Wiley & Sons (2011)
20. Kurtz, T.G.: The relationship between stochastic and deterministic models for chemical reactions. Journal of Chemical Physics 57(7), 2976–2978 (1972)
21. Kuwahara, H., Mura, I.: An efficient and exact stochastic simulation method to analyze rare events in biochemical systems. Journal of Chemical Physics 129, 165101 (2008)
22. McAdams, H.H., Arkin, A.: Stochastic mechanisms in gene expression. Proceedings of the National Academy of Science USA 94, 814–819 (1997)
23. McAdams, H.H., Arkin, A.: It's a noisy business? Trends in Genetics 15(2), 65–69 (1999)
24. McQuarrie, D.A.: Stochastic approach to chemical kinetics. Journal of Applied Probability 4, 413–478 (1967)
25. Mikeev, L., Sandmann, W., Wolf, V.: Efficient calculation of rare event probabilities in Markovian queueing networks. In: Proceedings of the 5th International Conference on Performance Evaluation Methodologies and Tools, VALUETOOLS, pp. 186–196 (2011)
26. Oppenheim, I., Shuler, K.E., Weiss, G.H.: Stochastic and deterministic formulation of chemical rate equations. Journal of Chemical Physics 50(1), 460–466 (1969)
27. Roh, M.K., Daigle Jr., B.J., Gillespie, D.T., Petzold, L.R.: State-dependent doubly weighted stochastic simulation algorithm for automatic characterization of stochastic biochemical rare events. Journal of Chemical Physics 135, 234108 (2011)
28. Rubino, G., Tuffin, B. (eds.): Rare Event Simulation Using Monte Carlo Methods. John Wiley & Sons (2009)
29. Rubinstein, R.Y., Kroese, D.P.: The Cross-Entropy Method. Springer (2004)
30. Samoilov, M., Plyasunov, S., Arkin, A.P.: Stochastic amplification and signaling in enzymatic futile cycles through noise-induced bistability with oscillations. Proc. Natl. Acad. Sci. USA 102(7), 2310–2315 (2005)
31. Sandmann, W.: Rare event simulation methodologies in systems biology. In: Rubino, G., Tuffin, B. (eds.) Rare Event Simulation Using Monte Carlo Methods, ch. 11, pp. 243–265. John Wiley & Sons (2009)
32. Sandmann, W.: Sequential estimation for prescribed statistical accuracy in stochastic simulation of biological systems. Mathematical Biosciences 221(1), 43–53 (2009)
33. Singer, K.: Application of the theory of stochastic processes to the study of irreproducible chemical reactions and nucleation processes. Journal of the Royal Statistical Society, Series B 15(1), 92–106 (1953)
34. Srivastava, R., You, L., Summers, J., Yin, J.: Stochastic vs. deterministic modeling of intracellular viral kinetics. Journal of Theoretical Biology 218, 309–321 (2002)
35. Stewart, W.J.: Introduction to the Numerical Solution of Markov Chains. Princeton University Press (1995)
36. Thattai, M., van Oudenaarden, A.: Intrinsic noise in gene regulatory networks. Proceedings of the National Academy of Science USA 98(15), 8614–8619 (2001)
37. van Kampen, N.: Stochastic Processes in Physics and Chemistry. Elsevier, North-Holland (1992)

An Approximate Execution of Rule-Based Multi-level Models

Tobias Helms, Martin Luboschik, Heidrun Schumann,
and Adelinde M. Uhrmacher

University of Rostock, Institute of Computer Science,
Albert-Einstein-Straße 22, 18059 Rostock, Germany
{tobias.helms,martin.luboschik,heidrun.schumann,
adelinde.uhrmacher}@uni-rostock.de
http://www.uni-rostock.de

Abstract. In cell biology, models increasingly capture dynamics at different organizational levels. Therefore, new modeling languages are developed, e.g., like ML-Rules, that allow a compact and concise description of these models. However, the more complex models become the more important is an efficient execution of these models. τ-leaping algorithms can speed up the execution of biochemical reaction models significantly by introducing acceptable inaccurate results. Whereas those approximate algorithms appear particularly promising to be applied to hierarchically structured models, the dynamic nested structures cause specific challenges. We present a τ-leaping algorithm for ML-Rules which tackles these specific challenges and evaluate the efficiency and accuracy of this adapted τ-leaping based on a recently developed visual analysis technique.

Keywords: computational biology, rule-based modeling, multi-level modeling, tau-leaping, efficient execution.

1 Introduction

The size and complexity of models in systems biology have steadily been increasing in the last decade, also promoted by the development of modeling languages whose syntax allows a more compact and succinct description of models [6,13]. Consequently, the need for an efficient execution has increased, so that many improvements of Gillespie's original stochastic simulation algorithm (SSA) [8] have been developed, e.g., by using more efficient data structures or performing tasks concurrently [7,17,5]. However, exact variants of the SSA still execute every single event that occurs inside the system. Thus, with large propensities (due to high numbers of molecules, large stochastic rate constants or diffusion constants, and multi-scale models), the time step between successive events might decrease drastically, rendering the simulation progress very slow [2].

Multi-level models describe a system at different levels, e.g., combining intracellular and intercellular dynamics. Multi-level rules (ML-Rules) [21] is a rule-based modeling formalism developed to model systems operating at different

A. Gupta and T.A. Henzinger (Eds.): CMSB 2013, LNBI 8130, pp. 19–32, 2013.

organizational levels. Therefore, an ML-Rules model can represent dynamically and arbitrarily nested biochemical reaction networks. So far, the simulation algorithm of ML-Rules bases on the exact SSA and thus, shares its drawbacks. In contrast, approximate algorithms trade accuracy for execution speed. One famous family of approximate algorithms are leap methods, e.g., τ-leaping [9], which abandon the idea of single event executions in favor of larger time jumps and an estimation of all events within the intervals. These methods can gain a significant performance improvement compared to exact SSA procedures [15]. Based on the τ-leaping variant presented in [3], we develop a τ-leaping approach to compute ML-Rules models and present methods to tackle the specific challenges caused by the dynamic nesting structure. As earlier experiments showed that the parameters of τ-leaping algorithms influence both speed and accuracy [15,18], we use visual analysis to illuminate this influence. As case studies serve a Wnt/β-catenin pathway model [22], a fission yeast model [21], and a lipid raft model [10].

2   Background

We use the τ-leaping variant of Cao [3] as the basis of our τ-leaping approach for ML-Rules. In the following, this algorithm is explained in more detail. Additionally, a brief description of ML-Rules is given.

2.1   Tau-Leaping

The τ-leaping algorithm was introduced by Gillespie et al. [9] to speed up the simulation of well-stirred biochemical reaction networks. Instead of simulating every single reaction that occurs inside the system, as done by exact algorithms, τ-leaping performs "leaps" along the time line. For each leap, τ-leaping calculates the number of firings for each reaction during this leap and executes all reaction firings simultaneously. The length of a single leap is denoted by τ. More formally, a τ-leap can be described by

$$\mathbf{X}(t + \tau) = \mathbf{X} + \sum_{r \in R(\mathbf{X})} K_r(\tau; \mathbf{X}, t) \cdot \mathbf{v}(r) \tag{1}$$

with $R(\mathbf{X})$ being the set of all potential reactions given the current state $\mathbf{X}$, $\mathbf{v}(r)$ as the state change map for the reaction r and $K_r(\tau; \mathbf{X}, t)$ representing the firings' number of r during τ for $\mathbf{X}$. Restricting the size of τ to a value sufficiently small that the propensity $a(r)$ remains nearly constant during the leap for each reaction $r \in R(\mathbf{X})$ allows an approximation of $K_r(\tau; \mathbf{X}, t)$ by a Poisson random variable $P(a(r), \tau)$ with mean and variance $a(r)\tau$. This condition on the selection of τ is called the *leap condition*. The degree of acceptable propensity changes is bounded by the error parameter ϵ.

   Many improvements of Gillespie's original approach have been developed recently [3,11]. Since we use the mechanism of Cao [3] as the basis of our algorithm, this method shall be described briefly. Initially, the reaction set $R(\mathbf{X})$ is computed. Afterwards, $R(\mathbf{X})$ is divided into a set of non critical reactions $R_{ncr}(\mathbf{X})$, and a set of critical reactions $R_{cr}(\mathbf{X}) = R(\mathbf{X}) \backslash R_{ncr}(\mathbf{X})$ using the parameter $n_c \in \mathbb{N}$. A reaction is assigned as critical if this reaction cannot be fired more than n_c times, i.e., at least one reactant would be completely consumed after n_c firings of this reaction. The separation is done to reduce the probability of negative populations caused by the unbounded Poisson distribution. Therefore, critical reactions are only allowed to fire at most once during a τ-leap.

Next, one candidate for τ, denoted τ', is computed based on $R_{ncr}(\mathbf{X})$. The set $RS_{ncr}(\mathbf{X})$ of reactant species of all $R_{ncr}(\mathbf{X})$ is determined initially, i.e., the set of species which are a reactant in at least one reaction of $R_{ncr}(\mathbf{X})$. For each reactant species $rs \in RS_{ncr}(\mathbf{X})$, three values are computed. Firstly, the function $g(rs)$, which is used to "guarantee that bounding the relative change of states is sufficient for bounding the relative change of propensity functions" [23] is computed by the equation from [23]:

$$g(rs) = h(rs) + \frac{h(rs)}{n(rs)} \sum_{i=1}^{n(rs)-1} \frac{i}{\mathbf{X}(rs) - i} \tag{2}$$

$h(rs)$ denotes the highest order of reactions in which rs is a reactant species. $n(rs)$ denotes the highest amount of rs which is consumed by any of the highest order reactions. The changes' mean and variance of rs of all reactions in $R_{ncr}(\mathbf{X})$ are computed afterwards by

$$\hat{\mu}(rs) = \sum_{r \in R_{ncr}(\mathbf{X})} v_r(rs) \cdot a(r) \qquad \hat{\sigma}^2(rs) = \sum_{r \in R_{ncr}(\mathbf{X})} v_r(rs)^2 \cdot a(r) \tag{3}$$

With the help of these equations, τ' is computed by

$$\tau' = \min_{rs \in RS_{ncr}(\mathbf{X})} \left\{ \frac{\max\{\epsilon \cdot \mathbf{X}(rs)/g(rs), 1\}}{|\hat{\mu}(rs)|}, \frac{\max\{\epsilon \cdot \mathbf{X}(rs)/g(rs), 1\}^2}{\hat{\sigma}^2(rs)} \right\} \tag{4}$$

If τ' is smaller than a multiple α of the propensity sum of all reactions, $1/a(R(\mathbf{X}))$, a number N_{SSA} of exact iterations is performed instead of a τ-leap. Very small τ values would cause many firing numbers being set to zero, so that the algorithm tries to overcome the critical region in the state space by falling back to a much more simpler and, in this case, often faster exact simulation.

If τ' is sufficiently large, a second τ candidate, denoted τ'', is sampled from $\mathrm{Exp}(1/a(R_{cr}(\mathbf{X})))$, i.e., τ'' represents the time interval until the next critical reaction will fire. The minimum of τ' and τ'' is used as the next τ value. If τ'' is smaller than τ', exactly one critical reaction is selected, which will fire once during this τ-leap. After computing τ, the number of reaction firings can be computed and these reactions can be executed simultaneously. If any negative population occurs after executing these reactions, all changes are discarded, τ' is halved and the procedure is repeated until a valid τ-leap is executed.

2.2 ML-Rules

Multi-level rules (ML-Rules) is a rule-based formalism which can be used to create hierarchical models, including models with downward and upward causation between different hierarchy levels [21]. It has been realized as part of the modeling and simulation framework JAMES II [14]. Models are described by species definitions, a start state, and rule schemata. A species definition declares a species type, i.e., a unique name (e.g., A, $Cell$) and a tuple of attributes. A concrete species is defined by its type, attribute values, and sub species. Furthermore, species are treated population-based, i.e., identical species are summarized and an amount value is added to the representative of them. Species are identical if they have the same type, the same attribute values, identical sub species, and are enclosed by the same species. A rule scheme comprises a set of reactant patterns, a set of products and a kinetic rate. Reactant patterns describe species by their names, their desired attributes (optionally expressed by variables), and by sub reactant patterns, i.e., nested patterns can reach across an arbitrary number of levels. For example, the reactant pattern $A[B]$ describes species of type A, which contain at least one species B. Products are defined analogously to reactant patterns. The kinetic rate of a rule scheme can be an arbitrary expression, i.e., ML-Rules is not fixed to mass action kinetics. Such expressions can comprise simple arithmetics, conditions, and functions. Mass action kinetics can be modeled by bounded variables. For example, the rule schema

$$B^b + C^c \xrightarrow{b \cdot c \cdot r_c} D$$

bounds the variable b to the amount of the selected species B, the variable c to the amount of the selected species C and uses these variables and a constant reaction rate r_c to compute the reaction propensity. Additionally, the hierarchy above the reactant species is considered to compute the propensity of a reaction (see figure 1).

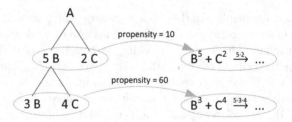

Fig. 1. Illustration of the rule schema instantiation of ML-Rules (adapted from [20, p. 143]). The rule scheme $B^b + C^c \xrightarrow{b \cdot c} \ldots$ is applied to the species on the left. Two reactions are instantiated. Due to the hierarchical multiplicity, a propensity of 60 $(12 \cdot 5)$ is assigned to the lower reaction.

So far, ML-Rules models have been executed by simulation algorithms based on the SSA [21]. These deal with dynamic hierarchical structures, complex rule schemata, and unbounded sets of species and reactions. To improve the performance of executing a model, a component-based simulation algorithm was developed recently to tailor the algorithm to specific requirements of concrete models. Furthermore, reinforcement learning techniques were used to adapt the algorithm due to changing model requirements during one simulation run [12].

3 τ-leaping for ML-Rules

The principle method of our τ-leaping approach for ML-Rules follows the one presented in [3] (see sec. 2.1). The algorithm is implemented inside JAMES II and will be part of the release 0.9.3. At first, all reactions are split into the sets of critical and non critical reactions. In contrast to the previous ML-Rules simulation algorithm, propensities of non critical reactions are computed locally, as the changes inside a context during a τ-leap depend neither on the number of context copies nor on the number of species higher up the hierarchy. Consequently, hierarchical multiplicities are not included in the propensity computation (see figure 2).

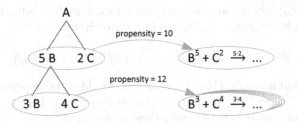

Fig. 2. Hierarchical multiplicities are not included in the propensity calculation of non critical reactions, because reactions are considered individually, i.e., for each of the five Bs, a reaction number of the lower reaction is computed and the corresponding number of reaction firings is applied to the specific B species

Next, each reaction context (rc) is considered separately and local τ'_l values are computed for each of these reaction contexts based on equation 4. Consequently, a τ'_l value reflects the relative changes inside its reaction context. τ' represents the minimum of all τ'_l values. After computing a suitable τ' value, it is checked whether τ' is too small and if so, a number N_{SSA} of SSA steps is performed. Otherwise, τ'' is computed for the critical reactions. If τ'' is smaller than τ', τ is set to τ'' and one critical reaction is selected which will fire exactly once. Otherwise, τ is set to τ' and no critical reaction will fire.

In contrast to flat chemical reaction networks, the computation of reaction firings and the simultaneous execution of all these reactions are complex tasks, because they have to be executed recursively along the hierarchy of the model. Thereby, each species is treated individually, because it is assumed that even identical species probably have different numbers of reaction firings during a τ-leap. The algorithm starts from the topmost reaction context of the model, i.e., the context which is not enclosed by a species:

1. Compute the number of reaction firings for the current reaction context rc and τ.
2. Remove all reaction reactants from rc.
3. For each remaining enclosed species of rc, execute this algorithm.
4. Add all reaction products to rc.

If the model state is invalid after a τ-leap, i.e., the amount of at least one species is negative, the changes are discarded, τ' is halved, and the algorithm is repeated.

3.1 Reaction Splitting

We applied the described τ-leaping algorithm to realistic ML-Rules models and observed that τ-leaping performed poorly. The number of executed reactions during a τ-leap was too small. The used models describe processes of cells. Typically, they comprise rules to diffuse species into or out of cells, rules to manipulate species inside a cell whose products and kinetic rates depend on the enclosing cell (e.g., on its volume), and rules to manipulate cells. For example, the following rule describes the diffusion of a protein into a cell:

$$Cell[sol?]^c + Protein^p \xrightarrow{c \cdot p \cdot r_c} Cell[Protein + sol?]$$

Multiple firings of a reaction based on this rule would add one protein to a number of cells. Additionally, such reactions tend to be critical, if the number of cells is small. The question is how to handle these reactions differently so that several proteins can enter the same cell within one τ-leap and the number of critical reactions is reduced. During calculating τ and the firing rates of reactions the cell in the above example can be ignored as its attributes do not change. Therefore, such reactions are split into two reactions, one describes the changes outside the cell and one describes the changes inside the cell. Referring to the example above, a concrete reaction would be split into the following two reactions:

$$1.\ \text{Inside the cell}\ :\ \ \xrightarrow{p \cdot r_c} Protein$$
$$2.\ \text{Outside the cell}:\ Protein \xrightarrow{p \cdot r_c}$$

The creation of these two reactions is possible, because all necessary information, e.g., the cell which is used and the values of the variables, are known after creating the basic reaction. The propensity of these reactions is reduced by the factor of the cells' amount, because in τ-leaping each cell is considered individually. These two reactions can now be handled as ordinary reactions, whereas the original reaction is not considered during the next τ-leap any more. Finally, the execution algorithm has only to ensure that both new reactions have the same firing number. Currently, we allow reaction splittings if the rule of a reaction contains exactly one pair of reactant and product pattern which satisfies the following conditions:

1. The species types and the attributes of the reactant and the product pattern are the same and the amount of both must be 1.
2. The amount of the top species of the reactant pattern is used as a factor for the reaction rate.
3. Both patterns comprise two hierarchy levels.

These conditions are a first rather restrictive approach, however, which facilitates automating reaction splitting, i.e., everything is done transparently to the user. Future approaches could generalize these conditions, so that more complex reactions could be split.

3.2 Population-Based τ-leaps

During the computation of the reaction firing numbers and the corresponding reactions, species are considered individually, i.e., the population-based approach cannot be used (see sec. 3). This individual consideration of species during a τ-leap can lead to a high computational effort. That is why we created a new parameter $\mu \in \mathbb{N}^+ \cup \infty$, which introduces the concept of populations to the execution of τ-leaps. Basically, during the computation of the reaction firing numbers and the corresponding reactions, it is used to partition a set of identical species into μ equally sized groups. Afterwards, all individual species of one group will evolve equally during the current τ-leap. If μ is greater than the amount of a species (guaranteed by $\mu = \infty$), one group is created for each individual of this species, i.e., all species are treated individually. Eventually, if $\mu = 1$, species are completely treated population-based, i.e., all identical species evolves equally. Consequently, reducing the value of μ on the one hand can decrease the accuracy of the simulation results but on the other hand can also decrease the computational effort of the τ-leaps.

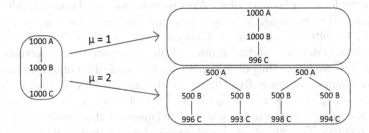

Fig. 3. Illustration of the impact of different μ values on one τ-leap. The rule scheme $C^c \xrightarrow{c \cdot r_c}$ is applied to the left species, i.e., one C is consumed after firing the corresponding reaction once. In this example, the concrete reaction firing numbers are chosen manually for illustration.

Figure 3 illustrates the concept of μ. The rule scheme $C^c \xrightarrow{c \cdot r_c}$ is applied to the species on the left, i.e., one reaction is created, which would consume exactly one C. If $\mu = 1$, species are treated completely population-based. Thus, one Poisson number is sampled for the reaction firing number and the amount of C is simply reduced according to that number, i.e., the reaction fires equally frequent inside all Bs. Afterwards, the whole species comprises again 1000 identical As, each enclose 1000 identical Bs. If $\mu = 2$, the As are separated into two groups, each comprises 500 As. For each group of As, the Bs are separated analogously. One reaction firing number is then computed for each group of Bs, and the amount of C is reduced accordingly. Thus, a total of four Poisson numbers are sampled. Finally, if μ would be set to ∞, 1000 groups of As would be created, each containing 1000 groups of Bs, i.e., one million Poisson numbers would be sampled.

4   Evaluation

We evaluate the developed τ-leaping approach for ML-Rules based on experiments with three different models, namely a realistic model of the Wnt/β pathway in neural progenitor cells [22], a cell cycle model [21], and a lipid raft model [10]. To analyze the effects of different parameter settings, we use a recently developed visual analysis technique [18].

4.1   The Wnt/β-Catenin Pathway Model

The Wnt/β-catenin pathway model described in [22] comprises five species types and twelve rule schemes. It defines one species type to represent cells and one species type to represent nuclei inside cells. The amount of cells can be defined by a parameter. Moreover, three other types are defined, representing Wnt proteins, β-catenin proteins, and Axin proteins. The Axin protein species type has one attribute reflecting the phosphorylation state of such species. Inside a cell, Axin proteins are phosphorylized and dephosphorylized (partially dependent on the number of Wnt proteins). β-catenin proteins are constantly moved inside and outside the nucleus of a cell. Depending on the number of β-catenin proteins inside the nucleus, dephosphorylized Axin proteins are synthesized inside a cell. Furthermore, Wnt proteins, Axin proteins, and β-catenin proteins are degraded constantly. Despite the number of β-catenin proteins, the degree of β-catenin degradation also depends on the number of phosphorylized Axin proteins.

We use 480 configurations of τ-leaping for the analysis, built from the cross product of twenty ϵ values (0.01, 0.02, ..., 0.2), four α values (5, 10, 15, 20), and six μ values (1, 2, 4, 6, 8, 10). n_c is always set to 10 and N_{SSA} is always set to 100. For each parameter setting, 100 replications with the simulation end time 200 are executed. The model state is observed after 0.4 time units have elapsed, i.e., 500 observations are made per simulation. Once we executed the experiment of the model with one cell, once with ten cells.

Referring to the performance, all configurations of τ-leaping need significantly less execution time per replication on average compared to the SSA. For one simulation of the model with one cell, the SSA need ≈ 39 s on average ($\sigma \approx 2.5$ s) on the used computer, the fastest τ-leaping configuration ($\epsilon = 0.2, \alpha = 5, \mu = 1$) need ≈ 0.5 s on average ($\sigma \approx 0.05$ s), and the slowest one ($\epsilon = 0.01, \alpha = 5, \mu = 10$) need ≈ 1.8 s on average ($\sigma \approx 1.3$ s). For one simulation of the model with ten cells, the execution times of the SSA and of most τ-leaping configurations nearly increase by thirty, e.g., the SSA need ≈ 1125 s on average ($\sigma \approx 76$ s). The ϵ parameter influences the performance significantly, i.e., the higher ϵ is chosen, the lower is the execution time. The α parameter influence the execution time negligibly. As excepted, for the model with one cell, the μ parameter has neither an effect on the simulation results nor on the execution time, because there is only one cell and one nucleus, i.e., population-based τ-leaps do not occur. For the model with ten cells, on the contrary, μ has an impact on the performance, i.e., τ-leaping performs better with $\mu = 1$ than with the other used values for μ (e.g., τ-leaping ($\epsilon = 0.2, \alpha = 5$) need ≈ 8 s with $\mu = 1$ and ≈ 17 s with $\mu \in \{2, 4, 6, 8, 10\}$). Interestingly, the difference depends on the value of ϵ, i.e., the higher ϵ is chosen, the smaller is the difference (if $\epsilon > 0.07$, the difference diminishes almost completely). We found out that only for $\mu = 1$, population-based τ-leaps are executed frequently due to identical cells. For the other values of μ, each cell differs permanently from each other after a short period of the simulation, so that only a few population-based τ-leaps are executed at the beginning. The executed SSA steps caused by small τ values are the main reason for this behavior. Nevertheless, the impact of these SSA steps should diminish with a higher number of cells, i.e., even with higher μ values several cells should be identical frequently during the simulation and thus these μ values should have an effect on the performance.

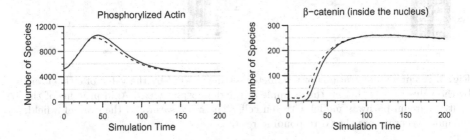

Fig. 4. The average results of the SSA (solid lines) compared to τ-leaping (dotted lines) with $\epsilon = 0.2$, $\alpha = 5$, and $\mu = 10$ (i.e., the least accurate configuration we used) of the species amount of phosphorylized Actin proteins and of β-catenin proteins inside the nucleus for the Wnt/β-catenin pathway model with one cell.

However, the interesting aspect is how much accuracy is traded for this gain of efficiency and especially due to their impact on the execution time how is the impact of the parameters ϵ and μ on the accuracy. Most importantly, all configurations of τ-leaping produce similar results compared to the results of the

SSA (e.g., see figure 4). Due to the high number of parameters, configurations, and observations, we use visual analysis to examine the impact of ϵ and μ on the accuracy, because it enables us to interactively explore data, e.g., to observe relations, correlations, and interdependencies between parameters and results. Precisely, we use the software developed in [18]. It visually connects the used parameter values and the computed accuracy measures with the corresponding simulation results of one species. Focusing on the first two, the software uses color coding and maps low values of the five parameters to white and high values to black, low accuracy values (in this case, represented by the p-value of the paired Wilcoxon rank sum test [24, p. 513]) of the observations to white and high values to saturated cyan (see figure 5). Each line corresponds to one parameter setting aligned to the according accuracy values evolving over time. The software now enables us to scroll trough, sort, and select parameter settings by concrete parameters and accuracy values of specific time intervals.

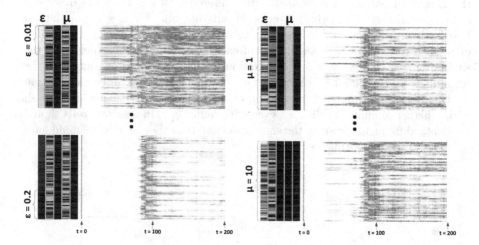

Fig. 5. Accuracy illustrations of parameter settings sorted by the value of ϵ (left) and by the value of μ (right) on the example of the dephosphorylized Axin proteins of the Wnt/β-catenin pathway model with ten cells. The more saturated the cyan of a field, the more accurate are the corresponding results.

To analyze the impact of ϵ on the accuracy, we started by sorting the parameter settings according to ϵ and scrolling through the configurations (see the left illustrations of figure 5). Characteristic for τ-leaping, the smaller ϵ is chosen, the more accurate the results get, which becomes visible by dark cyan regions in the top left illustration of figure 5. Vice versa, the accuracy decreases if ϵ is increased, especially in the beginning of the simulation (see the bottom left illustration in figure 5). Thus, referring to the impact of ϵ, the algorithm behaves as expected. Sorting the parameter settings according to μ, we observed another behavior: No relation between its value and the accuracy can be revealed (see the right illustrations of figure 5). As this might be specific for the chosen model, further experiments are needed to analyze the impact of μ on accuracy in more detail. Also, the parameter α does not influence the result accuracy in our model either, i.e., the computed τ' is usually chosen sufficiently high so that eventually only few SSA steps are executed. Interestingly, at several time points the accuracy of all configurations decreases slightly, which results in faint but noticeable vertical lines in the visualization (e.g., see figure 6). The reason for this is not yet clear, e.g., whether this is caused by large τ values in comparison to smaller observation rates or whether it is caused by the execution of SSA steps.

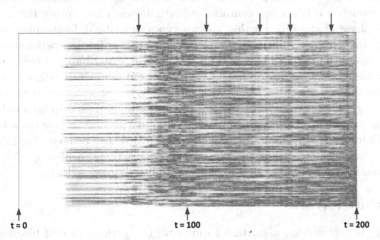

Fig. 6. The accuracy illustrations of some τ-leaping configurations again on the example of the dephosphorylized Axin proteins of the Wnt/β-catenin pathway model with ten cells. The illustrations show noticeable vertical lines, i.e., many configurations behave similar at these times. Some of these lines are marked by the arrows on top.

4.2 The Lipid Raft Model

The lipid raft model describes the synthesis, degradation and diffusion of lipid rafts in cell membranes [10]. It comprises receptor species, which diffuse inside and outside lipid rafts. While degrading a lipid raft, all containing receptors are removed. For this evaluation, the model was reimplemented in ML-Rules. We use the same 480 configurations of the experiments with the Wnt/β-catenin pathway model for the evaluation (see 4.1). Regarding the results, again ϵ is the most important parameter. However, if $\epsilon > 0.01$, the accuracy of the results decreases significantly. In contrast to the Wnt/β-catenin pathway model, the α parameter also significantly influences the results (if $\epsilon = 0.01$). For example, if $\alpha = 5$, many SSA steps are executed, so that the results are still relatively accurate. If $\alpha \in \{10, 15, 20\}$, SSA steps are executed rarely and the accuracy of the results decrease dramatically. Thus, accurate results are only achieved with $\epsilon = 0.01$ and $\alpha = 5$. The according configurations perform 16% better on average than the SSA, which needs ≈ 72 s. The μ parameter again only slightly influence the execution time of the simulations and does not noticeable influence the simulation results. The evaluation of the experiments with the lipid raft model again shows the importance of an appropriate configuration of all parameters of τ-leaping, i.e., using inappropriate configurations can lead to fast execution times but totally wrong results.

4.3   The Fission Yeast Model

The fission yeast model [21] represents the cell division and mating process of fission yeast cells. It comprises intracellular as well as intercellular reactions and contains a simple grid-based spatial level to describe specific diffusion reactions. In this model, a cell species contains two attributes, representing the volume and the phase of the cell. Further species types describe different pheromones and proteins. However, initial experiments show that τ-leaping performs worse for this model than the SSA. Reactions which change attributes of cells are the most frequent reactions in the model. Typically, the number of cells is small (1 - 100), so that either these reactions are declared as critical, or the computed τ' values are small. Consequently, τ-leaping only summarizes a small number of reactions per τ-leap, i.e., two reactions are summarized per τ-leap on average. All in all, these results show that τ-leaping for ML-Rules behaves like existing τ-leaping algorithms, i.e., if the most frequent reactions involve species with small amounts, τ-leaping cannot exploit its ability to execute leaps.

5   Related Work

τ-leaping is a promising algorithm to improve the performance of biochemical reaction networks [9]. Consequently, many improvements and extensions have been developed over the last years. For example, many τ-leaping variants deal with the problem of negative populations [1,25]. However, only a few variants focus on structured models. Related approaches are those τ-leaping variants that consider space. For example, the binomial spatial τ-leaping algorithm [19] and S-τ [16] are based on a grid-like structuring of space into subvolumes where diffusion events happen between and reaction events within those subvolumes. Similarly to our approach a τ candidate is calculated separately for each subvolume. Calculating τ candidates locally is also a strategy adopted for dynamical probabilistic P Systems [4]. However, those approaches do not have hierarchical contexts similar to those of ML-Rules: The former approaches due to not supporting hierarchies, the latter due to not supporting hierarchical rules and population-based membranes.

The performance of τ-leaping depends on the model and its configuration, i.e., the trade-off between execution time and accuracy [15,16]. To explore these dependencies in the multi-dimensional space of parameter settings and different accuracies, formulating and testing hypotheses iteratively would be the usual case. Instead, visual analysis enables the user to execute an exploratory investigation of the data helping him to get impressions of the data to formulate hypotheses specifically.

6   Conclusion

This paper presents a τ-leaping algorithm for the rule-based multi-level formalism ML-Rules, which supports dynamic nesting. It extends the traditional τ-leaping by treating identical individuals as populations (along with the parameter μ to control the grouping) and by calculating τ_l' values in the respective contexts (among which the minimum is selected as the overall τ'). The evaluation shows that the algorithm behaves like other τ-leaping variants, e.g., small ϵ values cause more accurate results than high ϵ values and the improvement of τ-leaping depends strongly on the used model. Depending on the model, the execution time can be decreased significantly, e.g., the execution time for one simulation of the used Wnt/β-catenin pathway model can be reduced by 2 orders on average. The new parameter μ, which configures the population-based execution of τ-leaps, rarely influences the execution time and does not affect the accuracy of the results. However, this relation is likely caused by the used models and thus, deserves further investigations - as do the areas of poor accuracy which have been revealed in our visual exploratory analysis. Currently, the developed τ-leaping approach only supports mass action kinetics. Since ML-Rules is not constrained to mass action kinetics, it has to be investigated which types of kinetics can be supported by τ-leaping.

Acknowledgments. This research is supported by the German Research Foundation, via research grants CoSA (UH 66/7-3) and VASSiB (part of SPP 1335).

References

1. Cao, Y., Gillespie, D.T., Petzold, L.R.: Avoiding negative populations in explicit Poisson tau-leaping. The Journal of Chemical Physics 123(5) (2005)
2. Cao, Y., Gillespie, D.T., Petzold, L.R.: The slow-scale stochastic simulation algorithm. The Journal of Chemical Physics 122(1), 014116 (2005)
3. Cao, Y., Gillespie, D.T., Petzold, L.R.: Efficient step size selection for the tau-leaping simulation method. The Journal of Chemical Physics 124(4) (2006)
4. Cazzaniga, P., Pescini, D., Besozzi, D., Mauri, G.: Tau Leaping Stochastic Simulation Method in P Systems. In: Hoogeboom, H.J., Păun, G., Rozenberg, G., Salomaa, A. (eds.) WMC 2006. LNCS, vol. 4361, pp. 298–313. Springer, Heidelberg (2006)
5. Dematté, L., Prandi, D.: GPU computing for systems biology. Briefings in Bioinformatics 11(3), 323–333 (2010)
6. Faeder, J.R.: Toward a comprehensive language for biological systems. BMC Systems Biology 9(68) (2011)
7. Gibson, M.A., Bruck, J.: Efficient Exact Stochastic Simulation of Chemical Systems with Many Species and Many Channels. The Journal of Chemical Physics 104(9), 1876–1889 (2000)
8. Gillespie, D.T.: Exact stochastic simulation of coupled chemical reactions. The Journal of Physical Chemistry 81(25), 2340–2361 (1977)
9. Gillespie, D.T.: Approximate accelerated stochastic simulation of chemically reacting system. The Journal of Chemical Physics 115(4), 1716–1733 (2001)

10. Haack, F., Burrage, K., Redmer, R., Uhrmacher, A.M.: Studying the role of lipid rafts on protein receptor bindings with Cellular Automata. IEEE/ACM Transactions on Computational Biology and Bioinformatics (accepted for publication, 2013)
11. Harris, L.A., Clancy, P.: A "partitioned leaping" approach for multiscale modeling of chemical reaction dynamics. The Journal of Chem. Physics 125(14) (2006)
12. Helms, T., Ewald, R., Rybacki, S., Uhrmacher, A.M.: A Generic Adaptive Simulation Algorithm for Component-based Simulation Systems. In: Proc. 27th Workshop on Principles of Adv. and Dist. Simulation, PADS 2013 (2013)
13. Henzinger, T.A., Jobstmann, B., Wolf, V.: Formalisms for Specifying Markovian Population Models. In: Bournez, O., Potapov, I. (eds.) RP 2009. LNCS, vol. 5797, pp. 3–23. Springer, Heidelberg (2009)
14. Himmelspach, J., Uhrmacher, A.M.: Plug'n simulate. In: Proc. 40th Annual Simulation Symposium (ANSS 2007), pp. 137–143 (2007)
15. Jeschke, M., Ewald, R.: Large-Scale Design Space Exploration of SSA. In: Heiner, M., Uhrmacher, A.M. (eds.) CMSB 2008. LNCS (LNBI), vol. 5307, pp. 211–230. Springer, Heidelberg (2008)
16. Jeschke, M., Ewald, R., Uhrmacher, A.M.: Exploring the Performance of Spatial Stochastic Simulation Algorithms. The Journal of Computational Physics 230(7), 2562–2574 (2011)
17. Li, H., Petzold, L.: Logarithmic Direct Method for Discrete Stochastic Simulation of Chemically Reacting Systems. Technical report, Department of Computer Science, University of California: Santa Barbara (2006)
18. Luboschik, M., Rybacki, S., Ewald, R., Schwarze, B., Schumann, H., Uhrmacher, A.M.: Interactive Visual Exploration of Simulator Accuracy: A Case Study for Stochastic Simulation Algorithms. In: Proc. 44th Winter Simulation Conference, WSC 2012 (2012)
19. Marquez-Lago, T.T., Burrage, K.: Binomial tau-leap spatial stochastic simulation algorithm for applications in chemical kinetics. The Journal of Chemical Physics 127(10) (2007)
20. Maus, C.: Toward Accessible Multilevel Modeling in Systems Biology - A Rule-based Language Concept. PhD thesis, University of Rostock, Germany (2013)
21. Maus, C., Rybacki, S., Uhrmacher, A.M.: Rule-based multi-level modeling of cell biological systems. BMC Systems Biology 5(166) (2011)
22. Mazemondet, O., John, M., Leye, S., Rolfs, A., Uhrmacher, A.M.: Elucidating the Sources of β-Catenin Dynamics in Human Neural Progenitor Cells. PLoS ONE 7(8), e42792 (2012)
23. Sandmann, W.: Streamlined formulation of adaptive explicit-implicit tau-leaping with automatic tau selection. In: Proc. 41st Winder Simulation Conference (WSC 2009), pp. 1104–1112 (2009)
24. Sheskin, D.J.: Handbook of Parametric and Nonparametric Statistical Procedures, 4th edn. Chapman & Hall/CRC (2007)
25. Tian, T., Burrage, K.: Binomial leap methods for simulating stochastic chemical kinetics. The Journal of Chemical Physics 121(21), 10356–10364 (2004)

Computing Cumulative Rewards
Using Fast Adaptive Uniformisation

Frits Dannenberg[1], Ernst Moritz Hahn[1,2], and Marta Kwiatkowska[1]

[1] University of Oxford, Department of Computer Science
[2] State Key Laboratory of Computer Science, ISCAS, China

Abstract. The computation of transient probabilities for continuous-time Markov chains often employs uniformisation, also known as the Jensen's method. The fast adaptive uniformisation method introduced by Mateescu approximates the probability by neglecting insignificant states, and has proven to be effective for quantitative analysis of stochastic models arising in chemical and biological applications. However, this method has only been formulated for the analysis of properties at a given point of time t. In this paper, we extend fast adaptive uniformisation to handle expected reward properties which reason about the model behaviour until time t, for example, the expected number of chemical reactions that have occurred until t. To show the feasibility of the approach, we integrate the method into the probabilistic model checker PRISM and apply it to a range of biological models, demonstrating superior performance compared to existing techniques.

1 Introduction

Model checking of continuous-time Markov chains (CTMCs) [3] is an established method that has been successfully used for quantitative analysis of a variety of models, ranging from biochemical reaction networks [9,17] to performance analysis of computer systems [2]. The analysis typically involves computing the *transient probability* of the system residing in a state at a given time t, or, for a model annotated with time-dependent rewards, the *expected reward* that can be obtained. Transient probabilities for finite-state CTMCs can be computed through the *uniformisation* method, also known as the Jensen's method. Uniformisation involves discretising the CTMC with respect to a fixed rate, which enables reduction of the transient probability calculation to an infinite summation of Poisson distributed steps of the derived discrete-time Markov chain, and approximating the probability by truncating to a finite summation. The number of terms required can be precomputed for a specified precision using the Fox-Glynn method [8].

Many biochemical reaction networks, however, induce CTMC models whose state space is potentially infinite. To handle such cases, [17] introduced *continuous-time propagation models*, a generalisation of continuous-time Markov chains. The idea of this model is to propagate the (probability or expectation) mass values along the system execution. In order to analyse propagation models,

A. Gupta and T.A. Henzinger (Eds.): CMSB 2013, LNBI 8130, pp. 33–49, 2013.
© Springer-Verlag Berlin Heidelberg 2013

the *fast adaptive uniformisation (FAU)* method was formulated [16]. Similarly to standard uniformisation, FAU applies discretisation, except that it does so dynamically, starting from some initial condition wrt to a sequence of rates (a birth process) rather than a single rate, and truncates the computation of the probability to a finite summation, although the number of summation terms cannot be precomputed. To deal with the unbounded state space, FAU explores only the relevant states, ignoring the probability of the insignificant states. Thus, the number of states to be maintained in memory can be kept small, at a cost of some loss of precision. Importantly, the FAU method can also speed up the analysis of very large finite models.

Fast adaptive uniformisation was implemented [6] and applied successfully on a variety of biological systems [6,17], but for transient probabilities only. Many useful quantitative analyses involve the computation of expected rewards, which can be *instantaneous* (incurred at time t) or *cumulated* (until time t). An example of an instantaneous reward property is the number of molecules of a given species at time $100s$, and of a cumulative property the expected number of reactions that occurred for the duration of $100s$. Although one can express cumulative reward properties by adding additional species to the model, for example by increasing the reward by 1 every time a reaction occurs, this has the disadvantage of introducing an additional dimension into the model and, as we show later, can severely affect the performance, resulting in higher memory requirement and a consequent loss of precision.

In this paper, we extend fast adaptive uniformisation for CTMCs to allow for efficient computation of cumulative reward properties, thus avoiding the overhead of adding the additional dimension. We cast our results in the framework of propagation models of [17]. We implement the method, including the reward extension, and integrate it into the probabilistic model checker PRISM. To show the practical applicability of FAU, we have applied it to a range of case studies from biology, demonstrating superior performance compared to existing techniques.

Related Work. FAU generalises adaptive uniformisation [18] by accelerating the discretisation and neglecting states with insignificant probability. Standard uniformisation is implemented in a number of tools, including PRISM, which we enhance with the FAU functionality in this paper. SABRE [6] is the first tool to implement FAU without cumulative rewards. Both PRISM and SABRE support models written in Systems Biology Markup Language (SBML) as input, in addition to their native modelling languages. SABRE is a stand-alone tool available for download or as a web interface; it additionally offers deterministic approximations using differential equations (by the Runge-Kutta fourth order method), which is faster and leads to accurate results for large numbers of molecules. PRISM does not support deterministic approximations, but provides extensive support for temporal logic model checking that is appropriate for molecular networks where some species occur in small numbers, or the encoding of spatial information is needed, as we consider in this paper. PRISM also provides statistical model checking and Gillespie simulation. The tool MARCIE

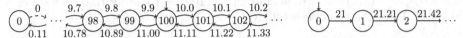

Fig. 1. Birth-death process   Fig. 2. Birth process

[19] implements FAU but does not support cumulative rewards. Further tools that support reward properties but not FAU include, for instance, MÖBIUS [4] and MRMC [11].

2 Preliminaries

We begin by giving an overview of the main definitions and results based on [13,6,17]. A *continuous-time Markov chain (CTMC)* is given by a set of discrete states S and the transition rate matrix $\mathcal{R} \colon S \times S \to \mathbb{R}_{\geq 0}$ where $\mathcal{R}(s, s) = 0$ for all $s \in S$. The rate $\mathcal{R}(s, s')$ determines the delay before the transition can occur, i.e. the probability of this transition being triggered within t time-units is $1 - e^{-\mathcal{R}(s,s') \cdot t}$. Let $E(s) \stackrel{\text{def}}{=} \sum_{s' \in S} \mathcal{R}(s, s')$ be the exit rate and define the generator matrix Q by $Q \stackrel{\text{def}}{=} R - diag(E)$, where $diag(E)$ is the $S \times S$ matrix with E on its diagonal and zero everywhere else. Then $\pi_t \colon S \to \mathbb{R}_{\geq 0}$, the transient probability vector at time t, can be expressed as $\pi_t = \pi \cdot e^{Qt}$ given the initial probability vector π.

We cast our method in the framework of *continuous-time (linear) propagation models* [17, Section 2.3.3] which generalise continuous-time Markov chains. We now recall the relevant results from [17].

Definition 1 (Continuous-time propagation model). *A* continuous-time propagation model (CTPM) *is a tuple* $\mathcal{M} = (S, \pi, \mathcal{R})$, *where*

- *S is a countable or finite set of* states,
- *$\pi \colon S \to \mathbb{R}_{\geq 0}$ where $|\{s \in S \mid \pi(s) > 0\}| < \infty$ is an* initialisation vector, *and*
- *$\mathcal{R} \colon S \times S \to \mathbb{R}_{\geq 0}$ is a* transition matrix, *such that for all $s \in S$ we have* $|\{s' \in S \mid \mathcal{R}(s, s') > 0\}| < \infty$.

The transition matrix $\mathcal{R}$ assigns a rate $\mathcal{R}(s, s')$ to each pair of states, as for CTMCs, and the initialisation vector π assigns an initial mass value $\pi(s)$ to each state $s \in S$. There are only finitely many states to which a positive mass is assigned initially. The models we consider are *finitely branching*, that is, for each state there are only finitely many states to which this state has a positive transition rate.

A CTPM is a CTMC if $\sum_{s \in S} \pi(s) = 1$ and $\mathcal{R}(s, s) = 0$ for all $s \in S$.

Example 1 (Continuous-time Markov chain). In Fig. 1, we depict a CTPM [7, 001-01], a so-called *birth-death process*. Each state s is a natural number describing the number s of molecules of a given species. In each state s, a new molecule can appear with rate $\lambda \cdot s$, and disappear with rate $\mu \cdot s$ for $\lambda \stackrel{\text{def}}{=} 0.1, \mu \stackrel{\text{def}}{=} 0.11$. We thus have $\mathcal{R}(s, s+1) \stackrel{\text{def}}{=} \lambda \cdot s$ for all $s \geq 0$, $\mathcal{R}(s, s-1) \stackrel{\text{def}}{=} \mu \cdot s$ for $s \geq 1$

and $\mathcal{R}(\cdot, \cdot) \overset{\text{def}}{=} 0$ otherwise. We assume that $\pi(100) \overset{\text{def}}{=} 1$ and for the other states $\pi(\cdot) \overset{\text{def}}{=} 0$. Thus, the model is a CTMC.

To reason about the timed behaviour of a CTPM, we now define its *generator matrix* which generalises that for CTMCs.

Definition 2 (Generator matrix). *The* generator matrix $Q(\mathcal{M})\colon S \times S \to \mathbb{R}$ *of a CTPM $\mathcal{M}$ is defined so that*

- $Q(\mathcal{M})(s, s') \overset{\text{def}}{=} \mathcal{R}(s, s')$ *for $s, s' \in S$ with $s \neq s'$, and*
- $Q(\mathcal{M})(s, s) \overset{\text{def}}{=} \mathcal{R}(s, s) - \sum_{\substack{s' \in S \\ s' \neq s}} \mathcal{R}(s, s')$.

The *propagation process*, which propagates probability mass or expectation values, is then defined as follows. Note that, for CTMCs, $\pi_t(s)$ is the (transient) probability that the model resides in state s at time t.

Definition 3 (Propagation process). *Given a CTPM $\mathcal{M} = (S, \pi, \mathcal{R})$, the propagation process at time t, $\pi_t(\mathcal{M})\colon S \to \mathbb{R}$, is defined as the solution of the differential equation*

$$\dot{\pi}(s') \overset{\text{def}}{=} \sum_{s \in S} \pi(s) \cdot Q(\mathcal{M})(s, s')$$

at time t, for $s' \in S$, given the initial value π.

The *standard uniformisation* [10] method for CTMCs splits the CTMC into a discrete-time Markov chain (DTMC) and a Poisson process as follows. Define the DTMC P by $P \overset{\text{def}}{=} I + \frac{1}{\Lambda} \cdot Q$ where Λ is the uniformisation rate such that $\Lambda \geq \max_{s \in S} E(s)$. Then π_t can be computed as $\sum_{n=0}^{\infty} \pi_t(\Psi^\Lambda)(n) \cdot \tau_n$ where $\pi_t(\Psi^\Lambda)(n)$ is the value of the Poisson distribution with rate $\Lambda \cdot t$ at point n, and $\tau_n = \tau_{n-1} \cdot P$ for $n > 0$, $\tau_0 = \pi_0$. For a given precision ϵ, the summation can be truncated using the Fox and Glynn method [8].

The *fast adaptive uniformisation (FAU)* [6,17] is a variant of the *adaptive uniformisation* [18] which splits the CTMC into a DTMC and a birth process. For an infinite sequence $\mathbf{\Lambda} = (\Lambda_0, \Lambda_1, \dots)$ of rates with $\Lambda_n \in \mathbb{R}_{\geq 0}$ for all $n \in \mathbb{N}$, the *birth process* is defined as the CTMC $\Phi^\mathbf{\Lambda} \overset{\text{def}}{=} (S, \pi, \mathcal{R})$, where

- $S \overset{\text{def}}{=} \mathbb{N}$,
- $\pi(0) \overset{\text{def}}{=} 1$ and $\pi(\cdot) \overset{\text{def}}{=} 0$ otherwise, and
- $\mathcal{R}(n, n+1) \overset{\text{def}}{=} \Lambda_n$ for $n \in \mathbb{N}$ and $\mathcal{R}(\cdot, \cdot) \overset{\text{def}}{=} 0$ otherwise.

Note that the Poisson process is a special case of the birth process with constant rates $\mathbf{\Lambda} = (\Lambda, \Lambda, \dots)$.

Transient probabilities of birth processes can be approximated efficiently using specialised techniques [17, Section 4.3.2, Solution of the birth process]. This is possible by applying *standard uniformisation* [10] in a way which takes advantage of the particular structure of the process. Finally, transient probabilities of general CTPMs can be computed as follows, where we reformulate P_n in terms of the rate matrix, rather than the generator matrix used in [17].

Theorem 1 (Solving propagation models using a birth process).
Consider

- *a CTPM $\mathcal{M} = (S, \pi, \mathcal{R})$,*
- *an infinite sequence of subsets $\mathbf{S} = (S_0, S_1, \ldots)$ with $S_n \subseteq S$ denoting active states,*
- *an infinite sequence $\mathbf{\Lambda} = (\Lambda_0, \Lambda_1, \ldots)$ with $\Lambda_n \geq \sup_{s \in S_n} \sum_{\substack{s' \in S, \\ s' \neq s}} \mathcal{R}(s, s')$ of uniformisation rates,*
- *probability matrices $P_n(\mathcal{M}) \colon S \times S \to \mathbb{R}_{\geq 0}$ for $n \in \mathbb{N}$, where for $s, s' \in S$ we have*

$$P_n(\mathcal{M})(s, s') \stackrel{\text{def}}{=} \begin{cases} \frac{\mathcal{R}(s, s')}{\Lambda_n} & \text{if } s \neq s', \text{ and} \\ \frac{\mathcal{R}(s, s')}{\Lambda_n} + 1 - \sum_{\substack{s'' \in S, \\ s'' \neq s}} \frac{\mathcal{R}(s, s'')}{\Lambda_n} & \text{otherwise,} \end{cases}$$

- *discrete-time distributions $\tau_n(\mathcal{M}) \colon S \to \mathbb{R}_{\geq 0}$ for $n \in \mathbb{N}$ with*

$$\tau_n(\mathcal{M})(s') \stackrel{\text{def}}{=} \begin{cases} \pi(s') & \text{if } n = 0, \text{ and} \\ \sum_{s \in S} \tau_{n-1}(\mathcal{M})(s) \cdot P_{n-1}(s, s') & \text{otherwise.} \end{cases}$$

We further require that $\{s \in S \mid \tau_n(\mathcal{M})(s) > 0\} \subseteq S_n$ for $n \in \mathbb{N}$.
Then we have that, at time t, for each $s \in S$:

$$\pi_t(\mathcal{M})(s) = \sum_{n=0}^{\infty} \pi_t(\Phi^{\Lambda})(n) \cdot \tau_n(\mathcal{M})(s).$$

The *fast adaptive uniformisation* method [17] builds on the result above and works as follows. Starting with the initial distribution at step $n = 0$, at each step n the FAU explores a subset $\hat{S}_n$ of the states S_n. The sets $\hat{S}_n$ are constructed by taking $\hat{S}_{n-1}$, adding the successor states $\{s' \in S \mid \exists s \in \hat{S}_{n-1}. \, \mathcal{R}(s, s') > 0\}$ of this set, and discarding states s with $\tau_n(\mathcal{M})(s) < \delta$, where δ is a fixed precision threshold. This process is repeated until step m, for instance so that $(1 - \sum_{n=0}^{m} \pi_t(\Phi^{\Lambda})(n)) < \varepsilon$. Thus, we add the probability from the birth process at each step, and stop the state space exploration as soon as the value obtained this way is at least $1 - \varepsilon$. In contrast to standard uniformisation, where the Fox and Glynn [8] algorithm can be utilised, we do not have an a priori step bound, but are still able to decide in a straightforward way when the infinite sum can be safely truncated.

Definition 4 (Fast Adaptive Uniformisation). *Let $\mathcal{M}$, $\mathbf{S} = (S_0, S_1, \ldots)$, and $\mathbf{\Lambda} = (\Lambda_0, \Lambda_1, \ldots)$ be as in Theorem 1. Further, consider*

- *a truncation point $m \in \mathbb{N}$,*
- *a finite sequence of subsets $\hat{\mathbf{S}} = (\hat{S}_0, \ldots, \hat{S}_m)$ with $\hat{S}_n \subseteq S_n$ denoting active states for $n \in \{1, \ldots, m\}$,*
- *probability matrices $\hat{P}_n(\mathcal{M}) \colon S \times S \to \mathbb{R}_{\geq 0}$ for $n \in \{0, \ldots, m\}$ where*

$$\hat{P}_n(\mathcal{M})(s, s') \stackrel{\text{def}}{=} \begin{cases} P_n(\mathcal{M})(s, s') & \text{if } s \in \hat{S}_n, \text{ and} \\ 0 & \text{otherwise,} \end{cases}$$

– discrete-time distributions $\hat{\tau}_n(\mathcal{M})\colon S \to \mathbb{R}_{\geq 0}$ *for $n \in \mathbb{N}$ with*

$$\hat{\tau}_n(\mathcal{M})(s') \stackrel{\text{def}}{=} \begin{cases} \pi(s') & \text{if } n = 0, \text{ and} \\ \sum_{s \in S} \hat{\tau}_{n-1}(\mathcal{M}) \cdot \hat{P}_{n-1}(s, s') & \text{otherwise.} \end{cases}$$

We define the fast adaptive uniformisation *(FAU)* value at time t for each $s \in S$ *as*

$$\hat{\pi}_t(\mathcal{M}, \hat{\mathbf{S}}, \mathbf{\Lambda})(s) \stackrel{\text{def}}{=} \sum_{n=0}^{m} \pi_t(\Phi^{\mathbf{\Lambda}})(n) \cdot \tau_n(\mathcal{M})(s).$$

Example 2 (Fast Adaptive Uniformisation). We sketch how one can perform FAU for the CTMC from Example 1 according to Definition 4 and Theorem 1: only for state $s = 100$ the initial distribution is positive, so we can use $S_0 \stackrel{\text{def}}{=} \{100\}$. Then, we use $S_n \stackrel{\text{def}}{=} \{\max\{0, 100 - n\}, \ldots, n\}$ and $\Lambda_n \stackrel{\text{def}}{=} (\lambda + \mu) \cdot n$. The corresponding birth process is sketched in Fig. 2. In Fig. 3, we depict S_n together with the relevant parts of the matrices P_n for $n = 0, 1, 2, \ldots, \infty$ (rounding-off the numbers). In addition, for $t = 0.1$ we provide the transient probabilities of being in the nth state of the birth process and the first n summands of $\pi_{0.1}(\mathcal{M})(100)$.

State $s = 0$ is absorbing, that is, once entered it cannot be left, and rates leading to a decrease in molecule count are higher than those leading to an increase. Thus, in the last line ("Step ∞") we see that, as n increases, the probability concentrates on the state $s = 0$. Thus, we can discard states with a high number of molecules from the reduced state sets $\hat{S}_n$, while retaining a sufficient amount of the total probability.

We now define instantaneous rewards, which can be used to express expected reward properties incurred at a given time. We annotate the models with state rewards.

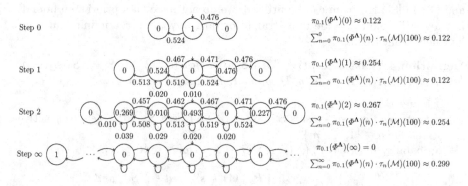

Fig. 3. Demonstration of Fast Adaptive Uniformisation

Definition 5 (State reward structure). *A* state reward structure *for a CTMC* $\mathcal{M} = (S, \pi, \mathcal{R})$ *is a function* $\mathbf{r}: S \to \mathbb{R}_{\geq 0}$.

Definition 6 (Instantaneous rewards). *Consider a CTMC* $\mathcal{M} = (S, \pi, \mathcal{R})$ *with state reward structure* $\mathbf{r}: S \to \mathbb{R}_{\geq 0}$ *and a time point* $t \in \mathbb{R}_{\geq 0}$. *The instantaneous reward is defined as*

$$\mathcal{I}_t(\mathcal{M}, \mathbf{r}) \stackrel{\text{def}}{=} \sum_{s \in S} \pi_t(\mathcal{M})(s) \cdot \mathbf{r}(s).$$

We show that instantaneous rewards can be easily accommodated within the FAU method, and we can approximate the expected mass value by terminating the state-space exploration using a criterion similar to the probability mass calculation in [17].

Definition 7 (Instantaneous reward approximation). *Let* $\mathcal{M}$, $\mathbf{S} = (S_0, S_1, \ldots)$, *and* $\mathbf{\Lambda} = (\Lambda_0, \Lambda_1, \ldots)$ *be as in Theorem 1 and let* $\hat{\mathbf{S}}$ *be as in Definition 4. Then we define*

$$\mathcal{I}_t(\mathcal{M}, \mathbf{r}, \hat{\mathbf{S}}, \mathbf{\Lambda}) \stackrel{\text{def}}{=} \sum_{s \in S} \hat{\pi}_t(\mathcal{M}, \mathbf{S}, \mathbf{\Lambda})(s) \cdot \mathbf{r}(s).$$

Corollary 1 (Error bounds for FAU). *Consider a CTMC* $\mathcal{M} = (S, \pi, \mathcal{R})$ *for which we have the uniformisation rates* $\mathbf{\Lambda}$ *(cf. Theorem 1) and a state reward structure* $\mathbf{r}$. *Consider* $m \in \mathbb{N}$, $\hat{\mathbf{S}}$, *and* $\hat{\tau}$ *as in Definition 4. Set* $\mathbf{\Lambda} \stackrel{\text{def}}{=} (\Lambda, \Lambda, \ldots)$. *Then if*

$$\max_{s \in S} \mathbf{r}(s) \cdot \left(1 - \sum_{n=0}^{m} \pi_t(\Phi^\Lambda)(n)\right) < \frac{\epsilon}{2} \text{ and } \max_{s \in S} \mathbf{r}(s) \cdot \left(1 - \sum_{s \in \hat{S}_m} \hat{\tau}_m(s)\right) < \frac{\epsilon}{2}$$

it follows that

$$\mathcal{I}_t(\mathcal{M}, \mathbf{r}) - \mathcal{I}_t(\mathcal{M}, \mathbf{r}, \hat{\mathbf{S}}, \mathbf{\Lambda}) < \epsilon.$$

Proof. Part of the expected reward value is lost due to the approximation of the infinite sum. This is accounted for by first inequality. By discarding states while exploring the state space, we lose further mass. This is accounted for by the second inequality. Adding up the maxima of the two errors, we can bound the error. □

Example 3 (FAU for Instantaneous Rewards). Consider the CTMC from Example 1 with a reward structure $\mathbf{r}$ assigning to each state s the number of molecules s. When using the transient probability values computed in Example 2, for the computation of the exact values we have $\mathcal{I}_{0.1}(\mathcal{M}, \mathbf{r}) = \sum_{s \in S} \hat{\pi}_t(\mathcal{M}, \mathbf{S}, \mathbf{\Lambda})(s) \cdot \mathbf{r}(s) \approx \ldots + 100 \cdot 0.299 + \ldots \approx 99.900$.

3   Cumulative Rewards

In this section, we extend the FAU method to reason about properties of the behaviour of a CTMC model cumulating the rewards *until* a given point of time. The correctness of the method is proved using the framework of CTPMs [17], where cumulative rewards were not considered.

For a given CTMC, we first extend its state space by adding *time-accumulating* states to remember how much time was spent in a specific state, and then, noting that the time-extended CTPM is not a CTMC, show how the expected reward computation can be approximated.

Definition 8 (Time-extended CTPM). *Given a CTMC* $M = (S, \pi, \mathcal{R})$, *the time-extended CTPM is defined as*

$$\mathrm{ext}(M) \stackrel{\mathrm{def}}{=} (S_{\mathrm{ext}}, \pi_{\mathrm{ext}}, \mathcal{R}_{\mathrm{ext}}), \quad \text{where}$$

- $S_{\mathrm{ext}} \stackrel{\mathrm{def}}{=} S \uplus S_{\mathrm{acc}}$, *where for each* $s \in S$ *we have exactly one corresponding time-accumulating* $s_{\mathrm{acc}} \in S_{\mathrm{acc}}$, *that is*, $S_{\mathrm{acc}} \stackrel{\mathrm{def}}{=} \{s_{\mathrm{acc}} \mid s \in S\}$,
- $\pi_{\mathrm{ext}}(s) \stackrel{\mathrm{def}}{=} \pi$ *for* $s \in S$ *and* $\pi_{\mathrm{ext}}(\cdot) \stackrel{\mathrm{def}}{=} 0$ *otherwise, and*
- *the transition matrix* $\mathcal{R}_{\mathrm{ext}} \colon S_{\mathrm{ext}} \times S_{\mathrm{ext}} \to \mathbb{R}_{\geq 0}$ *is defined such that, for* $s_{\mathrm{ext}}, s'_{\mathrm{ext}} \in S_{\mathrm{ext}}$, *we have*

$$\mathcal{R}_{\mathrm{ext}}(s_{\mathrm{ext}}, s'_{\mathrm{ext}}) \stackrel{\mathrm{def}}{=} \begin{cases} \mathcal{R}(s_{\mathrm{ext}}, s'_{\mathrm{ext}}) & \text{if } s_{\mathrm{ext}}, s'_{\mathrm{ext}} \in S \text{ and } s_{\mathrm{ext}} \neq s'_{\mathrm{ext}}, \\ 1 & \text{if } s_{\mathrm{ext}} = s \in S \text{ and } s'_{\mathrm{ext}} \in \{s_{\mathrm{ext}}, s_{\mathrm{acc}}\}, \\ 0 & \text{otherwise}. \end{cases}$$

We now use time-extended CTPMs to prove the central theorem of the paper. This is achieved by first approximating the residence time, and then the cumulative reward, by considering the reward per time unit of residing in a given state. We use the *mixed birth process probability* $\Psi^{\Lambda}(n) = \frac{1}{\Lambda_n} \cdot \sum_{i=n+1}^{\infty} \pi_t(\Phi^{\Lambda})(i)$, which denotes the probability that more than n state changes happen within time t in the birth process, divided by the nth uniformisation rate. This is used to collect the time *spent* in given state, as opposed to the probability $\pi_t(\Phi^{\Lambda})(i)$ to be in a state at a given point of time.

Theorem 2 (Residence time). *Consider a CTMC* $M = (S, \pi, \mathcal{R})$ *and let* $\rho_t(M) \colon S \to \mathbb{R}_{\geq 0}$ *be defined as*

$$\rho_t(M)(s) \stackrel{\mathrm{def}}{=} \int_0^t \pi_u(s) \, \mathrm{d}u$$

for $s \in S$. *Then we have*

$$\rho_t(M)(s) = \sum_{n=0}^{\infty} \Psi^{\Lambda}(n) \cdot \tau_n(M)(s),$$

for $s \in S$, τ *and* Λ *as in Theorem 1 and* $\Psi^{\Lambda}(n) \stackrel{\mathrm{def}}{=} \frac{1}{\Lambda_n} \cdot \sum_{i=n+1}^{\infty} \pi_t(\Phi^{\Lambda})(i)$.

Proof. Assume $\text{ext}(\mathcal{M}) = (S_{\text{ext}}, \pi_{\text{ext}}, \mathcal{R}_{\text{ext}})$. Then, by definition of the structure of the time-extended CTPM, we have for $s \in S$ that

$$\pi_t(\mathcal{M})(s) = \pi_t(\mathcal{M}_{\text{ext}})(s), \text{ and } \rho_t(\mathcal{M})(s) = \pi_t(\mathcal{M}_{\text{ext}})(s_{\text{acc}}).$$

By the structure of the time-extended CTPM, we have

$$\tau_0(\mathcal{M}_{\text{ext}})(s_{\text{acc}}) = 0,$$

$$\tau_{n+1}(\mathcal{M}_{\text{ext}})(s_{\text{acc}}) = \frac{1}{\Lambda_n} \cdot \tau_n(\mathcal{M}_{\text{ext}})(s) + \tau_n(\mathcal{M}_{\text{ext}})(s_{\text{acc}})$$

$$= \frac{1}{\Lambda_n} \cdot \tau_n(\mathcal{M})(s) + \tau_n(\mathcal{M}_{\text{ext}})(s_{\text{acc}})$$

and thus

$$\tau_n(\mathcal{M}_{\text{ext}})(s_{\text{acc}}) = \sum_{i=0}^{n-1} \frac{1}{\Lambda_i} \cdot \tau_i(\mathcal{M})(s).$$

From this and by Theorem 1 we have

$$\pi_t(\mathcal{M}_{\text{ext}})(s_{\text{acc}}) = \sum_{n=0}^{\infty} \pi_t(\Phi^{\Lambda})(n) \cdot \tau_n(\mathcal{M}_{\text{ext}})(s_{\text{acc}})$$

$$= \sum_{n=0}^{\infty} \pi_t(\Phi^{\Lambda})(n) \cdot \sum_{i=0}^{n-1} \frac{1}{\Lambda_i} \cdot \tau_i(\mathcal{M})(s) = \sum_{n=0}^{\infty} \sum_{i=0}^{n-1} \frac{1}{\Lambda_i} \cdot \tau_i(\mathcal{M})(s) \cdot \pi_t(\Phi^{\Lambda})(n)$$

$$= \sum_{i=0}^{\infty} \sum_{n=i+1}^{\infty} \frac{1}{\Lambda_i} \cdot \tau_i(\mathcal{M})(s) \cdot \pi_t(\Phi^{\Lambda})(n) = \sum_{i=0}^{\infty} \left(\frac{1}{\Lambda_i} \cdot \sum_{n=i+1}^{\infty} \pi_t(\Phi^{\Lambda})(n) \right) \cdot \tau_i(\mathcal{M})(s)$$

$$= \sum_{i=0}^{\infty} \Psi^{\Lambda}(i) \cdot \tau_i(\mathcal{M})(s) = \sum_{n=0}^{\infty} \Psi^{\Lambda}(n) \cdot \tau_n(\mathcal{M})(s). \qquad \square$$

The above theorem splits the behaviour of a CTMC into the birth process and a discrete-time process that determines the time spent in specific states of the CTMC. Thus, we can now apply the FAU to compute cumulative reward properties. To do this, we accumulate rewards for being in a state over time. Transition rewards $\mathbf{r}_t \colon S \times S \to \mathbb{R}_{\geq 0}$ are obtained for moving from one state to another.

We do not explicitly consider transition rewards for CTMCs here. However, given state rewards $\mathbf{r}$ and transition rewards $\mathbf{r}_t$, we can define cumulative reward rates $\mathbf{r}'$ as $\mathbf{r}'(s) \stackrel{\text{def}}{=} \mathbf{r}(s) + \sum_{s' \in S} \mathcal{R}(s, s') \cdot \mathbf{r}_t(s, s')$. For the properties under consideration, this new reward structure is equivalent to using transition rewards, as shown in [12, Equation 6].

We stress that time-extended CTPMs are used here only in the proof, and never constructed in our method.

Definition 9 (Cumulative rewards). *Consider a CTMC $\mathcal{M} = (S, \pi, \mathcal{R})$ with state reward structure $\mathbf{r} \colon S \to \mathbb{R}_{\geq 0}$ and a time duration $t \in \mathbb{R}_{\geq 0}$. The cumulative reward value is defined as*

$$\mathcal{C}_t(\mathcal{M}, \mathbf{r}) \stackrel{\text{def}}{=} \int_0^t \sum_{s \in S} \pi_u(\mathcal{M})(s) \cdot \mathbf{r}(s) \, du.$$

We now use the results from Theorem 2 to compute the reward obtained until a given point of time. The following corollary follows directly from Theorem 2 and can be used to approximate cumulative rewards.

Corollary 2 (Computing rewards). *For a CTMC $\mathcal{M} = (S, \pi, \mathcal{R})$ with state reward structure $\mathbf{r} \colon S \to \mathbb{R}_{\geq 0}$ and a time duration $t \in \mathbb{R}_{\geq 0}$, we have*

$$\mathcal{C}_t(\mathcal{M}, \mathbf{r}) = \sum_{s \in S} \rho_t(\mathcal{M})(s) \cdot \mathbf{r}(s) = \sum_{n=0}^{\infty} \sum_{s \in S} \Psi^{\Lambda}(n) \cdot \tau_n(\mathcal{M})(s) \cdot \mathbf{r}(s).$$

Definition 10 (Cumulative reward approximation). *Let $\mathcal{M}$, $\mathbf{S} = (S_0, S_1, \ldots)$, and $\Lambda = (\Lambda_0, \Lambda_1, \ldots)$ be as in Theorem 1 and let $m \in \mathbb{N}$ and $\hat{\mathbf{S}}$ be as in Definition 4. Then we define*

$$\mathcal{C}_t(\mathcal{M}, \mathbf{r}, t, \hat{\mathbf{S}}, \Lambda) \overset{\text{def}}{=} \sum_{n=0}^{m} \sum_{s \in S} \Psi^{\Lambda}(n) \cdot \hat{\tau}_n(\mathcal{M})(s) \cdot \mathbf{r}(s).$$

Calculating the cumulative rewards is of similar complexity to calculating the instantaneous rewards. After each step n, we multiply the probability in the discrete-time process by the corresponding cumulative reward and the value from Ψ, and then sum up the values obtained this way. The time overhead to compute the accumulated reward values is negligible. More importantly, it is not necessary to extend the state space, and hence the space complexity compared to FAU is not increased.

The corollary can be seen as a generalisation of a previous result [12, Theorem 1], where the computation of cumulative reward-based properties is also considered. However, the analysis in [12] relies on complete exploration of the state space and uses the special case $\Lambda = (\Lambda, \Lambda, \ldots)$, which reduces the birth process to a Poisson process.

The computation of the error and the bounds on the number of steps is more involved for cumulative rewards than for instantaneous rewards, as shown in Corollary 1. The precision which can be achieved depends on the structure of the CTMC and the state rewards. We often have models in which, for each state, the sum of the rates to new states (further away from initial states) is bounded. We remark that this does not restrict the rates back to previously visited states. For this class of models, which includes many realistic examples as shown below, we derive error bounds as follows.

Corollary 3 (Error bound cumulative). *Consider a CTMC $\mathcal{M} = (S, \pi, \mathcal{R})$ for which we have a fixed Λ so that for each $n \in \mathbb{N}$ and $s \in S_n$ (cf. Theorem 1) we have that $\sum_{s' \in S_{n+1}} \mathcal{R}(s, s') \leq \Lambda$. Further, consider a state reward structure $\mathbf{r}$ so that we have fixed constants $c, d \in \mathbb{R}_{\geq 0}$ where for all $n \in \mathbb{N}$ and $s \in S_n$ we have $\mathbf{r}(s) \leq c + dn$. Consider $m \in \mathbb{N}$, $\hat{\mathbf{S}}$, and $\hat{\tau}$ as in Definition 4. Set $\Lambda_b \overset{\text{def}}{=} (\Lambda, \Lambda, \ldots)$ and $B \overset{\text{def}}{=} t \cdot (c + d + d\Lambda t)$. If*

$$B - \sum_{n=0}^{m} (c + dn) \cdot \Psi^{\Lambda_b}(n) < \frac{\epsilon}{2} \quad \text{and} \quad B \cdot \left(1 - \sum_{s \in \hat{S}_m} \hat{\tau}_m(s) \right) < \frac{\epsilon}{2}$$

then we have

$$\mathcal{C}_t(\mathcal{M}, \mathbf{r}) - \mathcal{C}_t(\mathcal{M}, \mathbf{r}, \hat{\mathbf{S}}, \mathbf{\Lambda}) < \epsilon.$$

Proof. When applying the FAU method, the worst case of reward loss is when we have the birth process $\Phi^{\Lambda_b} = (\mathbb{N}, \pi, \mathcal{R})$ with reward structure $\mathbf{r}$, so that, for all $n \in \mathbb{N}$, we have $\mathbf{r}(n) \stackrel{\text{def}}{=} c + dn$. Denote the total accumulated reward until time t for this model by B. Thus, we lose no more reward than $B - \sum_{n=0}^{m}(c+dn) \cdot \Psi^{\Lambda_b}(n)$ in case we use $\hat{S}_n = S_n$ and perform m steps in the FAU.

To take into account the loss of rewards from using $\hat{S}_n \subseteq S_n$, we consider the total probability lost $(1 - \sum_{s \in \hat{S}_m} \hat{\tau}_m(s))$. In the worst case, this probability was already lost at the beginning. In this case, we lose up to $B \cdot (1 - \sum_{s \in \hat{S}_m} \hat{\tau}_m(s))$.

By adding up the two sources of error, we obtain the result.   □

If the rates or rewards are increasing more quickly, e.g., if we have a quadratic increase in the rewards, that is, $\mathbf{r}(s) \leq c + dn^2$ for $s \in \hat{S}_n$, the bounds on the error can be obtained using similar reasoning for a different value of B. Because of the simple structure of birth processes, it is possible to quickly approximate $\sum_{n=0}^{m}(c+dn) \cdot \Psi^{\Lambda}(n)$ to find the value m to terminate the approximation in the worst case.

Example 4 (FAU for Cumulative Rewards). We reconsider the CTMC from Example 1 for which we computed transient probabilities in Example 2. We are interested in the expected total number of changes to the number of molecules, and thus assign a reward of 1 to each state change. As discussed, we transform these transition rewards into a state reward structure $\mathbf{r}$. For instance, state $s = 100$ has two transitions with rates 10 and 11, both with reward of 1, so that the state reward here is $10 \cdot 1 + 11 \cdot 1 = 21$. We have $\Psi^{\Lambda}(0) \approx 0.042$, $\Psi^{\Lambda}(1) \approx 0.029$, $\Psi^{\Lambda}(2) \approx 0.017$. To compute cumulative rewards, according to Corollary 2 we can proceed as follows: after each step n of Example 2 and Fig. 3, for each $s \in S_n$ ($s \in \hat{S}_n$) we compute the product $\Psi^{\Lambda}(n) \cdot \tau_n(s) \cdot \mathbf{r}(s)$ and build the sum $v(n) = \sum_{s \in S} \Psi^{\Lambda}(n) \cdot \hat{\tau}_n(\mathcal{M})(s) \cdot \mathbf{r}(s)$ of these values. States $s \in S \setminus S_n$ need not be considered, because for those $\tau_n(s) = 0$. This value $v(n)$ is then added to the partially computed total cumulative reward. In the example, we have $v(0) \approx 0.042 \cdot 1 \cdot 100$, $v(1) \approx 0.029 \cdot (0.524 \cdot 99 + 0.476 \cdot 101)$, $v(2) = 0.017 \cdot (0.269 \cdot 98 + 0.010 \cdot 99 + 0.493 \cdot 100 + 0.227 \cdot 102)$. Finally, we obtain $\mathcal{C}_{0.1}(\mathcal{M}, \mathbf{r}) \approx 2.099$.

4   Case Studies and Implementation

We have integrated the fast adaptive uniformisation method in the probabilistic model checker PRISM [14] and intend to make it available in one of the next PRISM releases. Our implementation builds on top of the "explicit" engine, and is written in Java. Models can be input in the native language of PRISM or SBML. Properties are specified as non-nested continuous stochastic logic (CSL) [1] formulae extended with the reward operator [13], as either time-bounded until, or instantaneous or cumulative reward properties.

To show the practical applicability of our method, we apply it to three case studies. We terminate the state space exploration once we obtain $(1 - \sum_{n=0}^{m} \pi_t(\Phi^{\Lambda}(n))) < \varepsilon$ for an adequate ε, and discard states with probability of less than δ in the discrete-time process. Experiments were performed on a Linux computer with an Intel i7-3770 processor with 3.40GHz and 32GB of RAM.

Wherever possible, we have compared our results to the PRISM engine which performs best for that particular model. This includes comparison with the symbolic engines of PRISM ("mtbdd" and "hybrid"), which tended to perform worse than the "explicit" engine on our examples, likely due to loss of regularity. We note that symbolic engines cannot handle infinite-state models employed here, but the "explicit" engine is able to, provided that the reachable state space is finite. Conventional methods could perform better than FAU when the state space is sufficiently small, in view of the additional overhead necessary for FAU. We anticipate that the performance of PRISM is indicative of modern probabilistic model checkers, and therefore our conclusions are more generally applicable.

In this paper, we do not compare against simulation-based approaches, such as approximate probabilistic model checking available in PRISM (probability estimation and statistical hypothesis testing); while simulation has the advantage of not requiring the generator matrix to be constructed, and hence does not suffer from state-space explosion, it is sensitive to the size of time bounds and can only guarantee error bounds with a given confidence interval. FAU can provide guarantees for an arbitrary precision by controlling δ, although reducing δ will generally incur higher memory requirements. Investigating the trade-off between FAU and simulation-based techniques deserves further study.

Note that the performance figures given in the tables reflect the relative speeds of engines at the time of writing, and can change due to further optimisation.

4.1   Discrete Stochastic Model Test Suite

The Discrete Stochastic Model Test Suite [7] is a test suite of models encoded in the Systems Biology Markup Language (SBML), for which values of certain properties have been computed up to a given precision. It is aimed at stochastic simulator developers who can evaluate the accuracy of their tools against known results.

We used PRISM's SBML import functionality[1] to convert SBML to PRISM files. The models have infinitely many states, and so cannot be handled by existing PRISM engines (except "explicit", providing the reachable state space is finite). As the SBML import does not yet support the SBML feature of *events*, we were only able to analyse 35 out of the 39 test models. For this case study, we chose $\varepsilon = 10^{-9}$ and $\delta = 10^{-13}$ and apply analyses for a time bound of 50, which is the largest one for which results are included in the SBML models. The results for a selection of the models are given in Table 1. For each "Model", we give the "Time (s)" in seconds needed to perform the analysis, the maximal

---

[1] http://www.prismmodelchecker.org/manual/RunningPRISM/SupportForSBML

Table 1. Discrete Stochastic Model Test Suite Results

Model	Time (s)	States	Lost	Molecules	Reactions
001-01	1	321	8.7060E-09	60.6531	826.2856
001-03	2	347	9.6165E-09	0.6738	2,085.8503
001-04	1	163	8.1621E-09	6.0653	82.6286
001-05	22	2,999	1.1078E-08	6,065.3065	82,628.5611
001-06	1	321	8.7060E-09	60.6531	826.2856
001-07	51	161,617	2.0779E-08	60.6531	826.2856
001-08	2	321	8.7060E-09	60.6531	826.2856
001-18	1	277	8.5217E-09	77.8801	464.5184
001-19	1	321	8.7060E-09	60.6531	826.2856
002-01	2	44	5.9691E-09	9.9326	90.0674
002-02	2	151	7.7930E-09	99.3262	900.6738
002-03	2	107	8.3948E-09	49.6631	450.3369
002-04	19	1,377	1.0983E-08	9,932.6204	90,067.3790
002-05	1	151	7.7930E-09	99.3262	900.6738
002-06	5	36,255	9.9788E-09	99.3262	900.6738
002-07	1	151	7.7930E-09	99.3262	900.6738
002-08	2	44	5.9691E-09	9.9326	90.0674
003-01	2	48	5.3131E-09	28.5423	64.7560
003-02	2	156	7.5890E-09	144.9960	573.9888
003-05	2	48	5.3131E-09	35.7289	64.7560
004-01	1	124	8.0978E-09	24.9989	275.0011
004-02	1	173	9.0356E-09	25.0000	525.0000
004-03	4	773	3.5419E-08	25.0000	5,024.9999
ext. 001-01	66	161,617	2.0779E-08	60.6531	826.2856

number of "States" in memory, and the probability "Lost" through approximation. The column "Molecules" is an instantaneous reward property that returns the expected number of molecules of the first species of the model under consideration. The column "Reactions" is the expected number of reactions until time 50, which is a cumulative reward property. In the table, each row corresponds to two analyses; however, the computation time is the same for both since the same number of states had to be explored.

All analyses (with the exception of "ext. 001-01" not originally from the test suite, see below) took less than a minute. The results we obtain for "Molecules" agree with those provided by the test suite, for the number of decimal places given there (values for "Reactions" are not provided by the test suite). For the model "001-01", we attempted a naive approach to compute the number of reactions by adding a new species "Reactions", increasing the dimensionality. As can be seen from results given in the last row ("ext. 001-01") of Table 1, these performance figures were much worse than for our implementation. We remark that these figures are similar to those for the (unmodified) "001-07", in which also a species tracking a specific reaction is introduced.

4.2 DNA Strand Displacement

DNA strand displacement (DSD) [20] is a mechanism for performing computation with DNA molecules. A variety of logic circuits can be designed and implemented using DSD. Initial species of DNA are mixed together in a reaction tube, and then strand displacement reactions proceed autonomously, relying solely on hybridisation between complementary nucleotide sequences to perform computational steps. In this case study, we consider transducer gates modelled and analysed in [15, Section 2] (example `transducer_K=3.sm`). This model features the parameter N, which corresponds to the number of copies for initial species, and K, the number of transducers placed in series.

Table 2. DNA Strand Displacement Results

N	T	Fastest PRISM engine - explicit				FAU				
		Time (s)	States	Finished	Reactions	Time (s)	States	Lost	Finished	Reactions
1	10000	3	169	0.0593	15.3193	2	169	1.6882E-09	0.0593	15.3193
1	50000	1	169	0.9999	26.9997	33	169	1.7133E-09	0.9999	26.9997
1	100000	2	169	1.0000	27.0000	140	169	1.7892E-09	1.0000	27.0000
2	10000	1	5,748	0.0224	37.1224	7	5,299	1.3024E-08	0.0224	37.1224
2	50000	2	5,748	0.9999	51.2958	41	5,299	1.5028E-08	0.9999	51.2958
2	100000	2	5,748	1.0000	51.2963	155	5,299	1.5090E-08	1.0000	51.2963
3	10000	15	93,538	0.0138	59.7229	145	67,292	1.0059E-07	0.0138	59.7229
3	50000	52	93,538	0.9999	75.0530	179	67,292	1.0994E-07	0.9999	75.0530
3	100000	96	93,538	1.0000	75.0536	437	67,292	1.1002E-07	1.0000	75.0536
4	10000	268	970,539	0.0103	82.6250	835	514,414	5.6703E-07	0.0103	82.6250
4	50000	1,039	970,539	0.9998	98.6211	872	514,414	5.8001E-07	0.9998	98.6211
4	100000	1,976	970,539	1.0000	98.6218	1,209	514,414	5.8009E-07	1.0000	98.6218
5	10000	3,463	7,377,039	0.0085	105.6602	2,370	2,814,235	2.9759E-06	0.0085	105.6602
5	50000	-	-	-	-	2,416	2,814,235	2.9907E-06	0.9998	122.0891
5	100000	-	-	-	-	2,815	2,814,235	2.9907E-06	1.0000	122.0897
6	10000	-	-	-	-	5,453	12,163,811	1.3377E-05	0.0074	128.7586
6	50000	-	-	-	-	5,644	12,163,811	1.3393E-05	0.9998	145.4913
6	100000	-	-	-	-	5,960	12,163,811	1.3393E-05	1.0000	145.4920

We are interested in the probability that the computation is finished by time T ("Finished"), an instantaneous reward property, and the expected total number of reactions ("Reactions"), a cumulative reward property. We fix $K = 3$ and provide results for different N and T in Table 2. The state space of this case study is small enough to be compared against existing methods in PRISM. We included the results for the "explicit" engine because it was the fastest. In each row, the best performance in terms of state-space size or time is highlighted in boldface.

Note also that the FAU method is able to handle larger models than existing PRISM engines, and obtains better performance for larger model instances.

4.3   DNA Walkers

We consider models of *DNA walkers* [22], which can also be used to design logic circuits on the nanoscale. The main difference from DSD designs is that a DNA walker operates on a track of DNA strands (called anchorages) tethered to a surface, rather than in solution, and thus the model has to incorporate spatial information. An example of an XOR circuit is shown in Fig. 4. We assign True/False values to absorbing anchorages. The walker starts in the Initial position and can navigate down a series of junctions [21]. An enzyme cuts the anchorage when the walker is attached, allowing the walker to step onto the next anchorage. Depending on prior input, certain anchorages can be blocked, which in turn directs the walker at each junction.

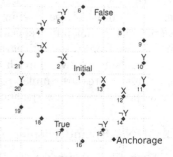

Fig. 4. Walker 'XOR' circuit. Adding the input X will block the anchorages labelled $\neg X$. Once the walker reaches True or False the computation ends.

Table 3. DNA Walkers

Model	Time (s)	States	Lost	Signal	Steps	Blocked (s)
xor(X, Y)	4126	228,803	1.9736E-02	0.6455	7.7696	606.2731
xor($\neg X, Y$)	4070	228,803	1.9736E-02	0.6455	7.7696	606.2731
xor($X, \neg Y$)	4684	239,680	2.2587E-02	0.5979	7.5610	659.3715
xor-($\neg X, \neg Y$)	4593	239,680	2.2587E-02	0.5979	7.5610	659.3715
xor-S-(X, Y)	2970	215,544	1.6719E-02	0.5374	8.8363	133.1672
xor-S-($\neg X, Y$)	3027	215,544	1.6719E-02	0.5374	8.8363	133.1672
xor-S-($X, \neg Y$)	3651	233,063	1.8775E-02	0.5473	8.4049	146.7377
xor-S-($\neg X, \neg Y$)	3630	233,063	1.8775E-02	0.5473	8.4049	146.7377
xor-large-(X, Y)	18382	443,584	5.1855E-02	0.5661	9.5020	577.2680
xor-large-($\neg X, Y$)	18142	443,584	5.1855E-02	0.5661	9.5020	577.2680
xor-large-($X, \neg Y$)	19418	455,685	5.2995E-02	0.5674	9.4983	567.3420
xor-large-($\neg X, \neg Y$)	18114	455,685	5.2995E-02	0.5674	9.4983	567.3420

A Markov chain model of the walker was developed [5] previously, and in this paper we apply model checking with FAU, using the parameter set $\varepsilon = 10^{-6}$ and $\delta = 10^{-8}$. We analyse three XOR-circuits, from Fig. 4 and two variants, summarising the results in Table 3. We model check the expected number of steps (column "Steps") and the probability of walkers reaching the desired anchorage (column "Signal") by time $T = 200$ min. The walker occasionally steps over blockades or between tracks, which may cause it to reach the wrong answer. Determining the size of the reachable state space appears to be a hard problem, not unlike determining the number self-avoiding walks on a lattice. We estimate the size to be around $1 \cdot 10^7$ and $9 \cdot 10^8$ reachable states for the normal and large tracks, respectively. This state space is too large to construct the models symbolically and compare against other PRISM engines.

The unmodified track, shown in Fig. 4, is "xor", and the suffix "-S" indicates that only one blocker is used instead of two consecutive ones, whereas suffix "-large" indicates a track with more anchorages. The expected number of steps correlates well with the track layout: when fewer anchorage are blocked ("-S"), the walker takes more steps on average. A larger track also results in more steps taken on average. Because the track has a point-symmetry, the results for inputs X, Y and $\neg X, Y$ are the same, as well as for inputs $\neg X, \neg Y$ and $X, \neg Y$. Occasionally, the blockade mechanism fails to block an anchorage. Column "Blocked" shows how much time the walker spends on anchorages that were supposed to be blocked, which is in line with expectations.

5   Conclusion

In this paper, we have extended fast adaptive uniformisation so that it can also be applied to cumulative reward properties. Cumulative measures allow one to express many important quantitative properties, such as the expected number of times a certain reaction happens and the average percentage of time the system spends in a given state. Our method does not introduce a significant overhead to the analysis, and in particular does not require the explicit construction of the extended state space of the underlying continuous-time propagation model. In contrast to simulation-based approaches, we can compute guaranteed error bounds for properties, as opposed to ensuring a given confidence interval. We have applied it to several case studies, obtaining superior performance in virtually all cases compared to existing methods.

Acknowledgements. The authors are supported in by the ERC Advanced Grant VERIWARE and a Microsoft Research PhD Studentship (FD). We would like to thank Taolue Chen and Andrew Phillips for useful discussion, and Chris Thachuk for his help in preparing the DSD models.

References

1. Aziz, A., Sanwal, K., Singhal, V., Brayton, R.K.: Model-checking continuous-time Markov chains. ACM TCS 1(1), 162–170 (2000)
2. Baier, C., Haverkort, B., Hermanns, H., Katoen, J.P.: Performance evaluation and model checking join forces. Commun. ACM 53(9), 76–85 (2010)
3. Baier, C., Haverkort, B.R., Hermanns, H., Katoen, J.P.: Model-checking algorithms for continuous-time Markov chains. IEEE TSE 29(6), 524–541 (2003)
4. Clark, G., Courtney, T., Daly, D., Deavours, D., Derisavi, S., Doyle, J.M., Sanders, W.H., Webster, P.: The Möbius modeling tool. In: PNPM, pp. 241–250 (2001)
5. Dannenberg, F., Kwiatkowska, M., Thachuk, C., Turberfield, A.J.: DNA walker circuits: Computational potential, design, and verification. In: Soloveichik, D., Yurke, B. (eds.) DNA 2013. LNCS, vol. 8141, pp. 31–45. Springer, Heidelberg (2013)
6. Didier, F., Henzinger, T.A., Mateescu, M., Wolf, V.: SABRE: A tool for stochastic analysis of biochemical reaction networks. In: QEST, pp. 193–194 (2010)
7. Evans, T.W., Gillespie, C.S., Wilkinson, D.J.: The SBML discrete stochastic models test suite. Bioinformatics 24(2), 285–286 (2008)
8. Fox, B.L., Glynn, P.W.: Computing Poisson probabilities. Comm. ACM 31(4), 440–445 (1988)
9. Heath, J., Kwiatkowska, M., Norman, G., Parker, D., Tymchyshyn, O.: Probabilistic model checking of complex biological pathways. Theoretical Computer Science 319(3), 239–257 (2008)
10. Jensen, A.: Markoff chains as an aid in the study of Markoff processes. Skand. Aktuarietidskr. 36, 87–91 (1953)
11. Katoen, J.P., Zapreev, I.S., Hahn, E.M., Hermanns, H., Jansen, D.N.: The ins and outs of the probabilistic model checker MRMC. PEVA 68(2), 90–104 (2011)
12. Kwiatkowska, M., Norman, G., Pacheco, A.: Model checking expected time and expected reward formulae with random time bounds. CMA 51, 305–316 (2006)
13. Kwiatkowska, M., Norman, G., Parker, D.: Stochastic model checking. In: Bernardo, M., Hillston, J. (eds.) SFM 2007. LNCS, vol. 4486, pp. 220–270. Springer, Heidelberg (2007)
14. Kwiatkowska, M., Norman, G., Parker, D.: PRISM 4.0: Verification of probabilistic real-time systems. In: Gopalakrishnan, G., Qadeer, S. (eds.) CAV 2011. LNCS, vol. 6806, pp. 585–591. Springer, Heidelberg (2011)
15. Lakin, M., Parker, D., Cardelli, L., Kwiatkowska, M., Phillips, A.: Design and analysis of DNA strand displacement devices using probabilistic model checking. Journal of the Royal Society Interface 9(72), 1470–1485 (2012)
16. Mateescu, M., Wolf, V., Didier, F., Henzinger, T.A.: Fast adaptive uniformisation of the chemical master equation. IET Syst. Biol. 4(6), 441–452 (2010)

17. Mateescu, M.: Propagation Models for Biochemical Reaction Networks. Ph.D. thesis, EPFL (2011)
18. van Moorsel, A.P.A., Sanders, W.H.: Adaptive uniformization. ORSA Communications in Statistics: Stochastic Models 10(3), 619–648 (1994)
19. Schwarick, M., Heiner, M., Rohr, C.: MARCIE - model checking and reachability analysis done efficiently. In: QEST, pp. 91–100 (2011)
20. Seelig, G., Soloveichik, D., Zhang, D.Y., Winfree, E.: Enzyme-free nucleic acid logic circuits. Science 314(5805), 1585–1588 (2006)
21. Wickham, S.F.J., Bath, J., Katsuda, Y., Endo, M., Hidaka, K., Sugiyama, H., Turberfield, A.J.: A DNA-based molecular motor that can navigate a network of tracks. Nature Nanotechnology 7(3), 169–173 (2012)
22. Wickham, S.F.J., Endo, M., Katsuda, Y., Hidaka, K., Bath, J., Sugiyama, H., Turberfield, A.J.: Direct observation of stepwise movement of a synthetic molecular transporter. Nature Nanotechnology 6(3), 166–169 (2011)

Linking Discrete and Stochastic Models: The Chemical Master Equation as a Bridge between Process Hitting and Proper Generalized Decomposition

Courtney Chancellor[1,2], Amine Ammar[4], Francisco Chinesta[2],
Morgan Magnin[1,3], and Olivier Roux[1]

[1] L'UNAM Université, École Centrale de Nantes, IRCCyN UMR CNRS 6597, France
[2] L'UNAM Université, École Centrale de Nantes, GeM UMR CNRS 6183, France
[3] National Institute of Informatics, Tokyo 101-8430, Japan
[4] Arts et Metiers ParisTech. Angers, France

Abstract. Modeling frameworks bring structure and analysis tools to large and non-intuitive systems but come with certain inherent assumptions and limitations, sometimes to an inhibitive extent. By building bridges in existing models, we can exploit the advantages of each, widening the range of analysis possible for larger, more detailed models of gene regulatory networks. In this paper, we create just such a link between Process Hitting [6,7,8], a recently introduced discrete framework, and the Chemical Master Equation in such a way that allows the application of powerful numerical techniques, namely Proper Generalized Decomposition [1,2,3], to overcome the curse of dimensionality. With these tools in hand, one can exploit the formal analysis of discrete models without sacrificing the ability to obtain a full space state solution, widening the scope of analysis and interpretation possible. As a demonstration of the utility of this methodology, we have applied it here to the p53-mdm2 network [4,5], a widely studied biological regulatory network.

1 Introduction

Our ability to gather data in the context of gene regulatory networks has skyrocketed in the past decades: technology has given scientists an unprecedented ability to take in large amounts of raw data on the genome and genomic expression. The scale of this newly available data is massive and uninterpretable without applying formal analysis. Computational tools are invaluable in this respect: to put an otherwise incomprehensible data set into the context of a modeling framework allows scientists to understand the behaviors of a system and make predictions thereof. Once one imposes a model, however, one is confined to the inherent assumptions and limitations of that method. The resulting compromise is why there exist many varieties of structures, each exploiting certain advantages while accepting certain limitations. There is a great interest in systems biology in how to best navigate this choice or, better yet, how to avoid it

A. Gupta and T.A. Henzinger (Eds.): CMSB 2013, LNBI 8130, pp. 50–63, 2013.

altogether: by building bridges in existing models, we can utilize the best aspects of each, widening the range of analysis possible for larger, more realistic models of gene regulatory networks.

In this paper, we begin what will become a body of work dedicated to this goal: approaching biological systems in inventive or innovative ways to maximize the accuracy and utility of modeling structures. We start with a recently introduced modeling structure called Process Hitting [6,7,8], a discrete model in which one tracks qualitative shifts of the system. Like other discrete frameworks, it possesses certain advantages in model fitting and model checking. In addition, Process Hitting offers straightforward ways of incorporating temporal and stochastic properties and, moreover, can be thought to contain the Generalized Logical Network in that it has been shown that the interaction graph and discrete Thomas parameters can be derived from the Process Hitting model. Simulation is required to obtain a full description of the local behaviors of the system, a computationally expensive and sometimes inhibitive aspect. Here, we propose a method to solve the system, obtaining a solution for all time, by translating Process Hitting to an equivalent differential equation form, namely the Chemical Master Equation (or CME), with the intention of applying a novel numerical technique,the Proper Generalized Decomposition (or PGD)[1,2,3], to overcome problems of dimensionality. Since this paper makes use of state-of-the art methods which readers may not yet be familiar with, we will briefly introduce Process Hitting, Proper Generalized Decomposition and the Chemical Master Equation in Section 2, outlining advantages, disadvantages and important features of each. In this section, we will also introduce the p53-mdm2 network [4,5,11], a biologically relevant model for which an extensive body of work exists and to which we will apply our methodology. Once these structures have been defined, the translation of Process Hitting to a special, discretized form of the Chemical Master Equation is outlined in Section 3. Application to the p53-mdm2 model and the analysis of results can be found in Section 4, along with a clearly defined path for future work. When fully realized, our proposed methodology has the potential to allow scientists to construct models without explicit knowledge of reaction kinetics, use a variety of analysis tools which exploit the discrete, formal structure of Process Hitting and, finally, efficiently solve this complete system for in-depth analysis.

2   Frameworks and Methods Used

The foundation of this paper is in patching together novel techniques to form a better, more complete modeling structure. While most readers may be familiar with discrete and stochastic modeling, for example, it may not be the case for Process Hitting or the Chemical Master Equation in particular. Here, we have included a brief description of all of the frameworks and methods used: although no one section gives neither a full nor formalized description, we hope to give an intuitive notion of each, as well as an understanding of their significance. We begin by introducing the p53-mdm2 system, followed by the relevant modeling frameworks, and finally by the numerical method of Proper Generalized Decomposition.

2.1   Introductory Example: The p53-mdm2 Network

We have chosen to apply our methodology to a proven and biologically relevant network, the p53-mdm2 regulatory system [4,5,11]. The protein p53 is a transcription factor for a variety of systems, particularly those relating to arrested cell growth, DNA repair and cell death. When a cell incurs DNA damage, p53 concentration levels rise, inducing cell repair or, if the damage persists, apoptosis, thus preventing the spread of genetically unstable cells. p53 is strictly regulated by ubiquitin ligase mdm2 via a negative circuit. Since this system is of such interest in cancer research, there exists an extensive body of experimental and modeling work dedicated to its study.

In the presence of DNA damage, high concentrations of p53 can promote cellular repair but can be lethal with long-term exposure. As a counter balance, these same high concentration levels of p53 up-regulate the transcription of gene mdm2, increasing the concentration of its protein in the cytoplasm of the cell. When this concentration is high enough, mdm2 moves into the nucleus, where it blocks further transcription of p53 and facilitates the degradation of existing proteins, preventing the cell from going into apoptosis. Since the ability to express p53 at sufficient levels is required for cellular repair, this inhibition cannot always prevail: byproducts of cellular damage help the degradation of mdm2 in the nucleus, and translocation itself is inhibited by p53 at a level much lower than the activation of mdm2. These interactions are summarized as a directed graph in Figure 1.

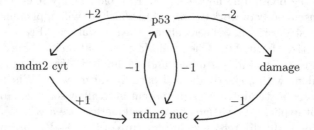

Fig. 1. Representation of p53-mdm2 network as a directed graph: interactions are summarized as activation or inhibition (+ or −), including discretized concentration levels ([0, 1, 2...]) defined by threshold values. Since mdm2 behaves differently in regards to the system depending on its location in the cell, cytoplasmic and nucleic mdm2 are represented separately.

2.2   A Brief Introduction to Process Hitting

Given the biological description of the system, we wish to construct a model for further analysis and study. Process Hitting is, more generally, a framework for modeling concurrent processes but is particularly apt for biological regulatory networks in that it conceals the kinetic mechanisms by which the system moves, describing instead the qualitative changes that may occur. In this fashion, it is possible to model a system with observational data and only partial knowledge of its inner workings. Although its structure seems simplistic, Process Hitting can capture complex dynamics and easily lends itself to model-checking, by which one can determine whether or not certain desired features are preserved by the model. In addition, temporal and stochastic properties can be naturally integrated into the Process Hitting structure. A full description of this framework including its implementation, can be found in [6,7,8].

In Process Hitting, here on referred to as PH, all interacting species —enzymes, genes, proteins, etc. —are abstracted as *sorts*. The sorts of the p53-mdm2 system are cellular damage, nuclear mdm2, cytoplasmic mdm2 and p53, which are given the labels Dam, Mn, Mc, and p53, respectively. These sorts are then subdivided into *processes*, which could represent concentration levels, spatial configuration, or any other form which has a distinct qualitative impact on the system. Dam, for example, has two processes, Dam 0 and Dam 1, the absence and the presence of cellular damage. Conversely, p53 contains three processes which represent the relevant concentration ranges of p53 in the cell. Processes interact with one another via *actions*, in which processes *hit* one another to create a *bounce* to some new level of the same sort, wherein we find the namesake of "Process Hitting". These actions move the state space one level at a time and are, therefore, asynchronous. For gene regulatory networks, processes are often abstractions of relevant concentration ranges, discretized domains of real numbers, and actions represent varying action and inhibition reactions. For instance, we know that, when at a very high level, p53 up-regulates the level of cytoplasmic mdm2. In PH action terms, this is demonstrated as

$$p53\_2 \rightarrow Mn\_0 \;\mathord{\uparrow}\; 1$$

which reads "p53_2 hits Mn_0 to bounce to Mn_1", as demonstrated in Figure 2.

In this structure, the absence of an activator is equivalent to inhibition and vice versa. Therefore, whenever p53 is below its activating threshold, it is effectively an inhibitor:

$$p53\_1 \rightarrow Mn\_1 \;\mathord{\uparrow}\; 0$$

$$p53\_0 \rightarrow Mn\_1 \;\mathord{\uparrow}\; 0$$

Whether or not this is a biologically valid assumption is subject to the modeler, who may remove any unwanted hits as suits the system in question. But what if a process can be influenced by more than one factor, as is the case with nucleic mdm2 which is influenced negatively by p53 and cellular damage but positively by cytoplasmic mdm2? PH does not enforce separability in its framework, that is, that effects be additive: the impact of two activators, for example, is not necessarily equivalent to the addition of their individual impacts. It is instead up to the modeler to define the results of combined influences. In order to remain consistent with the formalism of PH, we must introduce *cooperative sorts*, a representation of the collective influence of species. In our network, this leads to the creation of collective sort p53McDam which interacts with the normal sort Mn. In addition to defining the actions each combination of p53McDam will perform on Mn, we must also add actions between the normal sorts p53,

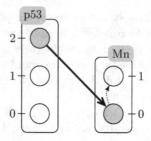

Fig. 2. Example of a Process Hitting action. Here, we show sorts p53 and Mn, boxes which contain processes. If the system is such that process p53_2 and Mn_0 are actice at the same time; p53_2 will have the chance to "hit" Mn_0, indicated by the solid arrow, which will then "bounce" to Mn_1, indicated by the light, dashed arrow.

Mc and Dam, and p53McDam which update the cooperative sort such that its state instantaneously reflects the current state of the system. Note that we do not need to indicate how an action is performed (the kinetics) nor dictate any particular behaviors as is the case in other discrete frameworks. Although we may abstractly incorporate kinetic concepts such as threshold values, as in the example of p53 via the discretization of concentration levels ($[0, 1, 2]$), we do not need to quantitatively define these thresholds. A list of the actions and the diagram of the Process Hitting of the p53 system can be found in the appendix.

2.3   The Chemical Master Equation

The Chemical Master Equation, or CME [1], is considered a canonical stochastic model in biological regulatory networks in which the modeler assumes full kinetic knowledge of the system. The vector $z = [z_i] \in \mathbb{Z}_{\geq 0}^{N_{sp}}$, $i = 1 \cdots N_{sp}$ contains the discretely valued counts of molecules for a given species i. Rather than tracking the state of the system as it varies in time according to an underlying deterministic process, we think about the system in terms of the probability $P(z, t|z_0, t_0)$ of existing at a certain state z at any given time t given some initial condition. From any state, reactions occur which move the system to a new configuration according to the reaction's known stoichiometry. This is a stark contrast to the Process Hitting framework, in which observational data and qualitative knowledge was sufficient. The Chemical Master Equation describes the evolution of the probability of the system existing at any given state z by considering the propensities, a, of all reactions r_j which leave z and those which enter z,

$$\frac{\partial P(z, t|z_0, t_0)}{\partial t} \equiv \sum_j^{N_{sp}} [a_j(z - v_j)P(z - v_j, t|z_0, t_0) - a_j(z)P(z, t|z_0, t_0)]$$

To simplify this representation, we may aggregate these terms to express the CME in matrix form, $\frac{\partial P}{\partial t} = \mathcal{A}P$ where $\mathcal{A}$ is known as the *connectivity matrix* of the regulatory network. $\mathcal{A}$ is sparse, with nonzero elements $A_{i,j}$ where a reaction links states i and j:

$$A_{ij} = \begin{cases} -\sum a_k(z), & i = j \text{ (reactions leaving state i)} \\ \sum a_k(z), & i \neq j \text{ (reactions moving from state i to state j)} \end{cases}$$

While the CME is a natural and rich description of the physical system in a biological regulatory network, it demonstrates what has come to be known as the curse of dimensionality, growing exponentially with the number of species, N_{sp}: that is, if, for each species, we limit the range of possible values to N, the total state space of the system will be $(N_{sp})^N$. Since biological regulation can have many subtly interacting factors, this is clearly an impasse in the application of the CME to gene regulatory networks, which can become impossible even for simplified models. Although simulation techniques are common, the structure of the CME permits the application of state-of-the-art numerical methods. To this end, PGD has already been shown to effectively and flexibly solve the Chemical Master Equation.

2.4   Proper Generalized Decomposition

Proper Generalized Decomposition [2,3] is an emerging numerical tool in the field of mechanical engineering, though it has been applied to a wide range of problems, including the CME [1]. The foundation of this method is to assume that the target, in this case, the probability, can be written as a sum of a product of separable functions.

$$P(z,t) \cong \sum_{j=1}^{M} F_1^j(z_1) \cdot F_2^j(z_2) \cdot \ldots \cdot F_N^j(z_N) \cdot F_t(t)$$

This is not an entirely novel idea, but its recent applications have proven promising in dimension reduction problems. Although the accuracy increases with every addition, only a limited number, M, of functions are needed to capture the behavior of the system. Note that, with each function of size N (that is, the state space of each variable is limited to size N), with N_{sp} functions, the resulting dimensionality is the M sum of N_{sp} functions of size N, or $M(N \times N_{sp})$ in contrast to the original N_{sp}^N. The inclusion of a time as a separated function means that the solution is not incremental but complete for all time.

PGD is performed iteratively, searching for each product of separable functions which will minimize the residual of the running sum. These functions are colloquially called "modes", although there is no underlying notion that they represent the greatest source of variance, as is the case with Principal Component Analysis [10] (PCA). At each step, one is searching for a single one-dimensional function, in this case a $N \times 1$ vector, with the remainder of the state space known. These sets of one dimensional functions are found until their sum, the resulting approximation, meets some stopping criterion. Since all operations can be performed by canonical techniques and are highly parallelizable, iterations are generally fast and computationally inexpensive. In addition, the form of PGD offers a natural way of handling unknown parameters by incorporating them as additional state space dimensions without changing the original algorithm. This, in particular, is a very desirable characteristic for application to biological regulatory networks since many parameters are often either unknown or come with some degree of uncertainty.

3   Translation of PH to CME

It is clear that Process Hitting contains certain desirable properties as a modeling framework for gene regulatory networks, such as the ability to construct a model from partial knowledge of the system, and that PGD is a fast, efficient way to numerically solve differential equations by breaking high dimensional systems into a search for one-dimensional modes. If we are able to move Process Hitting from its current format to that more like the CME, we have the potential to apply PGD and solve the system, not only for one desired configuration, but for the probability of existing at any configuration at any time given the initial state. An intuitive indication that such a link exists can be found in the simulation method used in Process Hitting: Gillespie's Next Reaction Algorithm.[13] This simulation technique allows for concurrent and competing processes and was developed by Gillespie as an *exact* simulation for the CME. That is, the Chemical Master Equation with any set of reactions can be simulated precisely by the Next Reaction Algorithm: if Process Hitting uses the Next Reaction Algorithm, there must exist some corresponding Chemical Master Equation that defines exactly the same behavior derived from the PH qualitative description of the system which does not require the addition of kinetic knowledge.

The key to translating the Process Hitting to its Chemical Master Equation structure is to re-imagine the kinetic reactions found in CME to be much more abstract, that is, not true physical reactions, but some collection of events which move the system from one state to another. These abstracted reactions cannot be interpreted in any greater detail than a PH action, though it is clear that, in reality, the system is being driven by a series of physical and chemical reactions. The hiding of kinetic processes was considered an advantage of PH, one which is conserved in translation to the CME. It is important that we concretize how one can derive these abstracted reactions, which we refer to as "faux-reactions", from any Process Hitting action. To illustrate this process, we will take a particular example

$$\text{p53McDam } 0 \to \text{Mn } 0 \, \uparrow \, 1 @ \left(4 h^{-1} \right)$$

That is, that when p53 AND cytoplasmic mdm2 AND cellular damage are at their lowest levels, the amount of nuclear mdm2 increases at the propensity of 4 per hour. The combination of those factors activate the uptake of mdm2 to the nucleus, bringing the system from state set $\{0, 0, 0, 0\}$ to state set $\{0, 0, 0, 1\}$ where it will then be susceptible to other actions. It is easy to write this in a more readable form

$$0p53 + 0Dam + 0Mc + 0Mn \xrightarrow{4h^{-1}} 0p53 + 0Dam + 0Mc + 1Mn$$

Granted, we have intentionally mimicked the structure of a stoichiometry equation, though it does not follow stoichiometric rules. Rather, this syntax merely states that there is some process (indicated by an arrow) which occurs at propensity $4h^{-1}$ that brings the system from state set $\{0,0,0,0\}$ to state set $\{0,0,0,1\}$. Any Process Hitting action can be written in these terms: for actions that do not dictate the full state set as our example has, undetermined variables give rise to multiple unique reactions with identical rates. So p53 0 → Dam 0 ↾ 1, for example, will result in four individual reactions wherein Mn and Mc take on values 0 and 1.

At this point, we must stretch our understanding of the Chemical Master Equation syntax to accommodate these new faux-reactions, though not by much. Recall that in the CME the system is described in terms of molecule counts, or whole numbers. In the same way that one discretizes the real number line in Generalized Logical Networks [12], so can we discretize the whole number line: $1p53$ does not signify one molecule of p53, but a quantity of p53 within some range. Thus, the faux-reaction equation effectively abstracts the mechanism by which the system moves as well as its actual quantitative contribution. The resulting set of equations can be but in the functional form of a discretized Chemical Master Equation which retains the syntax of its original counterpart, a sum of the effects of each reaction r_j which brings species to or away from a certain state, z at a given propensity $a(z)$.

4   Application and Results

We begin by constructing the Process Hitting for the p53 system as described in Section 1.1. This includes the creation of a cooperative sort p53McDam and all of the actions needed to update such that the state of the cooperative sort reflects instantaneously the state of the system. A full list of actions, as well as the PH diagram, can be found in the appendix. Reaction propensities were taken from [4,5]. At this early stage, we can do model-checking to find out if the fundamental structure of the graph supports certain desirable dynamics, exploiting the formal structure of the discrete model. Process Hitting has been implemented in a freely available software called PINT [9], which, among other things, can search for steady states, perform reachability analysis, and run simulation. In addition, PINT has the ability to import and export data from a number of other systems biology syntaxes, making it a flexible platform for newcomers to Process Hitting. From here, we can take the approach outlined in Section 3 to translate all actions into their corresponding faux-reactions.

In this paper, we begin by considering the most basic, intuitive representation of the Chemical Master Equation, that is, the case where the state space is represented by a one-dimensional vector of enumerate states, where the probability is a function of the state of the system and time, $P(Z, T)$. Here, Z is a vector of all possible states that can be occupied by the system; so, for example,

$\{Mc, Dam, p53, Mn\} = \{0, 0, 0, 0\}$ is state 1, $\{1, 0, 0, 0\}$ is state 2, and so on. This enumeration of states is acceptable for this particular problem but unsupportable as the number of species increases, making it impractical for most gene regulatory networks. However, by using this construction, we retain the potential to capture important emergent properties, to be addressed later on in the following subsection. In addition, the method by which we construct the discretized Chemical Master Equation is very straightforward in the case of a one dimensional state space. We can begin by considering the a matrix form of the CME as described in Section 2.3, $\frac{\partial P(z,t|z^0,t^0)}{\partial t} = \mathcal{A}P(z, t|z^0, t^0)$. As an example, the reaction

$$0p53 + 0Dam + 0Mc + 0Mn \xrightarrow{4h^{-1}} 0p53 + 0Dam + 0Mc + 1Mn$$

results in a nonzero element at $A_{\{0,0,0,0\},\{0,0,0,0\}}$ and $A_{\{0,0,0,1\},\{0,0,0,0\}}$, or, using arbitrary numbering for states, $A_{1,1}$ and $A_{7,1}$. The first element represents the system leaving state $1 \equiv \{0, 0, 0, 0\}$ and the second the system moving to state $7 \equiv \{0, 0, 0, 1\}$ from state $1 \equiv \{0, 0, 0, 0\}$. Once the connectivity matrix is constructed, we can apply PGD to find the decomposition of $P(Z, t)$ and solution of the Chemical Master Equation for all time.

4.1   Results of One-Dimensional Problem

In Figure 3, we have taken a snapshot in time to compare results of Process Hitting and our discretized version of the Chemical Master Equation. We begin the system in state $3 \equiv \{0, 1, 0, 0\}$, which represents the presence of damage ($Dam = 1$) without any other element ($p53 = Mn = Mc = 0$). Since the system is small, we can solve the resulting CME using implicit finite differences as a gold standard to which we can compare the PGD results. PH simulations were executed using PINT, averaging over 1000 runs. Although this particular graph only relates to a single instance in time, the PGD solution obtained is for all time, as can be seen in Figure 4. We stopped the iterative algorithm after reaching a precision of 10^{-3}. Although there is clearly good agreement in this solution, even better approximations can be made by continuing the iterative scheme, though only a limited number of modes are needed to obtain the basic information of the system.

4.2   Increased Depth of Analysis

By connecting two very different modeling frameworks, we are able to exploit the advantages of each and, potentially, fill in gaps of analysis. From Process Hitting, we can quickly and efficiently analyze the global behaviors of the system, using tools such as steady state or reachability analysis. The application of formal methods allows a modeler to ask fundamental questions of the system. However, there exist blind spots in this modeling type like any other. For example, the p53-mdm2 system as given in the appendix has no focal steady states and, although we can use model-checking to obtain further information, a complete

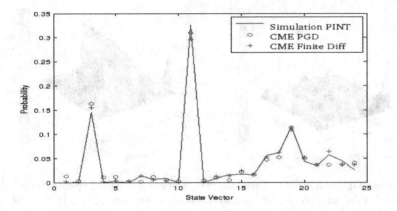

Fig. 3. Comparison of PH simulation with 1000 trials and translated CME, solved via Finite Differences and PGD

response requires simulation techniques. While manual investigation is possible in this small network, this ceases to be a viable strategy as the number of species increases. Moving to the probabilistic syntax of the Chemical Master Equation opens the potential to solve, rather than simulate, the complete system. In the p53-mdm2 network, we observe states whose probabilities do not change after a certain measure of time. This represents a basin of attraction, or a limit cycle, not yet capturable in PH analysis and predicted in [4,5]. In linear algebra terms, this is the null space of $\mathcal{A}$, all state vectors Z which satisfy $\mathcal{A}x = 0$. The PGD results for this particular problem can be seen on the left of Figure 4. Thus, even though the experimenter knows that the system does not settle into a particular steady state, he can know what states will be most or least prominent and the interactions (the faux-reactions) which connect them. Since the PGD can be implemented quickly, it facilitates experimentation with the model: one can easily test the results of adding or removing actions, working up to a model which demonstrates the correct dynamic behavior. Knowing that our system contains a limit cycle, we can induce a steady state by removing a single action in the Process Hitting. In this particular example, we have removed the action $Mn\_0 \rightarrow p53\_1 \upharpoonright 2$. That is, the absence of mdm2 in the nucleus is insufficient to bring p53 to its maximum level. As we can see on the right hand side of Figure 4, we do arrive at the desired focal steady state, which can be confirmed using PH analysis.

4.3 Separability, Emergence and Multidimensional Problems

We can effectively solve a PH system in the context of its one-dimensional state vector form, $P(Z, T)$, but the true goal of this work is to consider the multidimensional problem in which the probability is a function of all species of the system, $P(Mc, Dam, p53, Mn, T)$. The size of neither the solution nor the prob-

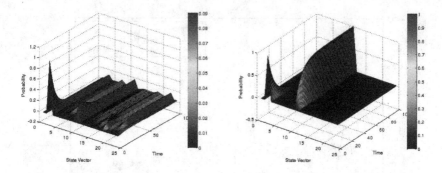

Fig. 4. The PGD solution of two systems. The original p53-mdm2 system as described in Section 2.1 (as seen on the left) contains a basin of attraction, or a limit cycle. The probabilities of the states in this cycle become constant as all trajectories from the initial condition enter the basin. By removing a single link in this cycle, we can induce a focus, or a steady state (as seen on the right). In this particular example, we have removed the action $Mn\_0 \to p53\_1$ Γ 2. That is, the absence of mdm2 in the nucleus is insufficient to bring p53 to its maximum level.

lem has changed, but the structure of the algorithm avoids the same problems of dimensionality. Where the connectivity matrix $\mathcal{A}$ was of size 24×24, it will now be $2 \times 2 \times 3 \times 2 \times 2 \times 2 \times 3 \times 2$, which contains the same number of elements. In this form, however, we do not directly enumerate the states and, at each step in the PGD algorithm, we search for a function of a very limited size corresponding to the number of processes in the sort, typically two or three. Without moving to a multi-dimensional structure, we cannot address larger and more interesting systems, however, moving into this new representation poses its own difficulties. A growing awareness of emergent properties in biological systems has developed from the fields of synthetic and systems biology. Though difficult to explain at a detailed level, the principle of emergence is, in effect, that the whole is somehow greater than the sum of its parts. The effect of activators and inhibitors are not additive but, rather, activators and inhibitors work synergistically to create amplified signals. For us, this means that we cannot assume that gene regulatory networks are separable, a feature we depend on for the PGD representation. However, that is not to say that it cannot be put into a *more or less* separated form: most likely, not all species demonstrate important emergent properties. The behavior of a system could very well be captured with the inclusion of a limited number of combined variables, perhaps $P(Dam, Mn, p53Mc, T)$, for example, in which $p53Mc$ captures the combined effects of p53 and cytoplasmic mdm2 while maintaining the individual influences of $p53$ and Mc. Developing a formal method for discovering these significant non-separable elements will be the topic of future work and brings the promise of opening up our methodology to the multidimensional formulation which, in turn, frees us from dimensionality.

5 Conclusion

The linking between very different modeling frameworks to arrive at a more complex, more powerful level of analysis promises to be an interesting vein of research in synthetic biology. In this paper, we have laid the foundation for just such a bridge between the discrete framework of Process Hitting and the Chemical Master Equation, in the hopes of overcoming the obstacle of dimensionality via the application of Proper Generalized Decomposition. This particular work considers the one dimensional "state vector" case, a form which conserves emergent– or non-separable– properties of the system. In order to broaden our scope to larger systems, we must be able to move to a multidimensional form. To do so, we must be able to re-introduce a limited number of non-separable, or combined, elements, which will be the topic of future works. The methodology introduced here has the potential to enrich analysis of gene regulatory networks and permit the study of larger, more complicated, and more realistic models, a necessity if we hope to make modeling frameworks fruitful in an applied setting.

References

1. Ammar, A., Cueto, E., Chinesta, F.: Reduction of the chemical master equation for gene regulatory networks using proper generalized decompositions. International Journal for Numerical Methods in Biomedical Engineering 28(9), 960–973 (2012)
2. Chinesta, F., Ammar, A., Leygue, A., Keunings, R.: An overview of the proper generalized decomposition with applications in computational rheology. Journal of Non-Newtonian Fluid Mechanics 166(11), 578–592 (2011)
3. Chinesta, F., et al.: PGD-Based computational vademecum for efficient design, optimization and control. Archives of Computational Methods in Engineering 20(1), 31–59 (2013)
4. Abou-Jaoudé, W., Ouattara, D., Kaufman, M.: Frequency tuning in the p53-mdm2 network. I. Logical approach. Journal of Theoretical Biology 258, 561–577 (2009)
5. Abou-Jaoudé, W., Ouattara, D., Kaufman, M.: Frequency tuning in the p53-mdm2 network. II. Differential and stochastic approaches. Journal of Theoretical Biology 264(4), 1177–1189 (2010)
6. Paulevé, L., Magnin, M., Roux, O.: Refining dynamics of gene regulatory networks in a stochastic π-calculus framework. In: Priami, C., Back, R.-J., Petre, I., de Vink, E. (eds.) Transactions on Computational Systems Biology XIII. LNCS (LNBI), vol. 6575, pp. 171–191. Springer, Heidelberg (2011)
7. Folschette, M., Paulevé, L., Inoue, K., Magnin, M., Roux, O.: Concretizing the process hitting into biological regulatory networks. In: Gilbert, D., Heiner, M. (eds.) CMSB 2012. LNCS, vol. 7605, pp. 166–186. Springer, Heidelberg (2012)
8. Paulevé, L., Magnin, M., Roux, O.: Tuning Temporal Features within the Stochastic-Calculus. IEEE Transactions on Software Engineering 37(6), 858–871 (2011)
9. Paulevé, L., Magnin, M., Roux, O.: Pint-Process Hitting Related Tools (October 10, 2010), http://processhitting.wordpress.com (April 16, 2013)
10. Jolliffe, I.: Principal component analysis, vol. 487. Springer, New York (1986)
11. Leenders, G., Tuszynski, J.: Stochastic and deterministic models of cellular p53 regulation. Frontiers in Molecular and Cellular Oncology 3(64) (2013)

12. Bernot, G., et al.: Application of formal methods to biological regulatory networks: extending Thomas asynchronous logical approach with temporal logic. Journal of Theoretical Biology 229(3), 339–347 (2004)
13. Paulevé, L., Youssef, S., Lakin, M., Phillips, A.: A Generic Abstract Machine for Stochastic Process Calculi. In: Proceedings of the 8th International Conference on Computational Methods in Systems Biology, pp. 43–54 (2010)

A The Process Hitting for p53-mdm2

In Section 2.2 we introduced the concept of sorts, processes and actions, as well as giving a brief introduction as to how actions are constructed based on a qualitative description of an activation or inhibition reaction. The actions constructed from the information outlined in Section 2.1 on the p53-mdm2 system are as follows:

$$Mn\_1 \to p53\_2 \ \textrm{↑} \ 1$$
$$p53\_0 \to Mc\_1 \ \textrm{↑} \ 0$$
$$p53\_1 \to Mc\_1 \ \textrm{↑} \ 0$$
$$Mn\_0 \to p53\_1 \ \textrm{↑} \ 2$$
$$Mn\_1 \to p53\_1 \ \textrm{↑} \ 0$$
$$p53\_0 \to Dam\_0 \ \textrm{↑} \ 1$$
$$p53\_1 \to Dam\_0 \ \textrm{↑} \ 1$$
$$Mn\_0 \to p53\_0 \ \textrm{↑} \ 1$$
$$p53\_2 \to Mc\_0 \ \textrm{↑} \ 1$$
$$p53\_2 \to Dam\_1 \ \textrm{↑} \ 0$$

$$p53McDam\_0 \to Mn\_1 \ \textrm{↑} \ 0$$
$$p53McDam\_1 \to Mn\_1 \ \textrm{↑} \ 0$$
$$p53McDam\_2 \to Mn\_1 \ \textrm{↑} \ 0$$
$$p53McDam\_3 \to Mn\_1 \ \textrm{↑} \ 0$$
$$p53McDam\_4 \to Mn\_1 \ \textrm{↑} \ 0$$
$$p53McDam\_5 \to Mn\_1 \ \textrm{↑} \ 0$$
$$p53McDam\_6 \to Mn\_0 \ \textrm{↑} \ 1$$
$$p53McDam\_7 \to Mn\_0 \ \textrm{↑} \ 1$$
$$p53McDam\_8 \to Mn\_0 \ \textrm{↑} \ 1$$
$$p53McDam\_9 \to Mn\_0 \ \textrm{↑} \ 1$$
$$p53McDam\_10 \to Mn\_0 \ \textrm{↑} \ 1$$
$$p53McDam\_11 \to Mn\_0 \ \textrm{↑} \ 1$$

Note that, for clarity and brevity, we have not listed those actions which update the cooperative sort p53McDam to reflect the current status of p53, Mc and Dam, nor have we included them in the graphical representation of the system shown in Figure 5. There are 48 such actions compared to the 22 listed here. As example, we have included two actions of Mc on the cooperative sort,

$$Mc\_1 \to p53McDam\_0 \ \textrm{↑} \ 2$$

$$Mc\_1 \to p53McDam\_1 \ \textrm{↑} \ 3$$

In both cases, p53McDam reflects the state space such that the sort Mc is at process Mc_0, which is no longer true if Mc_1 and must be updated. These examples are represented by blue, tightly dashed arrows in the figure, whereas the full updating action list is generally represented by the red, solid arrows linking the individual sorts p53, Mc and Dam, to their respective cooperative sort.

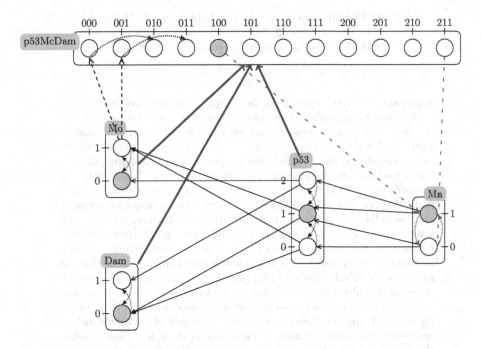

Fig. 5. A graphic representation of the Process Hitting. Each box represents a sort, which contains processes indicated by circles. Actions linking individual sorts are demonstrated by black arrows and can be found in the left hand column of the appendix chapter on PH. The p53-mdm2 system contains one cooperative sort, p53McDam. For the sake of clarity, not all actions linking the individual sorts and the cooperative sort have been shown; rather, red, solid arrows are drawn to indicate these updating actions, two examples of which are shown in tightly dashed blue lines. Likewise, not all actions linking p53McDam and Mn have been drawn, with two examples given in loosely dashed, green lines. As we have depicted it, our system is in state $\{0, 0, 1, 1\}$. There exists two actions which can be played, p53_1 → Dam_0 ⌐ 1 and p53McDam_4 → Mn_1 ⌐ 0. Process Hitting is asynchronous, thus, only one action can be played at any given time. Which action actually occurs depends on the rates, stochastic and temporal features assigned to each.

Coarse-Grained Brownian Dynamics Simulation of Rule-Based Models

Michael Klann[1], Loïc Paulevé[1], Tatjana Petrov[1], and Heinz Koeppl[1,2]

[1] BISON Group, Automatic Control Laboratory, ETH Zurich,
Physikstrasse 3, 8092 Zurich, Switzerland
[2] IBM Research - Zurich, Rueschlikon, Switzerland

Abstract. Studying spatial effects in signal transduction, such as co-localization along scaffold molecules, comes at a cost of complexity. In this paper, we propose a coarse-grained, particle-based spatial simulator, suited for large signal transduction models. Our approach is to combine the particle-based reaction and diffusion method, and (non-spatial) rule-based modeling: the location of each molecular complex is abstracted by a spheric particle, while its internal structure in terms of a site-graph is maintained explicit. The particles diffuse inside the cellular compartment and the colliding complexes stochastically interact according to a rule-based scheme. Since rules operate over molecular motifs (instead of full complexes), the rule set compactly describes a combinatorial or even infinite number of reactions. The method is tested on a model of Mitogen Activated Protein Kinase (MAPK) cascade of yeast pheromone response signaling. Results demonstrate that the molecules of the MAPK cascade co-localize along scaffold molecules, while the scaffold binds to a plasma membrane bound upstream component, localizing the whole signaling complex to the plasma membrane. Especially we show, how rings stabilize the resulting molecular complexes and derive the effective dissociation rate constant for it.

1 Introduction

Signal transduction pathways can contain several proteins and activation steps which give rise to complex spatiotemporal dynamics. Interactions between signaling molecules do not only transmit activations, but can also localize the molecules to certain structures or compartments in the cell [19]. This work aims at a simulator that is able to handle the complexity, and includes the localization of the molecules. Current modeling techniques include ordinary differential equations (ODE), or, if space is important, the corresponding partial differential equations (PDE). If the stochasticity is significant, the chemical or reaction-diffusion master equation (CME/RDME) describes the system dynamics. More specifically, each molecule can be tracked individually in the simulation, e.g. using Brownian/Smoluchowski or Green's function dynamics (see e.g. review [12]).

Transient complex formation of proteins and their post-translational modification in signaling can lead to a combinatorial number of distinct molecular

A. Gupta and T.A. Henzinger (Eds.): CMSB 2013, LNBI 8130, pp. 64–77, 2013.
© Springer-Verlag Berlin Heidelberg 2013

species. Rule-based languages, such as Kappa [6] or BioNetGen [8], provide a compact representation of such combinatorial processes. Nonspatial rule-based models can be efficiently simulated [3], and are amenable to further quantitative analysis, e.g. formal model reduction [9,10].

An extended Kappa framework was proposed in order to model the internal spatial structure of complexes that form during the process [4]. The Meredys simulator [20] in addition includes the molecule positions. Furthermore SRSim [11] provides a high resolution spatial rule-based extension, tracking the position and internal structure of all complexes. Indeed, the possibility to specify the binding angles sometimes naturally enforces a unique assembly path of a desired complex structure: for example, a polymer chain with angles between the bonds of $\pi - 2\pi/N$ in a plane will form a ring of N monomers, and local rules are sufficient to describe the global structure [4]. However, the exact molecular geometry of signaling molecules is often not known. In such cases, the simulation including the binding angles becomes more complex, without contributing additional insights.

Here we present a framework that supports a particle-based, spatial simulation, but omits the internal geometry. Still, as the internal structure in terms of a site-graph is maintained explicit, their cooperative effect on complex stability can be investigated effectively. In particular, we derive the dissociation rate constant for rings. The simulation is applied to MAPK (Mitogen activated protein kinase) signaling in yeast, where both localization and activation is mediated by a scaffold [19].

The paper is structured as follows. In Sect. 2, the general particle-based framework and the biophysical principles of complex formation are introduced. The formal framework underlying the simulator is outlined in Sect. 3. In Sect. 4, we show the application to MAPK signaling and we discuss the results.

2   Coarse-Grained Particle Diffusion and Reaction

The mobility of molecules in the cytoplasm is mainly governed by diffusion. Diffusion can be modeled efficiently by a random walk of the molecules of interest such that the myriad of solvent molecules can be omitted. We assume that the properties of the solvent allow us to use the Stokes-Einstein equation to obtain the (translational) diffusion coefficient $D_i \propto r_i^{-1}$ for a given molecular radius r_i of particle i. We also assume that rotational diffusion is much faster than translational diffusion such that the actual shape of the molecules averages to a sphere with radius r_i at the temporal resolution of the method. Therefore only the position but not the orientation of the molecules has to be tracked.

2.1   Remark on Diffusion-Controlled Reactions

Molecules can only react with each other if they are in contact/collide, and the collision process is governed by diffusion. Accordingly the observable bulk/macroscopic reaction rate constant k_{ij} between particles i and j in solution is determined both by the rate constant of collisions $k_D(i,j) = 4\pi(r_i + r_j)(D_i + D_j)$

(in 3D space) and the microscopic rate constant k'_{ij} that determines the reactive fraction of collisions [2]. For spherically symmetric molecules with isotropic reaction properties the microscopic and macroscopic rate constant are related in the form $k_{ij}^{-1} = k_D(i,j)^{-1} + k'_{ij}{}^{-1}$ (Collins-Kimball model). Reactions with very high $k_{ij} \approx k_D(i,j)$ require $k'_{ij} \to \infty$, i.e. every collision leads to a reaction such that they are diffusion-controlled. In contrast, reactions with $k_{ij} \ll k_D(i,j)$ are reaction-controlled; in this case $k'_{ij} \approx k_{ij}$. A diffusion factor/function can be introduced $f_D(i,j) := k_D(i,j)/(k_D(i,j) + k'_{ij})$ such that $k = f_D k'$ [2].

More in detail, signaling molecules have specific reaction sites, i.e. non-isotropic reaction properties. Such molecules have to be in contact (by translational diffusion) and correctly aligned with their reaction sites (by rotational diffusion). We define the corresponding nanoscopic rate constant as k'' and the conversion factor f_{DR} such that $k = f_{DR}k''$, however the derivation of f_{DR} is not straightforward and several approximations exist [2,21]. For completeness we also introduce $f_R = f_{DR}f_D^{-1}$ for the conversion $k' = f_R k''$, i.e. for integrating out the rotational diffusion effect only. In general: $k \le k' \le k''$.

Reversible reactions $A + B \rightleftarrows C$ with forward reaction at rate constant k_{AB} and backward k_C require to scale both rate constants with $f(A,B)$ in order to maintain the macroscopic reaction equilibrium (dissociation constant $K_d = k_C/k_{AB} = k'_C/k'_{AB} = k''_C/k''_{AB}$) [16,13]. For consistency we define the respective conversion factors $f = 1$ for unimolecular reactions that are not reversed by a bimolecular reaction, such that microscopic rate constants are always defined.

2.2   General Particle-Based Diffusion and Reaction Method

The present method implements the λ-ρ model [7], i.e. a discrete time continuous space random walk for diffusion. Each particle position x_i is updated by

$$x_i(t + \Delta t) = x_i(t) + \sqrt{2D_i \Delta t}\,\boldsymbol{\xi} \tag{1}$$

with diffusion coefficient D_i and standard normal random variable $\boldsymbol{\xi}$. Particles can overlap as discussed in [13], so collision testing is only needed for (static) reaction compartment boundaries. Reactions are executed with a probability that depends on their arity as follows.

Unimolecular Reactions. $A \to \ldots$ at rate constant k_A are executed in this method in every step with probability

$$P_A = 1 - exp(-k'_A \Delta t) \approx k'_A \Delta t \text{ if } \Delta t \to 0 \tag{2}$$

for each molecule which is of type A [7]. Note that the Bernoulli-trial scheme leads to a binomial distribution, which converges to the Poisson distribution for small probabilities in each step (law of rare events). The Poissonian reaction process has exponentially distributed inter event waiting times as expected.

Bimolecular Reactions. $A + B \to \ldots$ can only occur if two molecules are closer than their collision distance $(r_A + r_B)$. If so, then the reaction is executed with probability

$$P_{A,B} = \frac{k'_{AB} \Delta t}{4\pi(r_A + r_B)^3/3} \tag{3}$$

as derived in [15]. The accuracy constraint $P_{A,B} < 0.2$ gives an upper bound for Δt, if larger Δt are needed the reaction probability of [7] has to be used.

Higher Order and Hill-Type Reaction Schemes: for a particle-based simulation higher order reaction models have to be composed into their elementary uni- and bimolecular reaction steps (in which they also occur in nature).

2.3 Complexes

Complexes C that form out of $A + B \to C$ can still be modeled as spherical particles. The complex radius r_C is obtained e.g. under the assumption that the volume/mass of A and B is redistributed into C with constant mass density, i.e.

$$r_C = \left(\frac{3}{4\pi} \left[\frac{4\pi r_A^3}{3} + \frac{4\pi r_B^3}{3} \right] \right)^{1/3} = \left(\sum_{i \in C} r_i^3 \right)^{1/3} \tag{4}$$

as suggested in [22]. Alternatives for Eq. (4) are discussed and listed in Table 1 in [12]. The diffusion coefficient D_C is given by $D_C = D_0/r_0 r_C^{-1}$ based on a reference D_0 and r_0 in the Stokes-Einstein relation. Reactions between complexes α and β can occur when they are within their contact distance. The joint $k'_{\alpha\beta} = \sum_{A \in \alpha} \sum_{B \in \beta} k'_{AB}$ (assume $k'_{AB} = 0$ if no reaction between A and B is defined) could directly be used to calculate the binding probability by Eq. (3). However we decided to track each reaction individually as described in Sect. 3. Therefore also all resulting bonds in a complex are tracked individually. Such a bond in complex α between $A \in \alpha$ and $B \in \alpha$ can break with the corresponding dissociation rate constant/probability defined for the interaction. But if the two formerly directly connected molecules $A \in \alpha$ and $B \in \alpha$ are still connected by other bonds, they will stay together – and **aligned with their binding sites**. Therefore A and B can rebind in every step with P_{AB}, leading to bond recovery with rate constant

$$\bar{k}_{AB} = \frac{3k''_{AB}}{4\pi(r_A + r_B)^3} \tag{5}$$

effectively in a first order reaction (cf. Eq. 3 and Eq. 2) [14]. The recovery reaction does not take place in the (relatively large) reaction volume of the whole reaction compartment but in the (relatively tiny) interaction volume of the two agents. Therefore $\bar{k}$ is large compared to all other rates, and the lifetime of the open state of the bond $\tau \propto \bar{k}^{-1}$ is accordingly relatively small (cf. Test Case 2).

3 Spatial Stochastic Simulation of Rule-Based Model

Each particle of the simulation is an instantiation of a molecular species. A molecular species can be a protein, its post-translationally modified form or a protein complex that consists of proteins bound together. In order to reflect this internal structure of molecular species we represent them by *site-graphs*, in which modifications of protein residues and bonds are explicitly encoded, as introduced in Kappa [6].

Notations. We denote by $\{e_1 \mapsto v_1, \cdots, e_N \mapsto v_N\}$ the mapping from distinct elements e_i to values v_i. Given a mapping A, $\mathrm{dom}(A)$ denotes its domain, and $A(e)$ the value associated to e in A. We also write $A\{e \mapsto v\}$ for the mapping A updated so that e maps to v; and $A \setminus e$ for the mapping A updated so that the mapping from e is removed.

3.1   Site-Graphs

A site-graph is an undirected graph where typed nodes have sites, and edges are partial matchings on sites. Moreover, the sites which do not serve for forming edges are called *internal*, and they are assigned a value from a predefined set. The nodes of the site-graph can be interpreted as protein names, and sites of a node stand for protein binding domains. Internal states are used to encode post-translational modifications.

Let $\mathcal{S}$ denote the set of site labels, and $\mathcal{I}$ the set of internal values that can be assigned to sites. The function $I : \mathcal{S} \to \mathcal{P}(\mathcal{I})$ denotes the set of internal values that a site s can take. Let $\mathcal{A}$ be the set of node types. Each node type is being equipped with a set of sites, defined by a signature map $\Sigma : \mathcal{A} \to \mathcal{P}(\mathcal{S})$. Finally, the set of admissible bindings between sites is defined by the mapping $\mathcal{E} : \mathcal{A} \times \mathcal{S} \to \mathcal{P}(\mathcal{A} \times \mathcal{S})$ so that if $(a', s') \in \mathcal{E}(a, s)$ then necessarily $(a, s) \in \mathcal{E}(a', s')$. A rule-based model is defined over a fixed *contact map* defined by the tuple $(\mathcal{A}, \Sigma, \mathcal{E}, I)$ that we consider constant in the rest of this section.

Definition 1. A site-graph is a tuple $G = (V, T, F, E, \psi)$ with

1. a set of nodes V,
2. a node type function $T : V \to \mathcal{A}$,
3. a node interface function $F : V \to \mathcal{P}(\mathcal{S})$, such that for $v \in V$, $F(v) \subseteq \Sigma(T(v))$,
4. a set of edges between sites of different nodes, encoded by the function $E : V \times \mathcal{S} \to V \times \mathcal{S}$ such that if $E(v, s) = (v', s')$ then necessarily $v \neq v'$, $E(v', s') = (v, s)$, and $(T(v'), s') \in \mathcal{E}(T(v), s)$.
5. a site evaluation function $\psi : V \times \{s \in \mathcal{S} \mid I(s) \neq \emptyset\} \to \mathcal{I}$, so that $\psi(v, s) \in I(s)$.

Site-graphs will be used in two different contexts: (i) to model physically existing complexes, also termed *concrete site-graphs* (or reaction mixtures), and (ii) to specify the local interaction patterns (rewrite rules). The concrete site-graphs must have all interfaces complete, in the sense that, for all nodes $v \in V$, $F(v) = \Sigma(T(v))$.

Definition 2. Site-graph $G = (V, T, F, E, \psi)$ is a *union* of two site-graphs $G_1 = (V_1, T_1, F_1, E_1, \psi_1)$ and $G_2 = (V_2, T_2, F_2, E_2, \psi_2)$, denoted by $G = G_1 \oplus G_2$, if $V_1 \cap V_2 = \emptyset$, and $V = V_1 \cup V_2$, $F = F_1 \cup F_2$, $E = E_1 \cup E_2$, $\psi = \psi_1 \cup \psi_2$.

Definition 3. Given a site-graph $G = (V, T, F, E, \psi)$, a sequence of edges $(((v_1, s_1), E(v_1, s_1)), \ldots, ((v_k, s_k), E(v_k, s_k)))$ such that for $i = 1, \ldots, k$, $v_i \in V$,

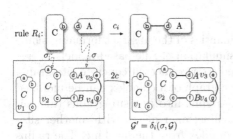

Fig. 1. Rule application. Rule R_i can be applied to a reaction mixture $\mathcal{G}$ via the embeding σ (indicated by the dotted arrows), and resulting in the reaction mixture $\mathcal{G}'$. The contact map is such that $\mathcal{A} = \{A, B, C\}$, $\mathcal{S} = \{a, b, c, d, e, f, g\}$, and $I(a) = \{u, p\}$. More details on the formalism can be found e.g. in [18].

$s_i \in F(v_i)$, and for $i = 1, \ldots, k - 1$, $E(v_i, s_i) = (v, s) \Rightarrow v_{i+1} = v$ and $s_{i+1} \neq s$, is called a *path* between nodes v_1 and v_k. A site-graph G is connected, denoted $cc(G)$ if there exists a path between every two nodes v and v'.

Two site-graphs can be related by an embedding function, which is important for defining the applicability of a rule to a reaction mixture (cf. Fig. 1).

Definition 4. The *embedding* σ between site-graphs $G = (V, T, F, E, \psi)$ and $G' = (V', T', F', E', \psi')$, is induced by a support function $\sigma^* : V \to V'$, if

1. σ^* is injective: for all $v, v' \in V$, $[\sigma^*(v) = \sigma^*(v') \implies v = v']$;
2. for all $v \in V$, $T(v) = T'(\sigma^*(v))$;
3. for all $v \in V$, $[s \in F(v) \implies s \in F'(\sigma^*(v))]$;
4. $\{(v, s) \mapsto (v', s')\} \subset E \implies \{(\sigma^*(v), s) \mapsto (\sigma^*(v'), s')\} \subset E'$;
5. $\{(v, s) \mapsto i\} \subset \psi \implies \{(\sigma^*(v), s) \mapsto i\} \subset \psi'$.

If σ^* is bijective, then σ is an *isomorphism*. The set of embeddings between the site-graph G and G' is denoted by $\mathsf{embed}(G, G')$. The set of all embeddings is denoted by $\mathbb{E}$.

3.2 Rule-Based Models

Definition 5. Consider three types of elementary transformations of site-graphs, denoted by $\delta_{\mathsf{ae}}, \delta_{\mathsf{de}}, \delta_{\mathsf{ci}} : \mathbb{E} \times \mathbb{G} \to \mathbb{G}$, with the following form:

1. $\delta_{\mathsf{ae}}^{v,s,v',s'}(\sigma, G) = (V, T, F, E\{((\sigma^*(v), s) \mapsto (\sigma^*(v'), s'), (\sigma^*(v'), s') \mapsto (\sigma^*(v), s)\}, \psi)$ (adding an edge)
2. $\delta_{\mathsf{de}}^{v,s}(\sigma, G) = (V, T, F, E \setminus E((\sigma^*(v), s)) \setminus (\sigma^*(v), s), \psi)$ (deleting an edge);
3. $\delta_{\mathsf{ci}}^{v,s,i}(\sigma, G) = (V, T, F, E, \psi\{(\sigma^*(v), s) \mapsto i\})$ (changing the state value),

where $G = (V, T, F, E, \psi) \in \mathbb{G}$, $\sigma \in \mathbb{E}$ induced by the suport function $\sigma^* : V_i \mapsto V$, V_i being a set of nodes such that $v, v' \in V_i$; $v \neq v'$, $(T(v'), s') \in \mathcal{E}(T(v), s)$, and $i \in I(s)$.

A *rule* R_i is a triple (G, δ, c), where G is a site-graph, δ is of type $\delta_{\mathsf{ae}}, \delta_{\mathsf{de}}$, or δ_{ci}, and c is a non-negative real number. Applying the rule to a site-graph $\mathcal{G}$ is unique for an embedding $\sigma \in \mathsf{embed}(G, \mathcal{G})$, and results in $\mathcal{G}' = \delta(\sigma, \mathcal{G})$ (for a rigorous explanation, see [5]). In particular, for the identity support function $\sigma^* = \mathbb{I}$, and for $\mathcal{G} = G$, we get $G' = \delta(\mathbb{I}, G)$, which is sometimes called the right-hand-side of the rule (cf. Fig. 1).

3.3   Stochastic Abstract Machine

This subsection defines the syntax and semantics of the abstract formal machine for our coarse-grained spatial stochastic simulation of rule-based models.

In our machine, a *complex* is associated to a connected concrete site-graph, denoted by $\mathcal{G}$, and a 3-D position. The site-graph of a complex can be modified according to the rules, and may undergo a split into two site-graphs (dissociation), adding a new complex in the machine; or a merge with another site-graph (association), removing the other complex from the machine. The radius of agents $a \in A$ is determined by the constant r_a, and the radius of the connected site-graph $\mathcal{G}$ is given by $\mathsf{rad}(\mathcal{G})$, according to Eq. (4) (Sect. 2.3):

$$\mathsf{rad}((V, T, F, E, \psi)) = \left(\sum_{v \in V} r_{T(v)}^3 \right)^{1/3} \tag{6}$$

The syntax of the abstract machine is given in Def. 6. A machine term M is a quadruple $(t, C, X, \mathcal{R})$ where t is the current time, C is a map from a complex i to its current site-graph $\mathcal{G}$, X is a map from a complex i to its position $\mathbf{x} \in \mathbb{R}^3$, and $\mathcal{R}$ the set of rules, as defined in Sect. 3.2.

The execution of the machine at time t follows Eq. (7). First, the position of each complex is updated according to a Brownian diffusion during a fixed Δt time (diffuse function, Def. 7). Second, the active rules at time t are applied according to the probabilities introduced in Sect. 2. The new site-graphs and positions maps give the new term of the machine at time $t + \Delta t$.

$$\frac{X' = \mathsf{diffuse}(X, C, \Delta t)}{(C', X'') = \mathsf{react}(C, X', \mathcal{R}, \Delta t)} \\ \frac{}{(t, C, X, \mathcal{R}) \to (t + \Delta t, C', X'', \mathcal{R})} \tag{7}$$

The react function (Def. 12) applies the rules embeddings that are active at time t in a random order. Each active rule embedding (Def. 11) is specified by the set of concerned complexes (either one or two), and the quadruple (σ, G, δ, k) where (G, δ, k) is a rule in $\mathcal{R}$, and σ is the actual embedding being a map from the nodes of the rule left hand-side G to the nodes of the concrete complex site-graphs. These rule embeddings gathers the embeddings from any rule in $\mathcal{R}$ to a single complex site-graph $C(i)$ (unary function) or to the union of two complex site-graphs $C(i_1) \oplus C(i_2)$, assuming that complexes i_1 and i_2 have their distance $|X(i_1) - X(i_2)|$ less than the sum of their radii (neighbors function). A rule with embedding σ is actually applied if (1) the embedding is still valid, i.e., previous rules application have not interfered with it; (2) the random number ζ_1 uniformly distributed in $[0; 1]$ is less than the rule probability.

The probability of applying a rule embedding (prob function, Def. 8) depends on the arity of the reaction and on the modification type, as described in Sect. 2. The application of rule embedding to the site-graphs and position maps is given by the do function (Def. 10). In the case of agent site values changes (δ_{ci}) or an internal bound creation (δ_{ae} within one complex i), the embedded rule modification is applied to the concerned site-graphs, without any side-effect. In the case

Stochastic Abstract Machine Definitions

Definition 6. Syntax of the abstract machine. $i_1, \cdots, i_N$ are the complexes identifiers, assumed all distinct.

$$M ::= (t, C, X, \mathcal{R}) \qquad \text{Time } t, \text{ complex maps } C \text{ and } X, \text{ rules } \mathcal{R}$$
$$C ::= \{i_1 \mapsto \mathcal{G}_1, \cdots, i_N \mapsto \mathcal{G}_N\} \quad \text{Map from a complex } i \text{ to its site-graph } \mathcal{G}$$
$$X ::= \{i_1 \mapsto \mathbf{x}_1, \cdots, i_N \mapsto \mathbf{x}_N\} \quad \text{Map from a complex } i \text{ to its position } \mathbf{x} \in \mathbb{R}^3$$

Definition 7. Euler-Maruyama integration of diffusion. *Returns the new position maps of complexes after a Brownian diffusion during Δt time, where ζ is a standard normal random variable, and D_0 and r_0 are constants (cf. Eq. (1), Sect. 2.2). Note: state dependent diffusion in different compartments is introduced in Appendix A.*

$$\mathsf{diffuse}(X, C, \Delta t) \triangleq \{i \mapsto | \ X(i) + \zeta\sqrt{2D_i\Delta t} \mid D_i = D_0\,r_0/\mathsf{rad}(C(i)), i \in \mathsf{dom}(X)\}$$

Definition 8. Rule probability. *The rates k', k'' and $\bar{k}$ refers to the rule rate k modified according to Sect. 2, $\mathsf{agents}(\delta_{ae})$ refers to the couple of agents concerned by the bound creation, and r_{a_i} is the radius of agent a_i.*

$$\mathsf{prob}(\{i\}, \delta_{ci}, \sigma, k, C, \Delta t) \triangleq k'\Delta t$$
$$\mathsf{prob}(\{i\}, \delta_{de}, \sigma, k, C, \Delta t) \triangleq k''\Delta t \text{ if } \mathsf{cc}(\delta_{de}(\sigma, C(i))) \text{ else } k'\Delta t$$
$$\mathsf{prob}(\{i\}, \delta_{ae}, \sigma, k, C, \Delta t) \triangleq \bar{k}\Delta t \triangleq \frac{3k''\Delta t}{4\pi(r_{a_1} + r_{a_2})^3} \text{ if } \mathsf{agents}(\delta_{ae}) = (a_1, a_2)$$
$$\mathsf{prob}(\{i_1, i_2\}, \delta, \sigma, k, C, \Delta t) \triangleq \frac{3k'\Delta t}{4\pi(\mathsf{rad}(C(i_1)) + \mathsf{rad}(C(i_2)))^3} \text{ if } i_1 \neq i_2$$

Definition 9. Complex formation and dissociation. *Returns the modified site-graphs and positions mappings after a merge or a split of complexes. The condition $i_2 \notin \mathsf{dom}(C)$ ensures that the new complex i_2 is a fresh identifier.*

$$\mathsf{merge}(i_1, i_2, \delta, \sigma, C, X) \triangleq ((C \setminus i_2)\{i_1 \mapsto \delta(\sigma, C(i_1) \oplus C(i_2))\}, X \setminus i_2)$$
$$\mathsf{split}(i_1, \delta, \sigma, C, X) \triangleq (C\{i_1 \mapsto \mathcal{G}_1, i_2 \mapsto \mathcal{G}_2\}, X\{i_2 \mapsto X(i_1)\})$$
$$\text{if } i_2 \notin \mathsf{dom}(C), \mathcal{G}_1 \oplus \mathcal{G}_2 = \delta(\sigma, C(i_1)), \mathsf{cc}(\mathcal{G}_1), \mathsf{cc}(\mathcal{G}_2)$$

Definition 10. Rule application. *Returns the modified site-graphs and positions mappings. Predicate $\mathsf{cc}(\mathcal{G})$ is true if and only if $\mathcal{G}$ is connected (Def. 3, Sect. 3.1).*

$$\mathsf{do}(I, \delta_{ci}, \sigma, C, X) \triangleq (C\{i \mapsto \delta_{ci}(\sigma, C(i)) \mid i \in I\}, X)$$
$$\mathsf{do}(\{i\}, \delta_{ae}, \sigma, C, X) \triangleq (C\{i \mapsto \delta_{ae}(\sigma, C(i))\}, X)$$
$$\mathsf{do}(\{i_1, i_2\}, \delta_{ae}, \sigma, C, X) \triangleq \mathsf{merge}(i_1, i_2, \delta_{ae}, \sigma, C, X) \text{ if } i_1 \neq i_2$$
$$\mathsf{do}(\{i\}, \delta_{de}, \sigma, C, X) \triangleq (C\{i \mapsto \delta_{de}(\sigma, C(i))\}, X) \text{ if } \mathsf{cc}(\delta_{de}(\sigma, C(i)))$$
$$\mathsf{do}(\{i\}, \delta_{de}, \sigma, C, X) \triangleq \mathsf{split}(i, \delta_{de}, \sigma, C, X) \text{ if not } \mathsf{cc}(\delta_{de}(\sigma, C(i)))$$

Stochastic Abstract Machine Definitions (Continued)

Definition 11. Active embeddings. unary *and* binary *returns the embedding specifications of rules* $\mathcal{R}$ *to single or couples of complexes, respectively;* neighbors *returns the couples of complexes close enough to react;* act_embeds *returns the* (I, E) *couples where* I *is the set of complexes concerned by the embedding specification* E.

$$\text{unary}(i, C, \mathcal{R}) \triangleq \{(\sigma, G, \delta, k) \mid (G, \delta, k) \in \mathcal{R}, \sigma \in \text{embed}(G, C(i))\}$$

$$\text{binary}(i_1, i_2, C, \mathcal{R}) \triangleq \{(\sigma, G, \delta, k) \mid (G, \delta, k) \in \mathcal{R}, \sigma \in \text{embed}(G, C(i_1) \oplus C(i_2)),$$
$$\sigma \notin \text{embed}(G, C(i_1)) \cup \text{embed}(G, C(i_2))\}$$

$$\text{neighbors}(C, X) \triangleq \{\{i_1, i_2\} \mid \text{rad}(C(i_1)) + \text{rad}(C(i_2)) \geq |X(i_1) - X(i_2)|,$$
$$i_1, i_2 \in \text{dom}(C), i_1 \neq i_2\}$$

$$\text{act\_embeds}(C, X, \mathcal{R}) \triangleq \{(\{i\}, E) \mid i \in \text{dom}(C), E \in \text{unary}(i, C, \mathcal{R})\}$$
$$\cup \{(\{i_1, i_2\}, E) \mid \{i_1, i_2\} \in \text{neighbors}(C, X),$$
$$E \in \text{binary}(i1, i2, C, \mathcal{R})\}$$

Definition 12. Sequential application of active rule embeddings, *order is assumed to be random; returns the modified site-graphs and position maps. The set of concerned complexes is denoted by* I, $\oplus_{i \in I} C(i)$ *denotes the union of site-graphs of complexes in* I, ζ_1 *is a random variable uniformly distributed within* 0 *and* 1, *and* embed *is the set of embeddings from rule left hand-side to reaction mixture (cf. Sect. 3.1).*

$$\text{apply}(\{(I, (\sigma, G, \delta, k))\} \cup Q, C, X, \Delta t) \triangleq \text{apply}(Q, C, X, \Delta t)$$
$$\textbf{if } \sigma \notin \text{embed}(G, \oplus_{i \in I} C(i))$$
$$\textbf{or } \zeta_1 \geq \text{prob}(I, \delta, \sigma, k, C, \Delta t)$$
$$\triangleq \text{apply}(Q, C', X', \Delta t)$$
$$\textbf{if } \sigma \in \text{embed}(G, \oplus_{i \in I} C(i)),$$
$$\zeta_1 < \text{prob}(I, \delta, \sigma, k, C, \Delta t),$$
$$(C', X') = \text{do}(I, \delta, \sigma, C, X)$$

$$\text{apply}(\emptyset, C, X, \Delta t) \triangleq (C, X)$$

$$\text{react}(C, X, \mathcal{R}, \Delta t) \triangleq \text{apply}(\text{act\_embeds}(C, X, \mathcal{R}), C, X, \Delta t)$$

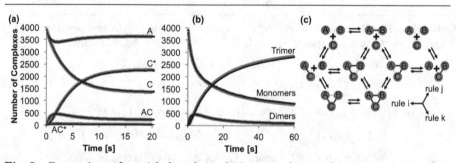

Fig. 2. Comparison of particle-based simulation according to the presented algorithm (black symbols) and ODE model (red line, see Appendix B) for two test cases: (a) enzymatic activation, Test Case 1 and (b) trimerization, Test Case 2. (c) Overview of the reactions in the ring formation process of example (b), binding sites are omitted here.

of a complex formation (δ_{ae} between two complexes i_1 and i_2), the two complexes are merged into one and receives the position of one of the two complexes in a non-deterministic manner (merge function, Def. 9). Finally, in the case of a bond deletion within a complex (δ_{de}), unless the site-graph is still connected, the complex is split in two complexes receiving the same position (split function, Def. 9).

4   Test Cases and Application

Test Cases: The accuracy of the simulation was tested on the principal reaction motifs of signaling models, where we assume the reaction-limited case for simplicity, i.e. $k = k' = k''$. The test cases are (in Kappa like syntax, see Fig. 1 for agents and sites):

1. Reversible enzymatic activation $A(d) + C(b) \rightleftharpoons A(d^1), C(b^1)$,
   $A(d^1), C(b^1, a_u) \rightarrow A(d^1), C(b^1, a_p)$ and $C(a_p) \rightarrow C(a_u)$.
2. Reversible trimerization by: $A(d) + C(b) \rightleftharpoons A(d^1), C(b^1)$,
   $A(e) + B(g) \rightleftharpoons A(e^2), B(g^2)$ and $C(c) + B(f) \rightleftharpoons C(c^3), B(f^3)$.

Fig. 2 depicts the simulation result which exactly match the ODE models derived in Appendix B, thus showing the correctness of the approach.

Test Case 2 exemplifies the formation of a ring and the cooperativity of the bonds in the ring. The ABC trimer can only dissociate if two bonds break and are open at the same time. For equal binding rate constants k_1, equal dissociation rate constants k_2 and under the quasi-steady-state assumption for complexes with one broken bond the effective dissociation rate constant becomes

$$k^* = 6k_2^2/(\bar{k}_1 + 2k_2) \tag{8}$$

as shown in Appendix B, Eq. (13). In a ring consisting of N agents the effective dissociation rate constant $k^* = \left(N(N-1)k_2^2\right)/\left(\bar{k}_1 + (N-1)k_2\right)$, will converge to Nk_2, i.e. with rising N the cooperativity vanishes. Especially small rings are therefore stabilized due to the high bond recovery rate constant $\bar{k}$. In the present example $k_1 = 5 \times 10^5 M^{-1}s^{-1}$, $k_2 = 0.2s^{-1}$, $r_A = r_B = r_C = 5nm$, $D_A = D_B = D_C = 1\mu m^2/s$. Then $\bar{k}_1 = 198s^{-1}$, and $k^* = 0.0012s^{-1} \ll k_2$, i.e. the ring structure is extremely stable. Diffusion limit $k_D = 1.51 \times 10^8 M^{-1}s^{-1} \gg k_1$ i.e. reaction-limited regime.

Signaling with Scaffolds in Space and Time: In order to exemplify how the presented method can be used to model signaling in space and time, we simulate the yeast pheromone response signaling model from Thomson et al. [19] (see Fig. 3a) up to the MAPK Fus3. In the model the actual signaling molecules Ste4, Ste11, Ste7 and Fus3 can bind to the scaffold Ste5. In contrast to Thomson et al. we assume that any activation by the upstream molecule also involves a possible binding interaction between these molecules – although that interaction might be weak. The resulting additional bonds, shown in red in Fig. 3b, enable the formation of three rings around Ste5, which can stabilize

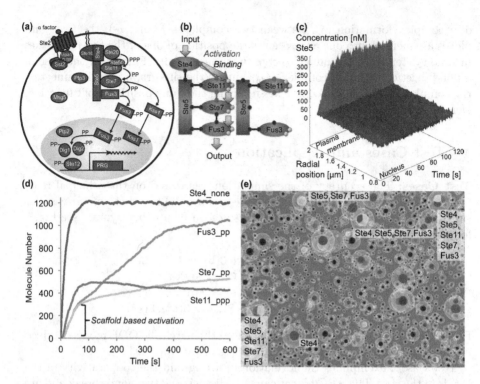

Fig. 3. (a) Yeast mating MAPK signaling pathway from [19]. (b) Binding and activation interactions around the scaffold Ste5. Incomplete signaling complexes cannot transmit the signal and do not include rings, which makes them less stable. (c) Localization of Ste5 to the plasma membrane. (d) Activation of the MAPK cascade. The initial fast activation phase of Fus3 takes place in scaffold based complexes that formed prior to activation. Further activations require that new Ste7-Fus3 pairs form (diffusion-reaction process) which gives rise to slower kinetics. (e) Visualization of the complexes in 3D space at $t = 184s$ using ZigCell3D from ScienceVisuals [23]. The transparent spheres with constant radius of 12.5 nm are used for all complexes. Nucleus and plasma membrane are not shown.

complete signaling complexes. The binding interactions Ste4-Ste11 and Ste11-Ste7 were omitted in [19] because pheromone response must only be activated if Ste5 is present. In the present model the binding interaction was weak enough such that likewise activation requires the presence of Ste5 (data not shown). Effectively, the bond only establishes along the scaffold, such that crosstalk with other signaling pathways can be prevented. The arising rings including Ste5 and the resulting low dissociation due to cooperativity of the bonds in the rings furthermore makes overexpression of Ste5 less harmful than in [19]. This is due to the fact that without the bonds parallel to Ste5, each Ste5-Ligand bond would arise independent of the other ligands. Too many Ste5 instances would therefore lead to complexes where most likely just 1 ligand is connected to Ste5 such that the signaling cascade is disrupted. In contrast, the additional bonds drive the equilibrium towards complete complexes.

The spatial and temporal dynamics of the activation process in the MAPK signaling cascade is depicted in Fig. 3c and d. Ste4 is a membrane bound protein, therefore all complexes containing Ste4 will likewise be membrane bound (cf. Appendix A). Fig. 3c shows how Ste5 accumulates at the plasma membrane accordingly. The 3D positions of molecules and complexes in the simulation are shown in Fig. 3e.

5 Discussion

The present work shows how a simulation for complex signal transduction models can be implemented, making use of a rule-based model description and including biophysical aspects as well as the spatial component. The present algorithm uses a coarse-grained description and simplified models for binding and dissociation rate constants on the macro-, micro- and nanoscopic level, thus refining the simulation method suggested in Meredys [20]. The applied rate constants and conversion factors as introduced in Sect. 2.1 could be further refined by more detailed models, which e.g. require to solve the reaction-diffusion or corresponding master equation [16,17].

An accuracy and performance test of the particle-based simulator core in [13] shows that the presented method performs at least as good as Smoldyn [1]. Smoldyn however can handle rule-based models only via the libMoleculizer plugin. Association and dissociation of complexes require costly graph traversal. However, associations require that two complexes are in contact, which is a rare event, and dissociation does not occur more frequently (at steady state). Given the computational cost of the random walk of the molecules and neighbor finding alone, the rule-based extension does not dramatically slow down simulation. The performance of the algorithm can be improved by using a Gillespie scheme (draw exponential waiting time for reaction event) instead of the Bernoulli trials for unimolecular reactions as in [14]. In that case also the order of unimolecular reactions is given by the (ordered) waiting times instead of the random order we proposed in order to execute all reactions in an unbiased manner. For bimolecular reactions it is extremely unlikely that more than two complexes are within the reactive distance such that in most cases there is no ambiguity which reaction is to be applied. Further improvements of the performance could come from multi-scale or mixed approaches for different domains of the simulation [12].

The coarse-grained rate constants enable the calculation of emerging properties like the cooperativity between bonds, that stabilizes rings. In the MAPK signaling example also rings are formed around the scaffold Ste5. Future work can analyze the stability of signaling complexes and the dose-response curve for Ste5 now including the cooperative effect of the bonds. The simulation already includes localization sites that determine the location of the agents (cytoplasmic or membrane based as shown in Appendix A). In the future we are planning to include the transport into the nucleus such that the complete signaling pathway can be analyzed. Furthermore formal model checking of spatial rule-based models has to be included to ensure meaningful models and simulation results.

Additional Information

The appendix is available from http://www.bison.ethz.ch/research/
spatial_simulations_si/CMSB2013_Appendix.pdf and the simulator
as well as example files from http://www.bison.ethz.ch/research/
spatial_simulations.

Acknowledgements. We thank Gerd Grünert for valuable discussions about spatial rule-based simulations and Pablo de Heras Ciechomski for the assistance with the visualization.

Funding: M.K.: Swiss Commission for Technology and Innovation (CTI) project 12532.1 PFLS-LS in the joint ZigCell3D software development project with ScienceVisuals, Lausanne, Switzerland. L.P. and T.P.: Swiss SystemsX.ch initiative. H.K.: Swiss National Science Foundation, grant no. PP00P2_128503.

References

1. Andrews, S.S., Addy, N.J., Brent, R., Arkin, A.P.: Detailed simulations of cell biology with smoldyn 2.1. PLoS Computational Biology 6(3), 1000705 (2010)
2. Berdnikov, V., Doktorov, A.: Steric factor in diffusion-controlled chemical reactions. Chemical Physics 69(1), 205–212 (1982)
3. Danos, V., Feret, J., Fontana, W., Krivine, J.: Scalable simulation of cellular signaling networks. In: Shao, Z. (ed.) APLAS 2007. LNCS, vol. 4807, pp. 139–157. Springer, Heidelberg (2007)
4. Danos, V., Honorato-Zimmer, R., Riveri, S., Stucki, S.: Rigid geometric constraints for Kappa models. Electronic Notes in Theoretical Computer Science (2012)
5. Danos, V., Feret, J., Fontana, W., Harmer, R., Krivine, J.: Abstracting the differential semantics of rule-based models: exact and automated model reduction. In: LICS 2010, pp. 362–381 (2010)
6. Danos, V., Laneve, C.: Formal molecular biology. Theoretical Computer Science 325(1), 69–110 (2004)
7. Erban, R., Chapman, S.: Stochastic modelling of reaction–diffusion processes: algorithms for bimolecular reactions. Phys. Biol. 6, 046001 (2009)
8. Faeder, J., Blinov, M., Hlavacek, W.: Rule-based modeling of biochemical systems with bionetgen. In: Systems Biology, pp. 113–167. Springer (2009)
9. Feret, J., Danos, V., Krivine, J., Harmer, R., Fontana, W.: Internal coarse-graining of molecular systems. Proceedings of the National Academy of Sciences 106(16), 6453–6458 (2009)
10. Feret, J., Henzinger, T.A., Koeppl, H., Petrov, T.: Lumpability abstractions of rule-based systems. In: Ciobanu, G., Koutny, M. (eds.) MeCBIC. EPTCS, vol. 40, pp. 142–161 (2010)
11. Gruenert, G., Ibrahim, B., Lenser, T., Lohel, M., Hinze, T., Dittrich, P.: Rule-based spatial modeling with diffusing, geometrically constrained molecules. BMC Bioinformatics 11(1), 307 (2010)
12. Klann, M., Koeppl, H.: Spatial simulations in systems biology: from molecules to cells. International Journal of Molecular Sciences 13, 7798–7827 (2012)

13. Klann, M., Koeppl, H.: Reaction schemes, escape times and geminate recombinations in particle-based spatial simulations of biochemical reactions. Physical Biology 10, 046005 (2013)
14. Klann, M., Ganguly, A., Koeppl, H.: Improved reaction scheme for spatial stochastic simulations with single molecule detail. In: Proceedings of the International Workshop on Computational Systems Biology, WCSB 2011, Zurich, pp. 93–96 (2011)
15. Klann, M., Lapin, A., Reuss, M.: Stochastic Simulation of Reactions in the Crowded and Structured Intracellular Environment: Influence of Mobility and Location of the Reactants. BMC Systems Biology 5(1), 71 (2011)
16. Morelli, M., Ten Wolde, P.: Reaction Brownian dynamics and the effect of spatial fluctuations on the gain of a push-pull network. J. Chem. Phys. 129, 054112 (2008)
17. Mugler, A., Tostevin, F., ten Wolde, P.: Spatial partitioning improves the reliability of biochemical signaling. Proceedings of the National Academy of Sciences 110(15), 5927–5932 (2013)
18. Petrov, T., Feret, J., Koeppl, H.: Reconstructing species-based dynamics from reduced stochastic rule-based models. In: Proceedings of the Winter Simulation Conference, p. 225. Winter Simulation Conference (2012)
19. Thomson, T., et al.: Scaffold number in yeast signaling system sets tradeoff between system output and dynamic range. Proceedings of the National Academy of Sciences 108(50), 20265–20270 (2011)
20. Tolle, D.P., Le Novère, N.: Meredys, a multi-compartment reaction-diffusion simulator using multistate realistic molecular complexes. BMC Systems Biology 4(1), 24 (2010)
21. Traytak, S.: Diffusion-controlled reaction rate to an active site. Chemical Physics 192(1), 1–7 (1995)
22. Weiss, M., Elsner, M., Kartberg, F., Nilsson, T.: Anomalous subdiffusion is a measure for cytoplasmic crowding in living cells. Biophysical Journal 87, 3518–3524 (2004)
23. ZigCell3D: from ScienceVisuals, `zigcell.sciencevisuals.com`

Modelling and Analysis of Phase Variation in Bacterial Colony Growth

Ovidiu Pârvu[1], David Gilbert[1], Monika Heiner[2], Fei Liu[3], and Nigel Saunders[1]

[1] School of Information Systems, Computing and Mathematics
Brunel University, Uxbridge, Middlesex UB8 3PH, UK
{ovidiu.parvu,david.gilbert,nigel.saunders}@brunel.ac.uk
[2] Computer Science Institute, Brandenburg University of Technology
Postbox 10 13 44, 03013 Cottbus, Germany
monika.heiner@informatik.tu-cottbus.de
[3] Harbin Institute of Technology
West Dazhi Street 92, 150001 Harbin, China
liufei@hit.edu.cn

Abstract. We describe an investigation into spatial modelling by means of an ongoing case study, namely phase variation patterning in bacterial colony growth, forming circular colonies on a flat medium. We explore the application of two different geometries, rectangular and circular, for modelling and analysing the colony growth in 2.5 dimensions. Our modelling paradigm is that of coloured stochastic Petri nets and we employ stochastic simulation in order to generate output which is then analysed for sector patterning. The analysis results are used to compare the two geometries, and our multidimensional approach is a precursor to more work on detailed multiscale modelling.

Keywords: Coloured stochastic Petri nets, spatial modelling, Systems Biology, pattern analysis, multidimensional, BioModel Engineering.

1 Motivation

This paper builds on [5], where we have introduced our methodology for the use of a structured family of Petri net classes which enables the investigation of biological systems using complementary modelling abstractions comprising the qualitative and quantitative, i.e., stochastic, continuous, and hybrid paradigms.

We extend our spatial modelling approach introduced in [2, 3] where we discretise space within a geometrical framework exploiting finite discrete colour sets embedded in coloured Petri nets. We motivate our work by describing an investigation into spatial modelling by means of an ongoing case study, namely phase variation patterning in bacterial colony growth, forming circular colonies on a flat medium. In order to illustrate the power and flexibility of our approach we explore the application of two different geometries, rectangular and circular, for modelling and analysing the colony growth in 2.5 dimensions: the 2 dimensions of the surface of the colony are modelled explicitly while the height

A. Gupta and T.A. Henzinger (Eds.): CMSB 2013, LNBI 8130, pp. 78–91, 2013.
© Springer-Verlag Berlin Heidelberg 2013

is modelled implicitly. In order to capture the stochastic properties of the case study, we have chosen coloured stochastic Petri nets ($\mathcal{SPN}^{\mathcal{C}}$) as our modelling paradigm and employ stochastic simulation in order to generate output which is then analysed for sector patterning. The analysis results are used to compare the two geometries, and our multidimensional approach is a precursor to more work on detailed multiscale modelling.

The main contributions of our paper are

- a detailed model of phase variation in bacterial colony growth, in two geometries (rectangular and circular),
- the development and application of techniques to analyse the properties of the patterns generated by phase variation,
- a comparison of the application of the two geometries.

This paper is organised as follows. The biological background and the basic model are described in Section 2. In Section 3 we explore the application of the two alternative geometries using Cartesian and polar coordinates, and the analysis is presented in Section 4. We conclude our paper with a brief summary in Section 5. Some additional data are given in the Appendix.

2   Phase Variation in Bacterial Colony Growth

Background. Microbial populations commonly use a stochastic gene switching process called phase variation, controlled by reversible genetic mutations, inversions, or epigenetic modification [14]. Understanding of its adaptive role has traditionally been within the context of "contingency gene theory" [10] in which populations will predictably include variants adapted to "foreseeable" frequently encountered environmental or selective conditions [12]. The mechanistically most common switches are mediated by random mutations in simple sequence repeats, as exemplified by H. influenzae [8], H. pylori [13, 15] and Neisseria [18]. Recent reconsideration suggests a different and additional role for phase variation in the generation of predictable functional diversity within multicellular microbial populations, providing differentiated sub-specializations within structured and predictable communities. Progress in this area requires the design of new models, moving from existing models of population proportions in freely competing populations to ones that include and address spatial and structural composition and interfaces.

The most readily observable compositional effect of phase variation in cultures grown in vitro is colonial sectoring. In this paper we present preliminary stochastic models that address colonial patterning including bi-directional reversible switching between two phenotypes, biologically relevant rates, and differences in the fitness of the two alternate phenotypes. We consider a colony of bacteria with two phenotypes A and B, which develop over time by cell division. Cell division may involve cell mutation, and back-mutation alternates phenotypes; see Fig. 1. We are interested in the proportion of phenotypes in the cell generations, and how their spatial distribution evolves over time.

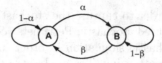

Fig. 1. Phase variation, basic scheme. α / β – forward/backward mutation rate.

Basic Model of Phase Variation. We start with the equations taken from the previous *deterministic* model of phase variation [16], which describe *synchronous* growth in cell colonies with two phenotypes A and B, but no spatial aspects. These equations include the assumption that "if phase variation occurs, the progeny consists of one A and one B." Previously [16], behaviour was explored by iterating the equations on a spreadsheet. We develop a stochastic Petri net ($\mathcal{SPN}$) that is directly executable by playing the token game which facilitates its comprehension, and permits the exploration of the behaviour by standard analysis and simulation techniques. Our initial $\mathcal{SPN}$ model, see Fig. 2, adopts an *asynchronous* modelling approach so that cells divide individually. The model parameters were taken from [16]; α and β represent the forward and backward mutation rates, and d_A, d_B the fitness of phenotype A and B, i.e. the proportions that survive to division.

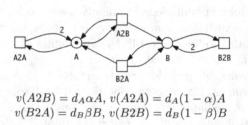

$$v(A2B) = d_A \alpha A, \quad v(A2A) = d_A(1 - \alpha)A$$
$$v(B2A) = d_B \beta B, \quad v(B2B) = d_B(1 - \beta)B$$

Fig. 2. $\mathcal{SPN}$ corresponding to Fig. 1; v – marking-dependent stochastic rates

Derived Measures of Interest. The n-th generation in a synchronous model yields 2^n bacteria. Vice versa, if we know the total number *total* of bacteria generated by asynchronous cell division, then we can obtain the corresponding synchronous generation counter n by

$$n = \log_2 total \tag{1}$$

For example, 26 synchronous generations (which may develop in about 24 hours) end up with a total population size of approximately $67 \cdot 10^6$. We obtain the proportion of phenotypes A and B modelled by the variables A and B by

$$propA = \frac{A}{A + B}; \quad propB = \frac{B}{A + B} \tag{2}$$

Simulating the stochastic model allows us to observe asynchronous population growth such that cells divide individually. Each event (firing of a transition) corresponds to the division of one cell. Consequently, the size of the population will grow in steps by 1, in contrast with the previous synchronous model.

Folding. To prepare for spatial modelling of cell colonies we fold our first (uncoloured) Petri net. We define two colour sets, *Phenotype* and *DivisionType*, see Appendix A provided in the supplementary materials, to fold the two places A and B into one coloured place *cell* with the colour set *Phenotype*, and to fold the four transitions into the coloured transition *division*. We obtain the model in Fig. 3. The derivation of our final model requires three further steps: adding space, controlling colony spreading, and controlling thickness, which we discuss in the next section.

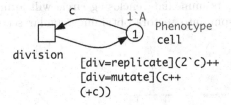

Fig. 3. $\mathcal{SPN}^{\mathcal{C}}$ as $\mathcal{SPN}$ short-hand notation; unfolding this $\mathcal{SPN}^{\mathcal{C}}$ generates the $\mathcal{SPN}$ in Fig. 2

3 Adding Space

The colony is represented in 2.5 dimensions by an explicit 2D grid with an implicit constant maximal height over all grid positions.

3.1 Alternative Geometries

Starting from a small initial population the colony spreads out as the number of bacteria increases maintaining a circular shape throughout its development. Thus, a circular geometry with polar coordinates for representing space seems to be most appropriate for this particular modelling task. However, previous attempts to model bacteria colony growth have represented space employing a rectangular geometry with Cartesian coordinates. Independently of the chosen spatial representation, the 2D space is discretised in compartments which are then mapped to a grid. Each position of the grid is referenced by a unique tuple (x, y), corresponding to a colour tuple in the model, where x is the index of the row and y of the column in the grid, respectively. Differences between modelling in these two coordinate systems will be highlighted next.

Cartesian Coordinate System. In the Cartesian coordinate system [19] approach, the 2D space is discretised by splitting it into equally sized rows and columns obtaining a 2D grid similar to a matrix as shown in Fig. 4. The mapping between this matrix and the compartments of the 2D grid is direct, because each position in the matrix corresponds to a compartment in the grid. The area of all the positions in the grid is equal. The volume of all grid positions is also equal because their maximal height is the same.

When division occurs, the parent remains in situ and the offspring can either stay with the parent or be displaced to a neighbouring position. The neighbourhood relation between different positions of the grid is represented as a function. The maximum number of neighbours for each position is eight depending on whether the considered position is in the interior of the grid, at the edge or in the corner.

Polar Coordinate System. On the other hand, when considering a polar coordinate system [19], the 2D space is discretised in a different manner. First of all, the space is divided into evenly spaced concentric circles. Each one of the concentric circles and its immediate enclosing circle will form an annulus [19]. All annuli are then split into sectors obtaining annular sectors like the ones presented in Fig. 4.

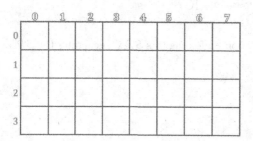 

Fig. 4. Discretising space considering Cartesian (left) and polar (right) coordinates. Each annulus in the polar case is mapped to a row in the grid and each sector to a column, such that a position in the grid (left) has one and only one corresponding annular sector (right) and vice versa.

When running a simulation from the centre of the discretised space, it is important that the offsprings are able to be displaced with equal chance in either of the directions identified by the sectors. For this purpose the origin of the space is considered as a position in the grid which has as neighbours all the immediate surrounding annular sectors. Therefore, the first row of the 2D grid will contain only one entry, the origin.

The number of neighbours for the origin is equal to the number of sectors. Similar to the neighbourhood relation in a Cartesian coordinate system, all other annular sectors have maximum eight neighbours, depending if their position is next to the origin, in the interior or at the edge.

Comparing the Geometries. One of the differences between the two geometries is that when using the rectangular geometry, the area and volume of all positions in the grid are constant while in the circular geometry the area and volume are variable. In case of the circular geometry the variability of the volume of each position in the grid has an effect on the transition rate function. Conversely, in case of the rectangular geometry, the transition rate function is not influenced by the volume of the positions since it is constant.

Another important aspect which sets the two geometries apart is the shape of the compartments due to the discretisation process. Let us compare one row from the grid obtained by discretising the space considering a Cartesian coordinate system and the sector obtained similarly by considering a polar coordinate system. The angle described by a row in the grid equals 0 degrees. Conversely, the sides of the sector form a sharp angle greater than 0 degrees (except when the number of sectors $\to \infty$).

For this particular case study, we are interested in the angle formed by the patterns of high intensity in the colony. Any sector in the circular geometry will automatically have a non-zero degrees angle associated. However, in the rectangular geometry a non-zero degrees angle is formed only if the colony spreads out on multiple rows and columns. In order to obtain comparable results we have removed the diagonal movement in the polar coordinates model such that the horizontal spreading of the colony is reduced.

Representing the Geometries Using Colour Sets. In spite of the multiple differences between the rectangular and circular geometries, the definition of the colour sets used for each Petri net is the same. The *Grid* colour set is equal to the Cartesian product of the *Grid2D* and *Phenotype* colour sets where *Grid2D* represents the two-dimensional grid and *Phenotype* the type of the bacteria; in our case either A or B.

Each Petri net place represents a subset of the discretised space. The maximum number of bacteria in each place is inversely proportional to the resolution of the grid. Increasing the resolution reduces the maximum capacity of the place, while decreasing it makes room for more bacteria.

One crucial difference between the geometries consists of the neighbourhood relation between two positions. This characteristic is captured by the neighbourhood functions *neighbourhood2D_rectangular* and *neighbourhood2D_circular* described in Appendix B, provided in the supplementary materials. They define all possible movements in the net. The neighbourhood function for polar coordinates may appear to be more complicated. However, its length is due to the need of separately considering the neighbours of the origin and not because of an increased complexity.

In this case study we are concerned with mutation rates and their influence on the system behaviour. Therefore, their total values for each position have to be kept constant irrespective of the number of neighbours. Introducing space means technically multiplying the number of transitions (one for each direction). To counterbalance this effect, we scale the transition rates by dividing them by N, where N is the number of neighbours.

3.2   Controlling the Spatial Dynamic Development of the Colony

Controlling Colony Spreading. The probability of staying with the parent or being displaced to a neighbouring position is modelled differently depending on the representation of space.

In the circular case the probability of a bacteria to be displaced to a neighbouring position has to take into account the size of the current position, because the area of the annular sectors is variable. We employ the interior-edge model described in Fig. 5 to capture this aspect. Considering a particular annular sector, the only bacteria which are able to be displaced from this sector to a neighbouring sector are the ones lying on the edge. Assuming that each bacteria can move in 8 directions (N, NW, W, SW, S, SE, E, NE) or remain in situ, only three out of the nine movements of the bacteria on the edge will be to a neighbouring position. The bacteria which lie in the corner are not treated separately in our approach. Thus, the probability of being displaced to a neighbouring position is:

$$P = \frac{3}{9} * \frac{\text{Area}_{\text{edge}}}{\text{Area}_{\text{grid position}}} \tag{3}$$

and the probability of staying with the parent is $1 - P$. $Area_{edge}$ is given by the maximum area which can be occupied by bacteria of size 1x1 μm located around the edge. The difference between the edges and interior of an annular sector is depicted in Fig. 5. $Area_{grid\ position}$ is computed as the total area of the annular sector. Both areas depend on the index i of the annulus to which the sector belongs. The value of i is set to 1 for the origin and is incremented with each enclosing annulus. Thus, the values of the areas are:

$$\text{Area}_{\text{edge}_i} = \frac{2rN + 2\pi r(2i + 1)}{MN}, \quad \text{Area}_{\text{grid position}_i} = \frac{\pi r^2(2i + 1)}{M^2N} \tag{4}$$

where M is the total number of annuli and N the total number of sectors. A step by step description of how the values of $Area_{edge}$ and $Area_{grid\ position}$ are computed is given in Appendix C provided in the supplementary materials.

As the area of annular sectors increases, the ratio between the area on the edge and the total area becomes smaller which means that the probability of a bacterium to be displaced to a neighbouring position decreases. On the other hand, in the rectangular case the area of the grid positions is constant which means that the model from the circular case would impose a constant probability for all positions in the grid. To add more flexibility to the model, the probability of staying with the parent or being displaced to a neighbouring position is modelled using two preference factors γ and ω without changing the total transition rate.

Increasing γ increases the preference to stay with the parent, while decreasing γ increases the preference to be displaced. Conversely, increasing ω increases the preference to be displaced, while decreasing ω increases the preference to stay

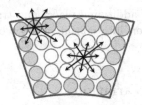

Fig. 5. Interior-edge model used for the circular geometry in order to represent the probability of a bacterium to be displaced to a neighbouring position. Bacteria lying on the edge are highlighted in yellow, bacteria lying in the interior in white and the annular sector boundary in blue.

with the parent. In the general case, the probabilities of staying with the parent or being displaced to a neighbouring position are:

$$P_{stay\ with\ parent} = \frac{\gamma}{\gamma + (\#neighbours * \omega)}$$
$$P_{displace\ to\ neighbour} = \frac{\omega}{\gamma + (\#neighbours * \omega)} \tag{5}$$

In the rectangular case, $\#neighbours$ is equal to 3 if the position on the grid is in the corner, 5 on the edge and 8 in the interior. Conversely in the circular case, $\#neighbours$ is equal to 5 on the edge, 8 in the interior, 6 in the annulus immediately enclosing the origin and "the number of sectors" for the origin.

All probabilities are encoded in the rate function of the transition *division*, irrespective of the employed geometry.

Controlling Thickness. The bacteria generated by cell division can pile up on top of each other and thus increase the colony thickness at that grid position. This thickness is limited because of the cells' requirements for access to oxygen and nutrients. In order to control the thickness we introduce a constant ρ, denoted as $POOLSIZE$ in the SPN^C model, which limits the maximum number of cells at a certain grid position. The constant ρ is set to give room for 26 generations. The entire set of colour-related definitions common to both circular and rectangular spatial representations and the final version of the models are given in Appendix A and D, provided in the supplementary materials. Rate functions are not described here due to space limitations, but they are defined in the computational models made available as supplementary materials.

The only structural difference between the models is that polar coordinates require additionally one Petri net place and two transitions, which are highlighted in green colour in the model (see Appendix D, Fig. 10, provided in the supplementary materials). The pre-transition of the place *pool* accounts for the variable pool size (volume) depending on the annulus to which each sector belongs. The extra place *src_index* and its pre-transition record to which annulus a given sector belongs, information which is used to adapt the rate of the transition *division*. A future version of our modelling tool will allow specifying a variable initial marking for a coloured place and accessing the index of a position in the grid without the need of additional places and transitions. Henceforth, this overhead should not be taken into consideration when comparing the spatial representations.

4   Analysing Phase Variation

4.1   Computational Experiments

The Petri nets were constructed using Snoopy [11], recently extended to support coloured Petri nets [6]. Simulations were run with Snoopy's built-in stochastic simulator and Marcie [7]. Simulation traces have been further processed by customized C++ programs, and finally visualised as images or mp4 movies.

All computational experiments were performed on automatically unfolded Petri nets. Unfolding the coloured Petri net for a 101×101 grid using a rectangular geometry yields an uncoloured Petri net with 30,605 places and 362,405 transitions with an unfolding time of 780 seconds on a regular desktop computer (Intel(R) Core(TM) i5-2500 CPU @ 3.30 GHz processor, 2 GB DDR3 RAM). Similarly, unfolding a coloured Petri net of the same dimensions using a circular geometry yields an uncoloured Petri net with 40,406 places and 382,191 transitions with an unfolding time of 2000 seconds. The number of places and transitions is higher in the circular case due to the overhead required by the current Snoopy version for recording to which annulus each sector belongs.

The unfolded Petri net is simulated using the Gillespie algorithm [4]. The output of the simulation comprises two traces for each grid position, corresponding to the two phenotypes A and B. The analysis follows the development over time of the proportion of the given phenotype in the total population, and the formation of the associated patterns. This requires converting the traces from the stochastic simulations into 2D representations, see Fig. 6, and analysing the development of the 2D sector-like patterns over time. We expect that the model will finally allow the prediction of mutation rates and fitness by counting and extracting information from the pattern segments, which in the future could give new insights into the population dynamics of mutation. Currently, the model predicts behaviour which has not been measured so far in the wet lab in the sense that it generates a time series description of the evolution of the patterns in the bacteria colony, while wet lab data just provide snapshots of final states.

4.2   Parameter Scanning

When the mutation rates are fixed, different combinations of values for parameters ρ and ω will result in different simulation outcomes. One batch of simulations was run for each parameter ρ and ω by choosing random values from the parameter space in order to observe how the behaviour is affected.

Changing ρ. In the first batch of simulations, all parameters were kept constant, except ρ, which had a different value for each run. The values for ρ were selected by starting with an initial value and linearly increasing it after each run.

Fig. 6. 2D representation of the final state of 4 stochastic simulations, 2 for rectangular (left) and 2 for circular (right), illustrating the development of sector-like patterns. Due to the stochastic nature of the simulations, the output is different in every run. The value of propB, see Equation 2, is encoded by colour. Yellow indicates patches with high density of phenotype B, dark purple patches of high density of phenotype A, red patches of approximately equal proportions. The black background shows the grid area not covered by phenotype B.

In the rectangular case the volume or capacity is constant throughout the grid, whereas in the circular case it is not. Therefore, ρ has a different interpretation depending on the chosen spatial representation. For comparison purposes, it is better to consider the *maximum height* of the colony which is constant throughout the entire grid for both geometries. Experiments with the same heights and corresponding ρ's were carried out for both geometries and two characteristic results for each one of the geometries are depicted in Fig. 7a-7d.

Increasing the value of the parameter ρ increases the maximum height of each grid position which implies that more bacteria can pile up onto each other. Since the number of generations is fixed and the maximum height limit of the colony was increased, it is to be expected that the final width of the colony is reduced; this can be observed in Fig. 7a-7d. The value of ρ was chosen for both geometries in such a way that the most outwards bacteria with respect to the centre do not reach the edge of the grid. The reason for this is that we expected some back-propagation of bacteria from the edge of the grid to affect the final outcome of the simulation.

Changing ω. The second batch of simulations changed only the value of ω for each run. Similar to the selection of values for ρ, the values for ω have been randomly selected from the entire search space. Images representing the final states of two simulations for each geometry are given in Fig. 7e-7h. The probability of the offspring to stay with the parent or be displaced to a neighbouring position depends on the dimensions of the grid position. All grid positions are equally-sized in the model using Cartesian coordinates, which means that the probabilities of staying/being displaced are constant. However, the area of the grid positions in the model using polar coordinates is different, which means that the probabilities are different as well. The value of ω specified as caption for the polar coordinates model in Fig. 7e-7h corresponds to the most outward annular sectors (i.e. annular sectors with the biggest area).

Considering that the value of γ is fixed, the preference of the offspring to be displaced to a neighbouring position is directly proportional to the value of ω. Increasing ω increases the chance of the offspring to be displaced which means that the clear cut between high and low density areas in the images fades away.

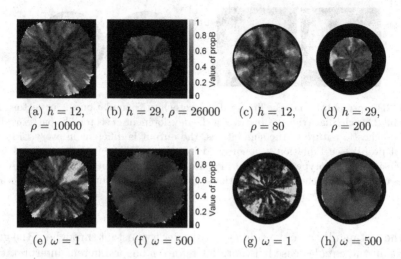

(a) $h = 12$, (b) $h = 29$, $\rho = 26000$ (c) $h = 12$,   (d) $h = 29$,
    $\rho = 10000$ $\rho = 80$ $\rho = 200$

(e) $\omega = 1$ (f) $\omega = 500$ (g) $\omega = 1$ (h) $\omega = 500$

Fig. 7. Different values of the parameter ρ, and implicitly maximum height (h), for the Cartesian coordinate system (a, b) and the polar coordinate system (c, d). Different values of the parameter ω for the Cartesian coordinate system (e, f) and the polar coordinate system (g, h).

Thus, in Fig. 7e-7h the images corresponding to a higher value of ω have a more uniform distribution of concentrations than the ones in which ω was smaller.

4.3   Sector Analysis

In the beginning, the analysis of the sectors was done by looking at the images of the colony at different time points and deciding if the sector-like patterns are similar to the ones in the wet lab. Unfortunately, only few images from the wet lab are available. New wet lab experiments are ongoing, but images of the colonies can not be provided yet.

For the purpose of improving the assessment of results, there was a need to formalise the analysis of sectors. The following set of measures was defined to describe the patterns from the final state of the simulation: *Area, angle* described by the sides, *distance* from the centre of the grid, and the *total number* of sectors.

Using specific image processing techniques from the open source computer vision library OpenCV [1], a sector detection module was implemented which takes images as input. The main steps of the algorithm are given in Appendix E.1, provided in the supplementary materials.

The advantage of the algorithm working directly with images and not with the raw output of the simulation is that the images can originate from either dry or wet lab. Thus, our analysis approach is generic. Since the experiments in the wet lab are still ongoing, the image processing procedure was validated only on in silico generated images, but our expectation is that the approach should work similarly well on images from the wet lab.

Results. One thousand stochastic simulations were run for both the rectangular and circular model with an average simulation time of 50 minutes. Images were generated from the final states of the simulations which were then provided as input to the sector analysis module. An example of the result of the sector detection procedure for each geometry is depicted in Appendix E.2, provided in the supplementary materials.

The output of the analysis procedure are csv files containing information about the area, angle, distance from the centre and number of detected sectors. The averaged results from all simulations for both the rectangular and circular case are described in Table 1. We employed a two-sample statistical test for comparing the results. The data corresponding to all measures and both geometries was tested for normality using the Shapiro-Wilk [17] and the Q-Q plot [21] methods. In all cases the null hypothesis, i.e. that the sample data is drawn from a normal distribution, was rejected. Thus, we tested if the sample data for both geometries is drawn from the same distribution using the Mann-Whitney [9, 20] non-parametric test. Similarly, the null hypothesis, i.e. that the sample data are drawn from the same distribution, was rejected. The p-values obtained for all tests are given in Table 2, Appendix E.3, provided in the supplementary materials.

Table 1. Rectangular ($\square$) and circular ($\bigcirc$) sector analysis with μ – mean, σ – standard deviation, c_v – coefficient of variation. Area and distance (from the centre) are given with respect to total grid area and maximum distance from the centre.

Measures	Area		Distance		Angle		Sectors	
	$\square$	$\bigcirc$	$\square$	$\bigcirc$	$\square$	$\bigcirc$	$\square$	$\bigcirc$
μ	3%	5%	41%	39%	56°	78°	1.47	1.78
σ	2%	2%	17%	16%	18°	25°	1.14	1.03
c_v	0.93	0.62	0.40	0.41	0.32	0.32	0.77	0.58

Both area and angle have higher values in the circular than in the rectangular case, which is to be expected due to the different 2D space discretisation. Sectors in the circular geometry inherently have a non-zero degree angle associated, while rows in a rectangular geometry do not. Moreover, the area of the annular sectors is increasing as they are farther away from the centre of the grid. Conversely, the area of all positions in the rectangular geometry is constant. The number of sectors is slightly bigger in the circular case because the bacteria from the starting position can be displaced in maximum "number of sectors" directions, while in the rectangular case only in maximum 8. Finally, the distance of the sectors from the grid centre is approximately equal for both geometries. Thus, according to these results the distance from the centre is the only reliable measure which has similar values for both geometries. Running batches of more simulations will increase the accuracy of the results and more fine-grained conclusions can be drawn. Histograms and corresponding normal distribution curves for all measures have been plotted and added to Appendix E.3 in order to complement the analytical comparison of the results described above.

5 Summary

In this paper we have described a methodology of modelling bacterial colonies which evolve in time and space using rectangular and circular geometries, and a procedure for sector-like patterns detection and analysis.

Currently it is not possible to state which geometry is more appropriate for the phase variation case study, because there are not sufficient images from the wet lab against which to validate our results. The emphasis of this paper is on the generic methodologies which we developed and which can be employed for different case studies modelled using coloured Petri nets. Work is ongoing in the wet lab to generate images of actual bacterial colonies which will then be used as targets for model fitting in order to generate more accurate computational models for describing bacterial colony growth under different conditions.

In the future we plan to extend our spatial modelling framework from 2.5D (i.e. 2D and implicitly modelling height) to full 3D representation which would allow the simulation and observation of more detailed aspects of bacterial colonies. We also want to extend our sector detection and analysis procedure from working with 2D sector-like patterns to linear and non-linear 3D surfaces.

All supplementary materials and appendices are made available at http:// people.brunel.ac.uk/~cspgoop/data/cmsb2013.

Acknowledgments. This research has been partially funded by the British EPSRC Research Grant EP I036168/1 and the German BMBF Research Grant 0315449H. Ovidiu Pârvu is supported by a scholarship from Brunel University. We would like to gratefully acknowledge the comments of the reviewers which have helped us to improve the quality of the paper.

References

1. Bradski, G., Kaehler, A.: Learning OpenCV: Computer Vision with the OpenCV Library. O'Reilly, Cambridge (2008)
2. Gao, Q., Gilbert, D., Heiner, M., Liu, F., Maccagnola, D., Tree, D.: Multiscale Modelling and Analysis of Planar Cell Polarity in the Drosophila Wing. IEEE/ACM Transactions on Computational Biology and Bioinformatics 99(PrePrints), 1 (2012)
3. Gilbert, D., Heiner, M., Liu, F., Saunders, N.: Colouring Space - A Coloured Framework for Spatial Modelling in Systems Biology. In: Colom, J.-M., Desel, J. (eds.) PETRI NETS 2013. LNCS, vol. 7927, pp. 230–249. Springer, Heidelberg (2013)
4. Gillespie, D.: Exact stochastic simulation of coupled chemical reactions. The Journal of Physical Chemistry 81(25), 2340–2361 (1977)
5. Heiner, M., Gilbert, D.: How Might Petri Nets Enhance Your Systems Biology Toolkit. In: Kristensen, L.M., Petrucci, L. (eds.) PETRI NETS 2011. LNCS, vol. 6709, pp. 17–37. Springer, Heidelberg (2011)

6. Heiner, M., Herajy, M., Liu, F., Rohr, C., Schwarick, M.: Snoopy – A unifying Petri net tool. In: Haddad, S., Pomello, L. (eds.) PETRI NETS 2012. LNCS, vol. 7347, pp. 398–407. Springer, Heidelberg (2012)

7. Heiner, M., Rohr, C., Schwarick, M.: MARCIE - Model checking and Reachability analysis done Efficiently. In: Colom, J.-M., Desel, J. (eds.) PETRI NETS 2013. LNCS, vol. 7927, pp. 389–399. Springer, Heidelberg (2013)

8. Hood, D.W., Deadman, M.E., Jennings, M.P., Bisercic, M., Fleischmann, R.D., Venter, J.C., Moxon, E.R.: DNA repeats identify novel virulence genes in haemophilus influenzae. Proceedings of the National Academy of Sciences of the United States of America 93(20), 11121–11125 (1996), PMID: 8855319

9. Mann, H.B., Whitney, D.R.: On a Test of Whether one of Two Random Variables is Stochastically Larger than the Other. The Annals of Mathematical Statistics 18(1), 50–60 (1947)

10. Moxon, E.R., Rainey, P.B., Nowak, M.A., Lenski, R.E.: Adaptive evolution of highly mutable loci in pathogenic bacteria. Current Biology: CB 4(1), 24–33 (1994), PMID: 7922307

11. Rohr, C., Marwan, W., Heiner, M.: Snoopy - a unifying Petri net framework to investigate biomolecular networks. Bioinformatics 26(7), 974–975 (2010)

12. Salaün, L., Ayraud, S., Saunders, N.J.: Phase variation mediated niche adaptation during prolonged experimental murine infection with helicobacter pylori. Microbiology 151(pt. 3), 917–923 (2005), PMID: 15758236

13. Salaün, L., Linz, B., Suerbaum, S., Saunders, N.J.: The diversity within an expanded and redefined repertoire of phase-variable genes in helicobacter pylori. Microbiology 150(pt. 4), 817–830 (2004), PMID: 15073292

14. Salaün, L., Snyder, L.A., Saunders, N.J.: Adaptation by phase variation in pathogenic bacteria. Advances in Applied Microbiology 52, 263–301 (2003), PMID: 12964248

15. Saunders, N.J., Peden, J.F., Hood, D.W., Moxon, E.R.: Simple sequence repeats in the helicobacter pylori genome. Molecular Microbiology 27(6), 1091–1098 (1998), PMID: 9570395

16. Saunders, N., Moxon, E., Gravenor, M.: Mutation rates: estimating phase variation rates when fitness differences are present and their impact on population structure. Microbiology 149(2), 485–495 (2003)

17. Shapiro, S.S., Wilk, M.B.: An analysis of variance test for normality (complete samples). Biometrika 3(52) (1965)

18. Snyder, L.A., Butcher, S.A., Saunders, N.J.: Comparative whole-genome analyses reveal over 100 putative phase-variable genes in the pathogenic neisseria spp. Microbiology 147(pt. 8), 2321–2332 (2001), PMID: 11496009

19. Weisstein, E.W.: Wolfram MathWorld, http://mathworld.wolfram.com

20. Wilcoxon, F.: Individual Comparisons by Ranking Methods. Biometrics Bulletin 1(6), 80–83 (1945)

21. Wilk, M.B., Gnanadesikan, R.: Probability plotting methods for the analysis of data. Biometrika 55(1), 1–17 (1968)

Using Probabilistic Strategies to Formalize and Compare α-Synuclein Aggregation and Propagation under Different Scenarios

Lucian Bentea[1], Peter Csaba Ölveczky[1], and Eduard Bentea[2]

[1] Department of Informatics, University of Oslo
[2] Center for Neurosciences, Vrije Universiteit Brussel

Abstract. We use PSMaude to define a formal real-time model of the aggregation and interneuronal propagation of the α-synuclein (α-syn) protein causing Parkinson's disease (PD). To the best of our knowledge, this is the first executable formal model of the propagation of α-syn aggregates through a neural network that is dynamically changing as a consequence of neuronal death. We then define different probabilistic strategies on top of our model to formalize the aggregation and propagation of α-syn in three different scenarios: (i) in a healthy person, (ii) in a person predisposed to PD, and (iii) in a predisposed person that is given some treatment with rapamycin. We use PSMaude to simulate our model in these different scenarios.

1 Introduction

The α-synuclein (α-syn) is a protein with unknown function that has been linked to the progression of Parkinson's disease (PD) [21]. It is necessary to understand how this process evolves under different pathobiological scenarios, e.g., in a healthy brain, or in a brain that is predisposed to PD, in order to find different ways to prevent or delay the onset and progression of PD. It would therefore be very valuable to have a model of the spread of this protein that would allow us to reason about the effect of different therapies, for example by realistic simulations and other forms of formal analysis. However, defining useful models of the spread of α-syn poses a number of challenges for formal modeling, including:

- The need to model *both aggregation* and *propagation* of α-syn in the brain.
- The need to model brain regions. Neurons behave differently in different regions; e.g., neurons in the midbrain can withstand less toxicity than neurons in other regions. Furthermore, α-syn propagates differently between neurons in the same brain region and between neurons in different brain regions.
- To model propagation, a model of the neural networks, that are changing dynamically due to the aggregation-induced death of neurons, is needed.
- It is also important to make a distinction between the different types of aggregates, since they have different dynamics and grow at different rates.
- Both *probabilistic* and *real-time* behaviors must be modeled.

A. Gupta and T.A. Henzinger (Eds.): CMSB 2013, LNBI 8130, pp. 92–105, 2013.

- The probabilities with which certain events take place depend on both the *age* and the *state* of the brain, leading to complex feedback loops.
- The probabilities differ in the three pathobiological scenarios that we would like to analyze (healthy person, person predisposed to PD, and PD patients receiving some treatment). It is desirable to have a modular way of modeling the different settings without redefining the entire model.

This is a tall order, and we are not aware of any previous executable formal model of the propagation of α-syn in (dynamically changing) neural networks.

In this paper, we use a recent extension of the rewriting-logic-based Maude system [9], called PSMaude [4], to formally model and simulate the aggregation and propagation of α-syn under the above scenarios. A novel feature of PSMaude is its expressive *probabilistic strategy language* that is used to quantify the nondeterminism in a probabilistic rewrite theory to obtain a fully probabilistic model that can be simulated and statistically model checked in PSMaude.

We model different types of α-syn aggregates, cell defenses against aggregation, the breakage of aggregates, the propagation of aggregates through the neural network, and the death of neurons due to high toxicity, that leads to a dynamically changing topology of the neural network. We then use PSMaude to formalize the different pathobiological conditions. In particular, we simulate the effect of rapamycin—a drug known to increase the efficiency of cell clearance mechanisms—in decreasing pathology in persons predisposed to PD.

We meet the above modeling challenges as follows:

- The expressiveness of rewriting logic allows us to define a hierarchical object-oriented model that models the brain as a network of different brain region objects, each of which contains its neural network, and where inside each neuron we have a multiset of α-syn species.
- We use well established techniques from Real-Time Maude [19] to model real-time behaviors in object-oriented PSMaude models.
- PSMaude's strategy language enables the following modeling methodology: We model all possible actions by rewrite rules. The different probabilities with which the various events in the three scenarios (healthy, predisposed, and taking medication) take place are then defined by three probabilistic strategies on top of our "base model." Furthermore, these probabilistic strategies can depend on both the age and the state of the brain.

A major obstacle to modeling α-syn aggregation and propagation is that important parameters, such as the different kinetic rates, are unknown. We have incorporated all the medical facts we are aware of into our model, but had to estimate a number of parameters. For example, the probabilities of certain actions change with age, and we have defined the probabilistic strategies as functions of the age of the person. However, the values of these probabilities are largely unknown. The point is that once the numbers become known, they can very easily be plugged into our model.

We then perform PSMaude simulations in the different settings. Since other formal approaches typically only perform Gillespie simulations, and since the size and complexity of reasonably realistic models and states make it hard to perform meaningful statistical model checking, we have only simulated our model.

Due to the space limitations, large parts of our model and strategies cannot be included in this paper. We refer to the longer report [5] for additional details and discussions; in particular, that report mentions and cites a lot of recent work in PD research that provides the basis for our model.

2　Preliminaries

Rewriting Logic and Maude. In rewriting logic the data types of the system are defined by an algebraic equational specification (Σ, E) (where Σ is a signature declaring sorts, subsorts, and function symbols, and E is a set of conditional equations), and the system's local state changes are defined by a set R of rewrite rules $(\forall \vec{x})\ l : t \longrightarrow t'$ **if** $cond$, where l is a label, t and t' are terms, $cond$ is a conjunction of equalities, and $\vec{x}$ is the set of all variables in the rule. In *probabilistic rewrite theories* [2] the righthand side t' may contain variables that do not occur in t, and that are instantiated according to a probability measure taken from a *family* π_l of probability measures—one for each instantiation of the variables in t.

Maude [9] is a high-performance simulation, reachability analysis, and LTL model checking tool for rewrite theories. Conditional rules are written `crl [l]: t => t' if cond`. In *object-oriented* Maude specifications, the state is a term of sort `Configuration` denoting a multiset of objects and messages, with multiset union denoted by juxtaposition. A declaration `class C | att₁:s₁, ..., attₙ:sₙ` declares a class C with attributes $att_1, \ldots, att_n$ of sorts $s_1, \ldots, s_n$, respectively. Objects are represented as terms $< o : C\ |\ att_1 : val_1,\ \ldots,\ att_n : val_n >$, where o is the object's identifier, C is the object's class, and $val_1, \ldots, val_n$ are the attribute values. If the sort s_i of an attribute att_i is `Configuration`, the attribute contains a subconfiguration, giving a *hierarchical* specification. A rule of the form

```
rl [1]: < O : C | a1 : x,      a2 : O', a3 : z > < O' : C' | b1 : w,      b2 : O >
     => < O ; C | a1 : x + w, a2 : O', a3 : z > < O' : C' | b1 : w + x, b2 : O >.
```

defines a family of transitions in which two objects O and O', of classes C and C', resp., synchronize and update their attributes; e.g., O adds the value w of the attribute b1 of O' to its attribute a1 whose current value is x. "Irrelevant" attributes (such as a3 and the righthand side occurrence of a2) may be omitted.

PSMaude. PSMaude [4] allows the probabilistic quantification of nondeterminism in probabilistic rewrite theories. The ways in which the nondeterminism is quantified are specified as *probabilistic strategies* on top of the "base" nondeterministic (and possibly probabilistic) model. PSMaude extends Maude by adding support for specifying both probabilistic rules and probabilistic strategies, and by providing a *probabilistic* rewrite command and a statistical model checker to analyze a probabilistic rewrite theory controlled by a probabilistic strategy.[1] Such a strategy consists of three parts: *(i)* a *rule strategy* that specifies the relative likelihood of applying a certain rewrite rule by giving each

---

[1] PSMaude is available at `http://heim.ifi.uio.no/~lucianb/prob-strat/`

rule (label) l a weight $w_l(s)$ that may depend on the state s; *(ii)* a *context strategy* that specifies the relative likelihoods of different components in the system to be involved in an application of a rule; *(iii)* a *substitution strategy* that specifies the relative likelihoods of the components involved in a rule application to have *different roles* in the event. A rule strategy is defined by `psdrule` *RuleStrat* `:=` `given state:` s `is:` $(l_1$ `->` $w_{l_1}(s)$ `;` $\ldots$ `;` l_n `->` $w_{l_n}(s))$, where $w_{l_i}(s)$ is the relative weight of the rule labeled l_i in state s. A context strategy `psdcontext` *CtxStrat* `:=` `given state:` s `rule:` l `is:` $C(s)$ `->` $w_{ctx}(s,l)$, defines the relative weight of applying the rule l in context $C(s)$ to be $w_{ctx}(s,l)$.

Parkinson's Disease. *Parkinson's disease* (PD) is a movement disorder characterized by the accumulation of a certain type of *protein aggregates* in the human brain. There are two main forms of PD: *(i)* sporadic PD, thought to be due to exposure to high levels of *environmental stress*, e.g., exposure to high levels of pesticides [22]; and *(ii)* familial PD, due to a genetic predisposition to PD.

The brain is divided into several *regions*, shown in Fig. 1, each containing *neurons* that are connected to one another and to neurons in neighboring regions. The space surrounding the neurons is called *extracellular space*. As shown in Fig. 2, a neuron is composed of a cell body, called *soma*, a number of *dendrites*, and an *axon*. A connection between an axon *terminal* and a dendrite of another neuron is called a *synaptic connection*.

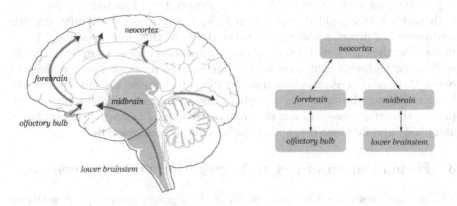

Fig. 1. Typical progression of PD (left) and connections between brain regions (right)

The aggregates in PD are mainly composed of a protein called *α-synuclein* (α-syn), that is synthesized in a neuron's soma. Single α-syn proteins, called *monomers*, may be in two different conformations: a *native* one, and a *misfolded* one that is believed to play a key role in PD development [6]; they may also bind to form large *aggregates*. Three main types of aggregates are linked to PD pathology: *(i)* small soluble aggregates, called *oligomers*, or *protofibrils*; *(ii)* moderately large insoluble aggregates called *fibrils*, that may fragment; and *(iii)* large insoluble aggregates, called *Lewy bodies* (LB), the pathological hallmark of PD.

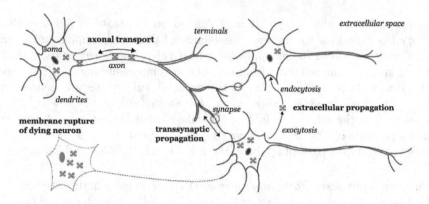

Fig. 2. The main means of propagation of pathogenic aggregates in PD

α-syn may spread through a neural network in several ways, shown in Fig. 2: *(i) extracellular propagation* between two neurons whose somas are close, but that are not connected synaptically; *(ii) transsynaptic propagation* between neurons in neighboring regions; *(iii) release into extracellular space* through the ruptured membrane of a dying neuron, from where α-syn may be uptaken by neighboring neurons.

A neuron has several defense mechanisms against protein aggregation, including: *(i)* the *proteasome* that regulates the concentration of proteins by degrading both native and misfolded proteins; and *(ii)* the *lysosomes* that may degrade any number of proteins and aggregates at the same time by *autophagy*. PD patients show deficient cell defenses against aggregation and an increase in certain pathogenic pathways that increase the aggregation propensity and toxicity of α-syn. A promising approach in treating PD is to increase the efficiency of the lysosomal system by enhancing autophagy inside each neuron. Following this approach, the drug *rapamycin* was shown in a preclinical PD model [20] to enhance autophagy and decrease α-syn aggregation and neurodegeneration.

3   Formalization of α-syn Aggregation and Propagation

This section presents a PSMaude model of the aggregation and propagation of α-syn in the brain. Our model is a hierarchical object-oriented one, in which one *brain* object contains a set of *brain region* objects, each containing the region's neural network, where each neuron object contains a multiset of α-syn monomers and aggregates. Our model describes *all possible behaviors* in the system and has 15 rewrite rules; we show a few of them and refer to [5] for more details.

The brain and the brain regions are modeled as objects of the classes

```
class Brain      | age : Nat, brainRegions : Configuration .
class BrainRegion | neuralNetwork : Configuration, neighbors : OidSet .
```

where the `age` attribute denotes the age of the brain in hours, and `brainRegions` contains a set of `BrainRegion` objects, whose `neighbors` attribute denotes the names of the neighboring brain regions, and `neuralNetwork` denotes the set of neurons and extracellular spaces in the brain region. Such neurons and extracellular spaces are instances of the classes `Neuron` and `ExtSpace`, respectively:

```
class ProteinSpace | proteins : ASynMSet .
class Neuron .  class ExtSpace .  subclasses Neuron ExtSpace < ProteinSpace .
```

where `proteins` is a multiset of α-syn monomers and aggregates, giving the contents of a neuron or extracellular space. Oligomers are represented by terms `oligomer(m)` and aggregation nuclei by `aggNucleus(m)`, with m a "++"-separated multiset of α-syn monomers. The term `dly(s, τ)` represents a *delayed species*, i.e., a species s which can only take part in further biochemical reactions after τ time units have elapsed.

We identify neurons and extracellular spaces by pairs `nid(g,n)` and `eid(g,n)`, respectively, where g is the name of the brain region in which they are located, and n is a number that identifies them inside region g. In this paper we assume a simplified arrangement of neurons within the same brain region in a square lattice, as shown in Fig. 3, that we formalize by a functions `reachable(n,n')` which is `true` iff the neurons n and n' are neighbors or are the same neuron. However, our model is parametric in the topology of the neural network, and can easily be adapted to a desired topology by redefining the function `reachable`.

An oligomer that has shrunk to a single monomer is a monomer. If instead its size is larger than a certain bound, it is an aggregation nucleus, and vice versa, and an aggregation nucleus can grow to a Lewy body:

```
var MN : ASynMonomer .  var MS : ASynMonomerMSet .
 eq    oligomer(MN) = MN .
ceq    oligomer(MS) = aggNucleus(MS) if noMonomers(MS) >  maxOligomerSize .
ceq aggNucleus(MS) = oligomer(MS)    if noMonomers(MS) <= maxOligomerSize .
ceq aggNucleus(MS) = LB(MS)          if noMonomers(MS) >  maxAggNucleusSize .
```

If the number of aggregates in a neuron exceeds its toxicity threshold, the neuron dies and all its content is released into the extracellular space in its vicinity:

```
ceq < nid(S, N) : Neuron  | proteins : ASYNMSET >
   < eid(S, N) : ExtSpace | proteins : ASYNMSET' >
 = < eid(S, N) : ExtSpace | proteins : ASYNMSET' ASYNMSET >
if toxicity(ASYNMSET) > toxicityThreshold(S) .
```

where `toxicityThreshold(S)` is the minimum number of aggregates to cause the death of a neuron in the brain region `S`, and `toxicity(ASYNMSET)` is a weighted sum of the number of different α-syn species in `ASYNMSET` [5].

α-syn Aggregation and Dissociation. The following rules model the formation of different types of aggregates inside a neuron's soma. Based on an *in vitro* study of α-syn aggregation in [14], we approximate the time it takes for a single monomer to bind as follows: 10 hours to bind to another monomer, 5 hours to bind to an oligomer, and 1 hour for a monomer to bind to an aggregation nucleus.

An oligomer inside a neuron's soma may recruit a (native or misfolded) α-syn monomer MN inside its soma, or it may lose one:

```
rl [oligomerElongation]:
    < O : Neuron | proteins : ASYNMSET oligomer(MS) MN >
=> < O : Neuron | proteins : ASYNMSET dly(oligomer(MS ++ MN), 5) > .


rl [oligomerDissociation]:
    < O : Neuron | proteins : ASYNMSET oligomer(MS ++ MN) >
=> < O : Neuron | proteins : ASYNMSET dly(oligomer(MS), 5) dly(MN, 5) > .
```

The following rule models how an aggregation nucleus inside a neuron's soma may break into two aggregates which may equal either aggregation nuclei or oligomers. We estimate the duration of a single such fragmentation to 10 hours [16]:

```
rl [aggNucleusFragmentation]:
    < O : Neuron | proteins : ASYNMSET aggNucleus(MN ++ NEMS ++ MN' ++ NEMS') >
=> < O : Neuron | proteins : ASYNMSET dly(aggNucleus(MN  ++ NEMS),  10)
                                      dly(aggNucleus(MN' ++ NEMS'), 10) > .
```

α-syn Propagation. The *transsynaptic* propagation of protein aggregates between two brain regions takes place along the thick edges represented in Fig. 3, denoting the axons of neurons. The *extracellular* propagation of α-syn (via exocytosis and endocytosis) may only take place between neighboring neurons within the same brain region. Based on the results in [12], we assume that any propagation event takes 16 hours for small aggregates, and is instantaneous for α-syn monomers.

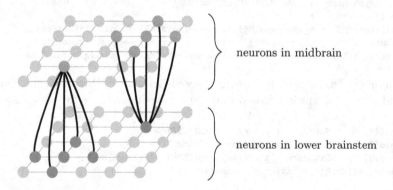

neurons in midbrain

neurons in lower brainstem

Fig. 3. Square lattice arrangement of neurons within each brain region, and possible means of transsynaptic propagation between neurons in neighboring brain regions

A small α-syn species SMALL can be transported transsynaptically[2] through the axon of a neuron nid(S, N) in the brain region S to some neuron nid(S', N') in a neighboring region S', if these neurons have the same or neighboring positions in their corresponding lattices (as shown by the thick lines in Fig. 3):

---

[2] We assume that Lewy bodies cannot be propagated this way due to their large size.

```
crl [synapticPropagation]:
   < S  : BrainRegion | neighbors : (S', OS),
       neuralNetwork : < nid(S, N) : Neuron | proteins : ASYNMSET SMALL > NNET >
   < S' : BrainRegion |
       neuralNetwork : < nid(S',N') : Neuron | proteins : ASYNMSET' >      NNET' >
=> < S  : BrainRegion |
       neuralNetwork : < nid(S, N) : Neuron | proteins : ASYNMSET >       NNET >
   < S' : BrainRegion |
       neuralNetwork : < nid(S',N') : Neuron | proteins : ASYNMSET'
                       dly(SMALL, if SMALL :: ASynMonomer then 0 else 16 fi)
                   > NNET' >
if reachable(N, N') .
```

Cell Defenses. A neuron's lysosomal system may degrade small α-syn species by autophagy:

```
rl [lysosomalAutophagy]: < O : Neuron | proteins : ASYNMSET SMALL >
                    => < O : Neuron | proteins : ASYNMSET > .
```

Time Elapse. We are only aware of one *in vitro* study of the kinetics of α-syn aggregation [14], which does not consider the stochastic fluctuations in the kinetics, but instead provides kinetic curves based upon which we estimate fixed durations of these reactions. We have therefore defined a real-time model with deterministic delays instead of stochastic delays, where the following rule models the elapse of time:

```
rl [tick]: < "Brain" : Brain | age : T,     brainRegions : REGIONS >
       => < "Brain" : Brain | age : T + 1, brainRegions : delta(REGIONS, 1) > .
```

where $\mathtt{delta}(m, \tau')$ defines the effect of time elapse by τ' units on the given multiset m, by decreasing the remaining delay τ of each delayed species $\mathtt{dly}(s, \tau)$ by τ' units in the neural network of each brain region [19]. This tick rule, advancing time by one hour, may also be applied when instantaneous rules are enabled.

4   Probabilistic Strategies

The model in Section 3 specifies all the possible actions that may happen in the aggregation and propagation of α-syn. However, it does not take into account that some events happen with higher probability than others. One could of course incorporate these probabilities directly into the model, but that solution has some drawbacks: *(i)* given the complex global dependencies between rules, it would be quite hard and ugly to try to encode such global properties into "local" probabilistic rewrite rules; and *(ii)* these probabilities are *different* for the different kinds of individuals that we want to analyze. Therefore, we follow a modeling and analysis methodology made possible by PSMaude: A simple and uncluttered rewrite theory defines all possible behaviors. The different global probabilities are then defined as different *probabilistic strategies* in PSMaude—on top of the "base" model—allowing us to probabilistically simulate our model.

In this section, we define three probabilistic strategies that formalize different probabilistic settings: *(i)* that of a healthy person, *(ii)* that of a person predisposed to sporadic PD, and *(iii)* that of a PD patient who is given a daily dose of rapamycin. These strategies are quite complex. For example, the efficiency of the body's defense mechanism decreases with age; therefore, the probability of applying a rule modeling such a defensive action decreases with the age of the brain. Likewise, the efficiency of a neuron's defense mechanisms is inversely proportional to the number of α-syn aggregates in its soma, i.e., to α-syn toxicity, which in the next stages promotes α-syn aggregation.

Since the kinetic rates for the different processes involved in the aggregation and propagation of α-syn are currently unknown, the weights that we set in the following strategies only approximate reality. We also made a series of assumptions, e.g., that the synthesis of α-syn and the propagation of α-syn monomers and aggregates take place at the same rate in all brain regions, etc. In our strategies they are encoded as *uniform distributions* over the corresponding choices. We refer to [5] for details.

Healthy Individual. We first set the likelihoods of different rules to be applied in the brain of a healthy person, who is *not* predisposed to PD:

```
psdrule HealthyRuleStrat :=
  given state: < "Brain" : Brain | age : AGE, ATTS >
          is: --- time elapse                        --- alpha-syn synthesis
              ( tick ) -> 20 ;                        ( synthesis ) -> 20 ;
              --- monomer folding dynamics
              ( toMisfolded ) -> (70 + AGE quo 200) ;    ( toNative ) -> 10 ;
              --- aggregation dynamics
              ( dimerization )           -> (20 + AGE quo 400) ;
              ( oligomerElongation )     -> (20 + AGE quo 400) ;
              ( aggNucleusElongation )   -> (20 + AGE quo 400) ;
              ( oligomerDissociation )   -> 2 ;
              ( aggNucleusFragmentation ) -> 5 ;
              --- cellular defense mechanisms
              ( proteasomalDegradation ) -> (50 / ((AGE quo 200) + 2)) ;
              ( lysosomalAutophagy )     -> (50 / ((AGE quo 200) + 2)) ;
              --- propagation dynamics
              ( exocytosis ) -> 8 ; ( endocytosis ) -> 4 ;
              ( synapticPropagation ) -> 8 ;
              --- extracellular dynamics
              ( dissolution ) -> 1 .
```

Since the weight of the `exocytosis` rule is 8, and the weight of `endocytosis` is 4, the probability of applying the `exocytosis` rule is twice that of applying the `endocytosis` rule in a state where both rules are enabled. The probabilities depend on the current age, e.g., the relative weight of rule `toMisfolded` is $70 + \text{AGE}/200$ (where `AGE` is the value of the `age` attribute in the `Brain` object in the current state); i.e., the probability of applying this rule increases with a person's age [10]. Likewise, the probability of selecting a rule modeling cell defense decreases with age [10].

A context strategy defines to which "components" a selected rule should be applied. We show the context strategy for the rule toMisfolded, which models how native α-syn monomers may become misfolded. The probability for a monomer to fold into a non-native conformation is different from one brain region to another [15], and is given by a function oxStressVulnerability that also depends on the number of oligomers in the neuron.

```
psdcontext HealthyContextStrat :=
  given state:
    < "Brain" : Brain | age : AGE,
      brainRegions : < S : BrainRegion | neighbors : OS,
                   neuralNetwork : < O : Neuron | proteins : ASYNMSET > NNET >
                REGIONS >
      rule: toMisfolded
        is: (< "Brain" : Brain | age : AGE,
               brainRegions :
                 < S : BrainRegion | neighbors : OS, neuralNetwork : [] NNET >
                 REGIONS >)
           -> ( oxStressVulnerability(S, noOligomers(ASYNMSET)) ) .
```

This context strategy sets the relative weight of applying the rule toMisfolded to neuron O in region S to oxStressVulnerability(S, noOligomers(ASYNMSET)) (the hole variable [] denotes the place of the rewrite).

Individual Predisposed to PD. In the *sporadic* form of PD, the likelihood of α-syn synthesis is the same in predisposed individuals as in healthy people. However, by assigning a larger weight $90 + \text{AGE}/200$ to rule toMisfolded we model that for persons predisposed to sporadic PD, different causes for oxidative stress are possible. Similarly, we set a larger weight of $40 + \text{AGE}/400$ to the rules dimerization, oligomerElongation and aggNucleusElongation, to model how oxidative stress due to aging, as well as other factors in individuals predisposed to sporadic PD, further promote α-syn aggregation. Finally, the cell defense mechanisms are also about 5 times less efficient than in a healthy individual; we therefore divide by 5 the weights associated with the rules proteasomalDegradation and lysosomalAutophagy. The context and substitution strategies are the same as for healthy individuals.

Predisposed Individual Treated with Rapamycin. In this scenario, the strategies are the same as in the previous setting, except for the rule weight associated with the rule lysosomalAutophagy, that is changed to formalize the administration of a daily dose of rapamycin at time 00:00, whose effects we estimate to be present for 6 hours [17]. As explained in Section 2, rapamycin enhances autophagy, and based on [11] we estimate that while rapamycin is active, it roughly doubles the efficiency of the lysosomal system:

```
( lysosomalAutophagy ) ->
    ((if AGE rem 24 < 6 then 100 else 50 fi) / ((AGE quo 200) + 2))
```

5   Probabilistic Simulation in PSMaude

We use PSMaude to simulate the system under the different scenarios mentioned in the previous section, from an initial state in which all neurons are empty, except for those in the lower brainstem, in which we set an initial insult in each neuron of 20 native and 20 misfolded α-syn monomers, as well as 20 oligomers made up of 3 monomers, and 10 aggregation nuclei made up of 6 monomers. Since we use approximate weights (due to missing kinetic data in the literature), the graphs below are also approximating reality. We simulate our system for a duration of a few months, that is enough to show some aspects of PD pathology, such as the propagation and accumulation of α-syn in the brain, the formation of Lewy bodies, and the effect of rapamycin. We refer to [5] for more details.

Healthy Individual. The results of the simulation in Fig. 4 show that α-syn also can accumulate in large amounts in healthy people, in agreement with experimental data [7], since aging influences α-syn aggregation. The graphs also show that the initial insult almost disappears, so that the progression of the disease seems to depend more on age than on the initial insult. The effect of aging is seen by the increase in total α-syn and misfolded monomers and oligomers after about 1000 hours. However, no neuronal death took place during the simulation, indicating that cell defenses are efficient in a healthy individual, at least for a few thousand hours.

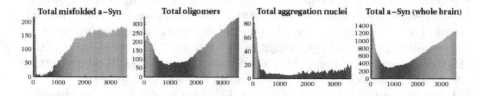

Fig. 4. The amount of different α-syn species over time in a healthy person

Sporadic PD. The results in Fig. 5 show the effect on the lower brainstem, the forebrain and the neocortex of taking rapamycin (the graph drawn in black). With rapamycin, the formation of Lewy bodies throughout the brain is delayed, and the total amount of aggregation nuclei is decreased. These graphs also show that the aggregates eventually reached the neocortex.

6   Related Work

Many formal stochastic approaches to biochemical reactions rely on *counting abstraction*, which represents a set of components by its number of elements (see, e.g., [13]). This abstraction does not allow us to keep track of the size of

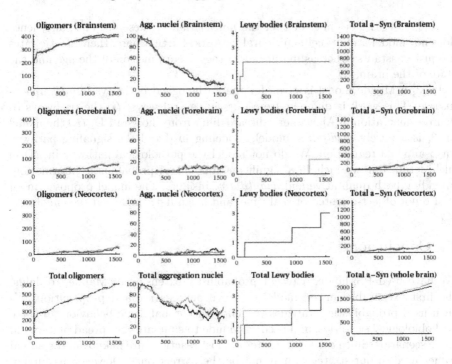

Fig. 5. Effect of rapamycin on patients with sporadic PD

single aggregates, which is important, since, e.g., the probability of degrading an aggregate is inversely proportional to its size. Our model gives a precise view of α-syn aggregation that takes stochastic fluctuations into account. We also model the dynamically changing neural networks and the intra-region and inter-region *propagation* of α-syn, which affects its aggregation. In contrast, [16] provides a *deterministic* differential equation model for protein aggregation in a *single* cell, and assumes a continuous change of concentrations for aggregates of different sizes. This model provides a view of the "average" dynamics of the aggregation process, and is not accurate for small amounts of aggregates. Furthermore, it does not consider the propagation of aggregates, which seems difficult to accomodate in a differential equation model.

The quantitative and probabilistic extension of Pathway Logic (PL) in [1] is also a rewriting-logic-based tool for modeling and analyzing biochemical pathways. PL focuses on a single cell, and does not support hierarchical models, nor does it support the complex probabilistic strategies in this paper.

Bio-PEPA [8] requires a static hierarchical model structure, and hence makes it hard to model dynamic aspects like the death of a neuron. It also does not support advanced probabilistic strategies, although it allows defining different kinetic laws for the reactions in the "base" model in a modular way.

In [18] the biochemical stochastic π-calculus has been used to model and simulate a detailed model of α-syn aggregation and degradation at the single-cell level. This work does not consider the *propagation* of these proteins, and hence does not model a multi-cellular neural network. Furthermore, their kinetic rates are just constants, whereas in our model they are a function of the age and the state of the brain.

In [3] a Maude model is given for the pathological pathways in *Alzheimer's* disease. This work is different from ours in several ways: *(i)* the pathways to neurodegeneration in Alzheimer's disease differ from those in PD; *(ii)* the model in [3] is a purely *qualitative* model, modeling in detail the signaling pathway for β-amyloid regulation. We do not model the pathological pathway in PD in such detail, but we include a significant amount of probabilistic information, providing both qualitative and *quantitative* insights. The model proposed in [3] is also not object-oriented or real-time, and has a flat, single-cell system state.

7   Conclusions

We have used PSMaude, a recent probabilistic extension of Maude, to define the first executable formal model of α-syn aggregation and propagation. We then used probabilistic strategies to define the probabilistic behaviors in three pathobiological scenarios, and used PSMaude to simulate the spread of α-syn in these scenarios. Since the kinetic rates of most reactions are unknown, the actual values used in our strategies may not be the correct ones. However, as further experimental data become available about the kinetics of different pathological processes, they could easily be incorporated into our model via the weights in the probabilistic strategies.

Since different neurodegenerative disorders have been shown to be prion-like, our model should be easily adaptable to model other diseases such as Alzheimer's and the Creutzfeldt-Jakob disease.

Due to the size and complexity of our model, we focus on simulation, and restrict ourselves to analyzing α-syn aggregation and propagation in fairly small neural networks. This is also the case with many existing approaches, which mainly use the Gillespie stochastic simulation algorithm to analyze different intracellular pathways, and rarely do a formal verification of such models. Nevertheless, our tool should be optimized to also make feasible nontrivial statistical model checking of α-syn aggregation and propagation in large neural networks.

Acknowledgments. We thank the anonymous reviewers for their very useful comments on a previous version of this paper.

References

1. Abate, A., Bai, Y., Sznajder, N., Talcott, C.L., Tiwari, A.: Quantitative and probabilistic modeling in Pathway Logic. In: BIBE, pp. 922–929. IEEE (2007)
2. Agha, G.A., Meseguer, J., Sen, K.: PMaude: Rewrite-based specification language for probabilistic object systems. ENTCS 153(2) (2006)

3. Anastasio, T.J.: Data-driven modeling of Alzheimer disease pathogenesis. Journal of Theoretical Biology 290, 60–72 (2011)
4. Bentea, L., Ölveczky, P.C.: A probabilistic strategy language for probabilistic rewrite theories and its application to cloud computing. In: Martí-Oliet, N., Palomino, M. (eds.) WADT 2012. LNCS, vol. 7841, pp. 77–94. Springer, Heidelberg (2013)
5. Bentea, L., Ölveczky, P.C., Bentea, E.: Formalization and simulation of α-synuclein aggregation and propagation under different pathobiological conditions in PS-Maude, http://folk.uio.no/lucianb/publications/asyn.pdf (manuscript)
6. Breydo, L., Wu, J.W., Uversky, V.N.: α-Synuclein misfolding and Parkinson's disease. Biochimica et Biophysica Acta - Molecular Basis of Disease 1822(2), 261–285 (2012)
7. Buchman, A.S., et al.: Nigral pathology and parkinsonian signs in elders without Parkinson disease. Annals of Neurology 71(2), 258–266 (2012)
8. Ciocchetta, F., Hillston, J.: Bio-PEPA: A framework for the modelling and analysis of biological systems. Theoretical Computer Science 410(33-34), 3065–3084 (2009)
9. Clavel, M., Durán, F., Eker, S., Lincoln, P., Martí-Oliet, N., Meseguer, J., Talcott, C.: All About Maude - A High-Performance Logical Framework. LNCS, vol. 4350. Springer, Heidelberg (2007)
10. Collier, T.J., Kanaan, N.M., Kordower, J.H.: Ageing as a primary risk factor for Parkinson's disease: evidence from studies of non-human primates. Nature Reviews Neuroscience 12(6), 359–366 (2011)
11. Crews, L., et al.: Selective molecular alterations in the autophagy pathway in patients with Lewy body disease and in models of α-synucleinopathy. PLoS ONE 5(2) (2010)
12. Hansen, C., et al.: α-synuclein propagates from mouse brain to grafted dopaminergic neurons and seeds aggregation in cultured human cells. The Journal of Clinical Investigation 121(2), 715–725 (2011)
13. Henzinger, T.A., Jobstmann, B., Wolf, V.: Formalisms for specifying Markovian population models. Int. J. Found. Comput. Sci. 22(4), 823–841 (2011)
14. Hoyer, W., Antony, T., Cherny, D., Heim, G., Jovin, T.M., Subramaniam, V.: Dependence of α-synuclein aggregate morphology on solution conditions. Journal of Molecular Biology 322(2), 383–393 (2002)
15. Kaul, T., et al.: Region-specific tauopathy and synucleinopathy in brain of the alpha-synuclein overexpressing mouse model of Parkinson's disease. BMC Neuroscience 12(1), 79 (2011)
16. Knowles, T.P.J., et al.: An analytical solution to the kinetics of breakable filament assembly. Science 326(5959), 1533–1537 (2009)
17. Korth-Bradley, J.M., et al.: Comparative sirolimus pharmacokinetics after single-dose administration of two prototype 0.5-mg tablets in healthy volunteers. Clinical Pharmacology in Drug Development 1(2), 52–56 (2012)
18. Lecca, P.: Stochastic pi-calculus models of the molecular bases of Parkinson's disease. In: BIOCOMP 2008, pp. 298–304. CSREA Press (2008)
19. Ölveczky, P.C., Meseguer, J.: Semantics and pragmatics of Real-Time Maude. Higher-Order and Symbolic Computation 20(1-2), 161–196 (2007)
20. Pan, T., et al.: Neuroprotection of rapamycin in lactacystin-induced neurodegeneration via autophagy enhancement. Neurobiology of Disease 32(1), 16–25 (2008)
21. Spillantini, M.G., Schmidt, M.L., Lee, V.M.Y., Trojanowski, J.Q., Jakes, R., Goedert, M.: α-synuclein in Lewy bodies. Nature 388(6645), 839–840 (1997)
22. Tanner, C.M., et al.: Rotenone, paraquat, and Parkinson's disease. Environmental Health Perspectives 119(6), 866–872 (2011)

Dynamic Image-Based Modelling
of Kidney Branching Morphogenesis*

Srivathsan Adivarahan[1], Denis Menshykau[1],
Odyssé Michos[2], and Dagmar Iber[1,3]

[1] Department for Biosystems Science and Engineering (D-BSSE), ETH Zurich,
Basel, Switzerland
[2] Wellcome Trust Sanger Institute, Cambridge, UK
[3] Swiss Institute of Bioinformatics
dagmar.iber@bsse.ethz.ch
http://www.bsse.ethz.ch/cobi

Abstract. Kidney branching morphogenesis has been studied exten-
sively, but the mechanism that defines the branch points is still elusive.
Here we obtained a 2D movie of kidney branching morphogenesis in
culture to test different models of branching morphogenesis with physiological growth dynamics. We carried out image segmentation and calculated the displacement fields between the frames. The models were
subsequently solved on the 2D domain, that was extracted from the
movie. We find that Turing patterns are sensitive to the initial conditions
when solved on the epithelial shapes. A previously proposed diffusion-
dependent geometry effect allowed us to reproduce the growth fields reasonably well, both for an inhibitor of branching that was produced in
the epithelium, and for an inducer of branching that was produced in
the mesenchyme. The latter could be represented by Glial-derived neurotrophic factor (GDNF), which is expressed in the mesenchyme and
induces outgrowth of ureteric branches. Considering that the Turing
model represents the interaction between the GDNF and its receptor
RET very well and that the model reproduces the relevant expression
patterns in developing wildtype and mutant kidneys, it is well possible
that a combination of the Turing mechanism and the geometry effect
control branching morphogenesis.

Keywords: image-based modelling, kidney, branching morphogenesis,
signaling networks, in silico organogenesis.

1 Introduction

Theoretical models have long been used to understand developmental patterning processes. Recent advances in imaging and computing allow us to develop

---

* S.A. is a Master Student. S.A. carried out the computational analysis under super-
vision of D.M. O.M. obtained the experimental data. D.I. conceived the project and
wrote the paper together with D.M. and S.A.

A. Gupta and T.A. Henzinger (Eds.): CMSB 2013, LNBI 8130, pp. 106–119, 2013.

increasingly realistic models of developmental pattern formation that can be validated with experimental data [1]. Such models open up new opportunities in that validated models can be used to clarify underlying mechanisms and to make predictions about further processes. The latter may enable a new field of *in silico* genetics where mutations are tested computationally before creating a mouse mutant. The advantage of such an approach is that models may predict a lack of phenotype because of compensating regulatory interactions that would otherwise have been overlooked. *In silico* genetics can thus help to avoid inconclusive experiments.

Most of the information about developmental processes are image-based and patterns typically evolve on growing domains (Figure 1). The geometry of the domain, in turn, can greatly affect model predictions. It is therefore important to simulate models on such physiological, growing domains [2]. This requires the development and combination of suitable techniques. In this paper we describe the methodology to obtain the geometries and displacement fields of developing kidneys that are undergoing branching morphogenesis. These can then be used to test models that describe the processes that regulate branching by simulating

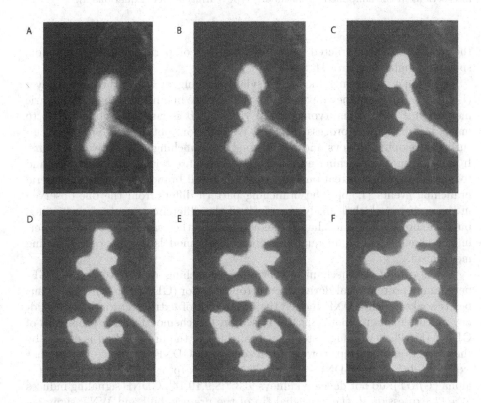

Fig. 1. Time course of kidney branching morphogenesis. The figure shows six out of 48 frames of a movie of kidney branching morphogenesis *in vitro*.

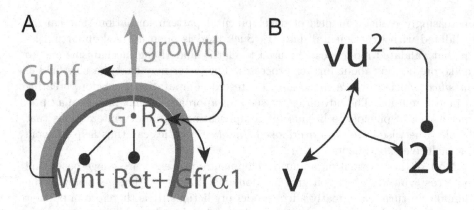

Fig. 2. The core Network regulating Kidney Branching Morphogenesis. (A) The dimer GDNF (G) binds GFRα1 and RET receptor (R) to form the GDNF-receptor complex, $G \cdot R^2$. The complex induces the expression of the receptor, *Ret*, and of *Wnt11* (*W*). Moreover, signaling by the GDNF-RET receptor complex triggers bud outgrowth. Adapted from Figure 1A in [15]. (B) Graphical representation of the ligand-receptor interactions in the simplified Schakenberg-type Turing model (Equations 2).

the models on the extracted geometries and by comparing predicted signaling spots and embryonic growth field.

The kidney collecting ducts form via branching of an epithelial cell layer (Figure 1). During kidney development the ureteric bud invades the metanephric mesenchyme around embryonic day (E)10.5 [3]. It is currently not possible to image this branching process *in utero*. We therefore obtained the data by culturing developing kidneys and by imaging the branching process over 48 hours. In culture, most branching events in the kidney are terminal bifurcations and to a lesser extent trifurcations, and only 6% of all branching events are lateral branching events [4,5,6]. The branching pattern differs from the one observed in the embryo, which is likely the result of the different geometric constraints, but the core signaling mechanism should still be the same. The culture experiments should thus be adequately suited to test models for this core signaling mechanism.

At the core of the mechanism controlling branching appears to be the TGF-beta family protein Glial-derived neurotrophic factor (GDNF) (Figure 2A). Thus beads soaked with GDNF induce the outgrowth of extra ureteric buds in kidney culture explants [6,3,7,8,9,10]. Based on the chemoattractive properties of GDNF [11,12], it was suggested that branching of the ureteric bud is caused by the attraction of the tips toward local sources of GDNF [13]. Mice that do not express *Gdnf*, or the GDNF receptor *Ret*, or co-receptor GDNF family receptor alpha (*Gfrα1*), do not develop kidneys [6,3,7,8,9,10,14]. GDNF signaling induces *Wnt11* expression in the epithelial tip of the ureteric bud and WNT signaling up-regulates expression of *Gdnf* in the mesenchyme, which results in the establishment of an autoregulatory epithelial-mesenchymal feedback signaling loop.

We have recently developed a model for the core network (Figure 2A) of GDNF (G), RET (R), and WNT (W) [15], which in non-dimensional form reads

$$\dot{G} = \underbrace{\Delta G}_{\text{diffusion}} + \underbrace{\rho_{G0} + \rho_G \frac{W^2}{W^2+1}}_{\text{production}} - \underbrace{\delta_G G}_{\text{degradation}} - \underbrace{\delta_C R^2 G}_{\text{complex formation}}$$

$$\dot{R} = \underbrace{D_R \Delta R}_{\text{diffusion}} + \underbrace{\rho_R + \nu R^2 G}_{\text{producation}} - \underbrace{\delta_R R}_{\text{degradation}} - \underbrace{2\delta_C R^2 G}_{\text{complex formation}}$$

$$\dot{W} = \underbrace{D_W \Delta W}_{\text{diffusion}} + \underbrace{\rho_{W0} + \rho_W \frac{R^2 G}{R^2 G+1}}_{\text{production}} - \underbrace{\delta_W W}_{\text{degradation}} . \tag{1}$$

When solved on a idealized 3D bud-shaped domain the model gives rise to GDNF-RET patterns that are reminiscent of, lateral branching events, bifurcations, and trifurcations. Much as reported for the embryo, the split concentration patterns as characteristic for bifurcations and trifurcations dominate in the model for physiological parameter values, while elongation and subsequent lateral branching are rather rare. Further simulations on deforming domains showed that the split concentration profiles can support bifurcating and trifurcating outgrowth [15].

We previously noticed in a model for lung branching morphogenesis that the interaction between Sonic Hedgehog (SHH) and its receptor PTCH1 results in Schnakenberg-type reaction kinetics [16]. Similarly, we notice that the model for the biochemical interactions between GDNF and its receptors (Figure 2A) reduces to Schnakenberg-type reaction kinetics of the form

$$\frac{\partial u}{\partial \tau} = \Delta u + \gamma(a - u + u^2 v)$$
$$\frac{\partial v}{\partial \tau} = D\Delta v + \gamma(b - u^2 v). \tag{2}$$

if we assume large concentrations of WNT, i.e. $W \gg 1$, a negligible receptor-independent decay rate, δ_G, for GDNF, and $\nu \sim 3\delta_C$. u and v then correspond to the receptor RET and its ligand GDNF respectively (Figure 2B). Schnakenberg reaction kinetics [17] can result in Turing pattern [18], i.e. in the emergence of stable pattern from noisy homogenous initial conditions, as a result of a diffusion-driven instability [19].

Alternatively, it has been proposed that outgrowth of branches in the lung and mammary gland may be controlled by a diffusion-based geometry effect [20,21]. If ligand is produced only in part of a tissue, then diffusion will result in a higher concentration at the centre of the ligand-producing domain. If the ligand supports outgrowth then this could support budding. When analysed on epithelial shapes of developing chicken lungs, it was concluded that, due to the same diffusion effect, the lowest concentration was observed at the highly curved tips. The branching controlling factor would thus have to be an inhibitor of branching [20].

We tested both mechanisms by obtaining a 2D movie of cultured ureteric buds and by following their epithelial dynamics over time. We extracted the shapes and displacement fields and simulated our model on these physiological domains. We find that the Turing type pattern is unstable to the noise in the initial conditions when solved on the epithelial shapes. A Turing mechanism alone can thus not control branching morphogenesis in the kidney. The diffusion-based geometry effect allowed us to reproduce the measured growth fields reasonably well, as long as it was based on an inhibitor of branching, which was expressed in the epithelium, or on an inducer of branching that was expressed in the mesenchyme. It is well possible that a combination of the Turing mechanism and the geometry effect control branching morphogenesis.

2 Results

To obtain the shapes of the ureteric bud during branching morphogenesis, E11.5 kidney rudiments were dissected and imaged as previously described [22]. Kidneys were imaged every 60 minutes using the epifluorescence inverted microscope Nikon TE300. We obtained a total of 49 frames, six of which are shown in Figure 1. To solve our computational models on these dynamic geometries we first segmented the images to obtain the boundary of the epithelium and calculated the displacement field between subsequent stages. The initial geometry and displacement fields were imported into the commercial FEM solver COMSOL Multiphysics to perform the simulations and parameter optimization.

2.1 Image Segmentation and Border Extraction

The images were segmented in MATLAB with a threshold based filter (Figure 3A,B). Prior to segmentation, the contrast of the image was increased with the built in MATLAB function imadjust. Next the images were segmented with a threshold filter. Threshold filters group pixels according to their intensity - pixels with intensities higher than a threshold value are assigned to the epithelium and those with intensity below a threshold value are assigned to the exterior. To apply threshold filters we used the MATLAB function imb2bw, which normalizes the intensity of each pixel prior to the application of the threshold filter. Threshold filters can wrongly assign islands of bright pixels to the kidney epithelium. To eliminate such small islands, we first labelled all separate objects with the MATLAB function bwlabeln and the object with the largest area was selected. (Figure 3C). We next extracted the border of the epithelium with the MATLAB function bwboundaries (Figure 3D).

The extracted boundaries had to be smoothened before they could be used for simulations and further calculations. The smoothening was done using the

MATLAB function `smooth` which uses a moving average method to smooth over the entire curve. Visual inspection confirmed that the extracted smoothened shape identifies the boundary of the kidney epithelium correctly (Figure 3E). The number of points in the extracted boundary was large, and were reduced to a set number using the interpolation function `interparc` [23].

2.2   Calculation of a Displacement Field

To simulate the signaling models on growing domains we needed to determine the displacement fields between the different stages. The displacement field between two consecutive stages was calculated by determining the minimum distance from each point on the curve at time t to the curve at time $t + \Delta t$ using the MATLAB function `distance2curve` [24].

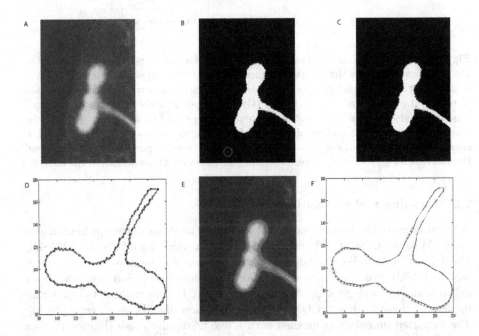

Fig. 3. Image Segmentation and Calculation of the Displacement Fields. (*A*) An original image from the movie shown in Figure 1. (*B*) Segmented image with a threshold filter. Isolated points are still present; one such group of points has been marked by red circle. (C) Segmented image after the removal of isolated points. (*D*) Extraction and smoothing of the boundary; the black dotted line shows the boundary obtained from C and the grey line shows the smoothed boundary. (*E*) The smoothed boundary (in red) superimposed on the original image. (*F*) Calculated displacement field between two images - the blue lines represent the displacement vectors; the black and the grey lines represent the contours at two subsequent stages.

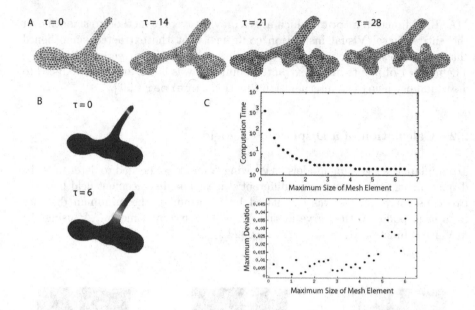

Fig. 4. Mesh Generation. (A) The meshed growing domain as generated in COMSOL Multiphysics at various time steps as indicated; the maximum size of the element was 3, minimum element quality 0.4 (B) Solution of a traveling wave equation (Equation 3) on a static domain at $\tau=0$ and $\tau=6$. (C) The upper plot shows the computational time for varying maximum size of mesh element (from 0.2 to 6.4) for $\tau = 6$ while the lower plot shows the maximum deviation of the solution at $\tau = 6$ for different mesh sizes. The deviation was calculated between a solution for a particular mesh size and the solution for a mesh with a mesh element size that was 0.2 greater than the current.

2.3   Meshing and Simulations

We next imported the curve that describes the initial shape of the epithelium into the FEM-solver COMSOL Multiphysics, using the ASCII file format. COMSOL Multiphysics is a well-established software package and several studies confirm that COMSOL provides accurate solutions to reaction-diffusion equations both on constant[25] and growing domains [26,27,28]. Details of how to efficiently implement these models in COMSOL have been described by us recently [29,30]. The imported domain was meshed with a free triangular mesh (Figure 4). The quality of the mesh can be assessed according to the following two parameters: mesh size and the ratio of the sides of the mesh elements. The linear size of the mesh should be much smaller than any feature of interest in the computational solution, i.e. if the gradient length scale in the model is 50 μm then the linear size of the mesh should be at least several times less than 50 μm. Additionally, the ratio of the length of the shortest side to the longest side should be 0.1 or higher.

Next the displacement field was imported into COMSOL and the domain was deformed accordingly. COMSOL Multiphysics uses the Arbitrary Lagrangian-

Eulerian (ALE) formalism to solve PDEs on a deforming domain. Figure 4A shows a sequence of meshes generated on a deforming domain. To confirm the convergence of the simulation, we solved a traveling wave equation of the form

$$\frac{\partial u}{\partial \tau} = \Delta u + u(1 - u) \tag{3}$$

on a series of refined meshes (Figure 4B). As the maximum mesh size decreased the maximum deviation in the solution decreased initially without greatly increasing the computation time (Figure 4C). As the mesh size was further decreased, the maximum deviation remained about constant while the computation time sharply increased. There is thus an optimal mesh size that needs to be defined for each particular model that is simulated on the domain.

2.4   Kidney Branching Morphogenesis

Depending on the choice of parameters Turing pattern can reproduce almost any pattern. We were interested whether we could find a parameter combination that would allow the model to reproduce the measured displacement field over time, while respecting all biological costraints. We started with a single frame of the extracted shape of the kidney epithelium from the movie (Figure 5). When we simulated the Schnakenberg Turing model (Equations 2) on this shape we noticed that the emerging pattern depended on the initial conditions. Given the noise in these initial conditions many different patterns emerged for the same parameter set. We therefore conclude that the Turing-based mechanism alone cannot explain the stereotyped, reliable pattern observed in the embryo.

A number of alternative mechanisms have previously been proposed to control branching morphogenesis in the lung. Most of these are based on the distance between the mesothelium and the epithelium, and thus cannot apply to the kidney. However, one mechanism relies on the particular tissue geometry [20,21], and we decided to test this one also for the kidney.

Fig. 5. Turing pattern on a static domain in the shape of embryonic kidney epithelium depends on noise in initial conditions. The three panels show the steady-state pattern of the receptor-ligand complex, u^2v (rainbow color code: red - highest level, blue - lowest level). The three panels were computed with the same parameter set, but with different random initial conditions The parameter values used for generating the figure are: $a = 0.2$, $b = 1.5$ and $\gamma = 0.04$, $D = 100$.

The mechanism requires that expression of the signalling factor is restricted to part of the tissue, and diffuses from there into the surrounding tissue. If ligand expression is restricted to the epithelium (and receptors to the mesenchyme) the model reads

$$\text{Epithelium:} \quad \frac{\partial L}{\partial \tau} = D\Delta L + 1$$

$$\text{Mesenchyme:} \quad \frac{\partial L}{\partial \tau} = D\Delta L - L. \tag{4}$$

Gdnf is expressed in the mesenchyme, and we therefore also wanted to study this case. The shape of the mesenchyme could not be extracted from the movies, and we therefore added an idealized domain in the shape of an ellipse to approximate the real shape of the mesenchyme. If ligand expression is restricted to the mesenchyme (and receptors to the epithelium), then the model reads

$$\text{Epithelium:} \quad \frac{\partial L}{\partial \tau} = D\Delta L - L$$

$$\text{Mesenchyme:} \quad \frac{\partial L}{\partial \tau} = D\Delta L + 1. \tag{5}$$

Next we tested if the model could predict the areas of growth that were observed during kidney branching morphogenesis. To that end we adjusted the only parameter value in the model, D, to minimize the deviation, Δ, between the computed signaling field and the registered displacement field based on the L2 distance (Euclidean distance), i.e.

$$\Delta = \sqrt{\int_L (|\overline{v}| - S)^2}, \tag{6}$$

$|\overline{v}|$ refers to normalized length of vectors of the displacement field, S refers to the normalized computational signal. We used $S = L$ to model a ligand that induces branch outgrowth, and $S = 1/L$ to model a ligand that inhibits branch outgrowth. The PDE models were solved on the kidney shapes of four separate stages (Figure 6B) for a wide range of the non-dimensional diffusion coefficient $D \in [10, 10^5]$. For each stage, 1000 parameter sets were sampled randomly from a logarithmic uniform distribution within these ranges.

The lowest deviation was obtained for the model where the ligand L was expressed in the epithelium and acted as an inhibitor of branch outgrowth (Figure 6A, black, solid line). The best fitting pattern matches the observed growth field quite well, though not perfectly (Figure 6B). The second closest match was obtained when an activator of branching was expressed in the mesenchyme, which could be represented by GDNF. The other two cases did not provide a good match. The observed displacement of the stalk could not be captured by the model. However, we note that according to experimental observations (at least a later stages) the receptor *Ret* is not expressed in the stalk [31]; this displacement must thus be the result of other processes than GDNF/RET signaling.

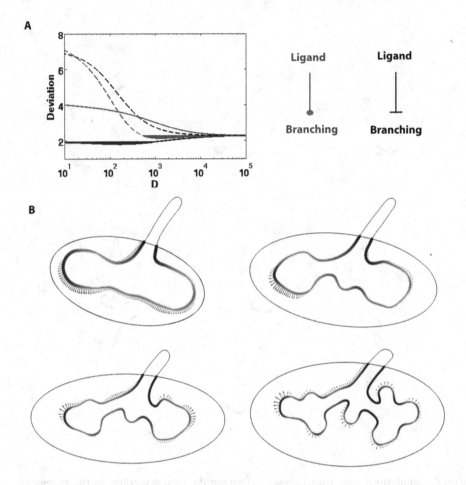

Fig. 6. A diffusion-based geometry effect results in patterns similar to those of the displacement field. (A) The deviation Δ (Eq. 6) between signalling pattern and displacement field if L is expressed either in the epithelium (solid lines) or in the mesenchyme (broken lines), and if L acts either as activator (green lines) or inhibitor (black lines) of branch outgrowth. (B) The correspondence of the signalling effect and the displacement field for the best matching case, i.e. for L expressed in the epithelium, acting as an inhibitor of branching, and $D = 200$. The solid colours represent the value of $1/L$ (red - high to blue -low). The arrows mark the displacement field, with the length of the arrows indicating the strength of the displacement.

Due to high computational cost we were unable to perform the optimization on the deforming domain; thus all results discussed so far were obtained on a series of static frames. To test whether the resulting pattern would be stable on a growing, deforming domain, we ran a simulation with the best fitting parameter value ($D = 200$) on the recorded kidney movie, i.e. we started with the first frame and deformed the domain according to the measured displacement field of

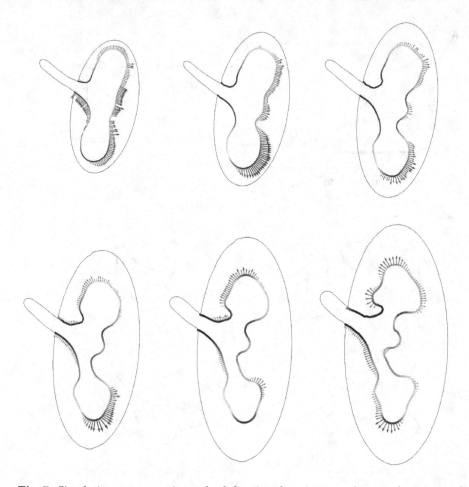

Fig. 7. Simulations on a continuously deforming domain, according to the measured displacement field. The correspondence of the signalling effect and the displacement field for the case of L expressed in the epithelium, acting as an inhibitor of branching, and $D = 200$. The solid colours represent the value of $1/L$ (red - high to blue -low). The arrows mark the displacement field, with the length of the arrows indicating the strength of the displacement.

the kidney explants (Figure 7). The distribution of the signalling activity that we obtained on a series of static frames (Figure 6B) is indeed similar to the one obtained on a growing domain (Figure 7).

3   Discussion

Mathematical models can help with the understanding of biological complexity if thoroughly rooted in experimental data. Most data in developmental biology is image based. In this contribution we used 2D movies that document branching of cultured kidneys to test a mathematical model for branching morphogenesis. We used MATLAB-based functions to extract the shapes and to calculate the displacement fields between frames. The shapes and displacement fields were subsequently imported into COMSOL to simulate the model on physiological geometries. Our previously proposed model for the core signaling mechanism is based on Schnakenberg reaction kinetics that can give rise to Turing pattern [15]. When simulating the model on the extracted epithelial shapes we noticed that the pattern were sensitive to the noisy initial conditions, which rules out such mechanism for robust pattern formation.

We next tested a diffusion-based geometry effect [20,21], which reproduced the growth fields of the cultured kidneys reasonably well. Much as in previous studies in the lung and mammary gland [20,21], when the ligand was produced only in the epithelium, it needed to be an inhibitor of branching to explain the observed growth fields. On the other hand, if produced in the mesenchyme, it had to be an inducer of outgrowth. The latter could be represented by GDNF, an inducer of the outgrowth of the ureteric bud.

While the Turing mechanism was unstable when solved on the epithelial domain, we note that it is well conceivable that both the Turing mechanism and the geometry effect act together during branching morphogenesis. After all, the Turing mechanism allowed us to reproduce the phenotype of all relevant, published mutants when solved on idealized domains [15]. Further analysis of movies that include both the epithelium and the mesenchyme will be important to address this.

The here-described methods permit the analysis of 2D movies. To further enhance the power of the analysis it would be valuable to obtain image frames in 3D rather than 2D. The calculation of the displacement field is more difficult in 3D. Several softwares are available to support morphing between 3D structures. The software package AMIRA employs the landmark-based Bookstein algorithm [32], which uses paired thin-plate splines to interpolate surfaces over landmarks defined on a pair of surfaces. The landmark points need to be placed by hand on the 3D geometries to identify corresponding points on the pair of surfaces. This is both time consuming and limits the accuracy of the reconstructed 4D series, in particular if the geometries are complex as is the case during kidney branching morphogenesis and if single frames are further apart. Further developments are clearly needed to enable a faster and more accurate reconstruction of 4D datasets. Similarly, computational methods need to be improved to facilitate the solution of computational models on complex, growing domains, that comprise several subdomains (tissue layers). While this is feasible, it is currently computationally very expensive, which makes it difficult to screen larger parameter spaces.

Acknowledgments. The authors acknowledge funding from the SNF Sinergia grant "Developmental engineering of endochondral ossification from mesenchymal stem cells", and a SystemsX RTD on Forebrain Development.

References

1. Iber, D., Zeller, R.: Making sense-data-based simulations of vertebrate limb development. Curr. Opin. Genet. Dev. 22, 570–577 (2012)
2. Iber, D., Tanaka, S., Fried, P., Germann, P., Menshykau, D.: Simulating Tissue Morphogenesis and Signaling. In: Nelson, C.M. (ed.) Tissue Morphogenesis: Methods and Protocols, Methods in Molecular Biology. Springer (2013)
3. Majumdar, A., Vainio, S., Kispert, A., McMahon, J., McMahon, A.P.: Wnt11 and Ret/Gdnf pathways cooperate in regulating ureteric branching during metanephric kidney development. Development 130, 3175–3185 (2003)
4. Watanabe, T., Costantini, F.: Real-time analysis of ureteric bud branching morphogenesis in vitro. Developmental Biology 271, 98–108 (2004)
5. Meyer, T.N., Schwesinger, C., Bush, K.T., Stuart, R.O., Rose, D.W., et al.: Spatiotemporal regulation of morphogenetic molecules during in vitro branching of the isolated ureteric bud: toward a model of branching through budding in the developing kidney. Developmental Biology 275, 44–67 (2004)
6. Costantini, F., Kopan, R.: Patterning a complex organ: branching morphogenesis and nephron segmentation in kidney development. Dev. Cell. 18, 698–712 (2010)
7. Treanor, J.J., Goodman, L., de Sauvage, F., Stone, D.M., Poulsen, K.T., et al.: Characterization of a multicomponent receptor for GDNF. Nature 382, 80–83 (1996)
8. Pichel, J.G., Shen, L., Sheng, H.Z., Granholm, A.C., Drago, J., et al.: Defects in enteric innervation and kidney development in mice lacking GDNF. Nature 382, 73–76 (1996)
9. Pepicelli, C.V., Kispert, A., Rowitch, D.H., McMahon, A.P.: GDNF induces branching and increased cell proliferation in the ureter of the mouse. Developmental Biology 192, 193–198 (1997)
10. Sánchez, M.P., Silos-Santiago, I., Frisén, J., He, B., Lira, S.A., et al.: Renal agenesis and the absence of enteric neurons in mice lacking GDNF. Nature 382, 70–73 (1996)
11. Tang, M.J., Cai, Y., Tsai, S.J., Wang, Y.K., Dressler, G.R.: Ureteric bud outgrowth in response to RET activation is mediated by phosphatidylinositol 3-kinase. Developmental Biology 243, 128–136 (2002)
12. Tang, M.J., Worley, D., Sanicola, M., Dressler, G.R.: The RET-glial cell-derived neurotrophic factor (GDNF) pathway stimulates migration and chemoattraction of epithelial cells. J. Cell. Biol. 142, 1337–1345 (1998)
13. Sariola, H., Saarma, M.: Novel functions and signalling pathways for GDNF. Journal of Cell Science 116, 3855–3862 (2003)
14. Schuchardt, A., D'Agati, V., Larsson-Blomberg, L., Costantini, F., Pachnis, V.: Defects in the kidney and enteric nervous system of mice lacking the tyrosine kinase receptor Ret. Nature 367, 380–383 (1994)
15. Menshykau, D., Iber, D.: Kidney branching morphogenesis under the control of a ligand-receptor-based Turing mechanism. Physical Biology 10, 046003 (2013)
16. Menshykau, D., Kraemer, C., Iber, D.: Branch Mode Selection during Early Lung Development. Plos Computational Biology 8, e1002377 (2012)
17. Schnakenberg, J.: Simple chemical reaction systems with limit cycle behaviour. Journal of Theoretical Biology 81, 389–400 (1979)
18. Gierer, A., Meinhardt, H.: A theory of biological pattern formation. Kybernetik 12, 30–39 (1972)
19. Turing, A.: The chemical basis of morphogenesis. Phil. Trans. Roy. Soc. Lond. B 237, 37–72 (1952)

20. Gleghorn, J.P., Kwak, J., Pavlovich, A.L., Nelson, C.M.: Inhibitory morphogens and monopodial branching of the embryonic chicken lung. Developmental dynamics: An official publication of the American Association of Anatomists (2012)
21. Nelson, C.M., Vanduijn, M.M., Inman, J.L., Fletcher, D.A., Bissell, M.J.: Tissue geometry determines sites of mammary branching morphogenesis in organotypic cultures. Science 314, 298–300 (2006)
22. Riccio, P.N., Michos, O.: Dissecting and culturing and imaging the mouse urogenital system. Methods in Molecular Biology 886, 3–11 (2012)
23. D'Errico, R.: Interparc function (2012), http://www.mathworks.in/matlabcentral/fileexchange/34874-interparc
24. D'Errico, R.: Normal distance function (2012), http://www.mathworks.com/matlabcentral/fileexchange/34869-distance2curve
25. Cutress, I.J., Dickinson, E.J.F., Compton, R.G.: Analysis of commercial general engineering finite element software in electrochemical simulations. J. Electroanal. Chem. 638, 76–83 (2010)
26. Carin, M.: Numerical Simulation of Moving Boundary Problems with the ALE Method: Validation in the Case of a Free Surface and a Moving Solidification Front. In: Excert from the Proceedings of the COMSOL Conference (2006)
27. Thummler, V., Weddemann, A.: Computation of Space-Time Patterns via ALE Methods. In: Excert from the Proceedings of the COMSOL Conference (2007)
28. Weddemann, A., Thummler, V.: Stability Analysis of ALE-Methods for Advection-Diffusion Problems. In: Excert from the Proceedings of the COMSOL Conference (2008)
29. Menshykau, D., Iber, D.: Simulating Organogenesis with Comsol: Interacting and Deforming Domains. In: Proceedings of COMSOL Conference 2012 (2012)
30. Germann, P., Menshykau, D., Tanaka, S., Iber, D.: Simulating Organogensis in COMSOL. In: Proceedings of COMSOL Conference 2011 (2011)
31. Pachnis, V., Mankoo, B., Costantini, F.: Expression of the c-ret proto-oncogene during mouse embryogenesis. Development 119, 1005–1017 (1993)
32. Bookstein, F.L.: Principal warps: Thin-plate splines and the decomposition of deformations. Pattern Analysis and Machine Intelligence (1989)

Statistical Model Checking Based Calibration and Analysis of Bio-pathway Models*

Sucheendra K. Palaniappan[1], Benjamin M. Gyori[2], Bing Liu[3],
David Hsu[1,2], and P.S. Thiagarajan[1,2]

[1] School of Computing, National University of Singapore, 117417, Singapore
[2] NUS Graduate School for Integrative Sciences and Engineering,
National University of Singapore, 117417, Singapore
[3] Computer Science Department, Carnegie Mellon University, Pittsburgh, PA 15213, USA

Abstract. We present a statistical model checking (SMC) based framework for studying ordinary differential equation (ODE) models of bio-pathways. We address cell-to-cell variability explicitly by using probability distributions to model initial concentrations and kinetic rate values. This implicitly defines a distribution over a set of ODE trajectories, the properties of which are to be characterized. The core component of our framework is an SMC procedure for verifying the dynamical properties of an ODE system accompanied by such prior distributions. To cope with the imprecise nature of biological data, we use a formal specification logic that allows us to encode both qualitative properties and experimental data. Using SMC, we verify such specifications in a tractable way, independent of the system size. This further enables us to develop SMC based parameter estimation and sensitivity analysis procedures. We have evaluated our method on two large pathway models, namely, the segmentation clock network and the thrombin-dependent MLC phosphorylation pathway. The results show that our method scales well and yields good parameter estimates that are robust. Our sensitivity analysis framework leads to interesting insights about the underlying dynamics of these systems.

1 Introduction

Biochemical networks–often called bio-pathways–govern a variety of cellular functions. Their malfunctioning can lead to major diseases [1]. Thus it is important to understand their dynamics using mathematical models [2]. However, building and analyzing such models poses considerable challenges. In this paper, we address the particular challenge of accounting for variable behavior across individual cells. A natural way to cater for this is to use a probabilistic system model such as continuous time Markov chains (CTMCs) [3]. However, such models typically track the occurrences of individual reactions. Hence for pathways of realistic size, calibrating these models using experimental data and analyzing them using stochastic simulations is very difficult. The alternative is to use ordinary differential equations (ODEs) to capture the dynamics. This approach is often computationally more tractable, although it requires that the number of molecules of each type involved in the pathway be abundantly present [4]. In this paper our focus

---

* This research was partially supported by the Singapore MOE ARC grant MOE2011-T2- 2-012.

A. Gupta and T.A. Henzinger (Eds.): CMSB 2013, LNBI 8130, pp. 120–134, 2013.

is on accounting for cell-to-cell variability in the setting of ODE based models. Specifically, our main contribution is a statistical model checking (SMC) based framework, using which a system with such variability can be efficiently calibrated and analyzed.

Variability in a population of cells has at least two major causes. First, as shown in [5], differences in the initial concentrations of proteins are the primary source of variability in response to external stimuli. Second, due to differing internal and external conditions among cells, the values of kinetic rate constants also vary across cells [6,7]. In our ODE setting the variables will represent the concentrations of the biochemical species (typically proteins) in the pathway, and hence the initial concentrations of these species will constitute the initial values of the variables. Further, the parameters appearing in the equations will consist of the kinetic rate constants governing the reactions. Thus we can capture cell-to-cell variability in the behavior of the bio-pathway by studying the ODE dynamics across a range of values for the initial concentrations and kinetic rate constant values. We do this in a probabilistic setting by assuming initial probability distributions (usually uniform) over an interval of values for the initial concentrations and rate constants. We then show that the resulting space of trajectories can be used to construct a natural probability measure space if the vector field defined by the ODE system is continuously differentiable. In our setting this requirement is easily met.

To analyze the ODE system, we first formalize properties using our specification logic and decide a corresponding confidence level (probability) with which we wish to assess them. Consequently, an SMC procedure –which poses the problem as a hypothesis test– is used to decide approximately, but with statistical guarantees, whether the properties are satisfied with the desired probability. SMC continues to sample and verify trajectories from the ODE system until a decision can be made. It is well-established that SMC is efficient since its complexity does not depend on the size of the system. Moreover, posing the problem as a sequential hypothesis test reduces the overall number of samples needed to make a decision [8]. These components form a principled method for analyzing the dynamics of a bio-pathway in the presence of dynamic variability across a population of cells.

To demonstrate the applicability of our approach, we develop an SMC based parameter estimation method. The unknown model parameters usually consist of initial concentrations and kinetic rate constants. Here, for convenience, we shall assume all the initial concentrations are known but that their nominal values can vary over a cell population. The parameter estimation procedure searches through the value space of the unknown parameters to determine the "best" combination of values that can explain the given data and predict new behaviors [9]. The key step in this procedure is to determine the fitness-to-data of the current set of parameter values. We use our specification logic to encode both experimental time series data and known qualitative trends concerning the dynamics of the pathway. We then use our SMC procedure to determine the goodness of the given set of parameter values, while taking into account that these values can fluctuate across the population of cells that the data is based on. Subsequently, we use a global optimization strategy known as SRES [10] to choose a new set of candidate parameter values according to the SMC based score assigned to the current set.

An important analysis task to be performed on the model is quantifying the influence of different parameters on the model dynamics. The information gained from such a sensitivity analysis procedure can help in robustness analysis, optimal experimental design and drug target selection [11]. We show how SMC can be used to generate the statistics needed by the global sensitivity analysis method MPSA [12]. Consequently, one can incorporate a rich class of dynamic behaviors–encoded as formulas in our specification logic–to drive our sensitivity analysis method.

We evaluated our method on two pathway models taken from the BioModels database [13]. For both case studies, we assumed that noisy experimental data and qualitative dynamic traits of a few species were known. This data was separated into training and test components. A subset of the rate constants were assumed to be unknown and estimated using our parameter estimation procedure. The first model, the segmentation clock pathway, consists of 16 differential equations and 75 rate constants, out of which 39 were fixed to be unknown. The second model, the thrombin-dependent myosin light chain (MLC) pathway consists of 105 differential equations and 197 rate constants, out of which 100 were fixed to be unknown. Our results (Section 5) show that our SMC based technique is efficient and scales well. We also applied our sensitivity analysis method to obtain interesting insights into the dynamics of these two bio-pathways.

1.1 Related Work

Probabilistic model checking of stochastic models is an active field of research [14–17]. Of particular interest in our context are sampling based methods such as [18, 19], which verify probabilistic properties using a fixed number of sampled trajectories. In contrast, SMC based methods such as [14, 20] adaptively generate a sufficient number of trajectories to determine if the property is satisfied while meeting the strengths of the statistical test specified by the user. Characterizing the behavior of dynamical systems where the initial conditions and the rate parameters are under-determined has also been discussed in [21] with a focus on sampling methods and computing reachable sets.

Turning to parameter estimation using temporal logic constraints, a brute force search of the parameter space is employed in [16] for Petri nets. In the ODE context, parameter estimation combined with model checking appears in [22] using again a brute force sampling based parameter search approach, and in [23], using an evolutionary strategy to guide the search. However, both these techniques only generate a single simulation trace of the ODE to evaluate a proposed set of parameters. A symbolic model checking approach is explored for the restricted class of multi-affine ODEs in [24, 25]. The work reported in [19] deploys a genetic algorithm to search for the best set of parameters. A fixed number of samples–this number is fixed in an ad hoc manner–is generated, and the probability of satisfying a property is calculated to be the fraction of the samples which satisfy the property. In all these studies, the quality of the estimated parameters is not validated using test data (i.e. data that was not used as training data). While [19] does mention identifying critical parameters, we believe that our approach is the first systematic attempt to develop a property-based sensitivity analysis framework using statistical model checking.

In the next section, we introduce ODE models and their dynamics. In Section 3, we discuss our specification logic and the statistical model checking procedure. Subsequently, we present our parameter estimation and sensitivity analysis framework.

Experimental results are reported in Section 5. Additional experimental results are reported in the supplementary material [26].

2   ODE Based Models and Their Behaviors

A popular formalism for describing the dynamics of a biochemical network is a system of ODEs. For each molecular species x_i in the pathway, there will be an equation of the form $dx_i/dt = f_i(\mathbf{x}, \Theta_i)$. Here f_i describes the kinetics of the reactions that produce and consume x_i, $\mathbf{x}$ denotes the concentrations of the molecular species taking part in these reactions, while the vector Θ_i gives the rate constants governing these reactions.

Each x_i is a real-valued function of $t \in \mathbb{R}_+$, where $\mathbb{R}_+$ denotes the set of non-negative reals. We shall realistically assume that $x_i(t)$ takes values in the interval $[L_i, U_i]$, where L_i and U_i are non-negative rationals with $L_i < U_i$. Hence the state space of the system is $\mathbf{V} = [L_1, U_1] \times \ldots \times [L_n, U_n]$, a bounded subset of $\mathbb{R}_+^n$. Let $\Theta = \bigcup_i \Theta_i = \{\theta_1, \theta_2, \ldots, \theta_m\}$ be the set of all rate constants. We again assume that the range of values for each θ_j is $[L^j, U^j]$ for $1 \le j \le m$. We shall present the SMC procedure while assuming that all the rate constants are known. In Section 4, it will become clear how unknown rate constants are handled.

An implicit assumption in what follows is that the value of a rate constant, when fixed initially, does not change during the time evolution of the dynamics, although this value can be different for different cells. To capture the cell-to-cell variability regarding the initial states, we define for each variable x_i an interval $[L_i^{init}, U_i^{init}]$ with $L_i \le L_i^{init} < U_i^{init} \le U_i$. The actual value of the initial concentration of x_i is assumed to fall in this interval. Similarly, we shall assume that the nominal value of the rate constant θ_j falls in the interval $[L_{init}^j, U_{init}^j]$ with $L^j \le L_{init}^j < U_{init}^j \le U^j$. We set $INIT = (\prod_i [L_i^{init}, U_i^{init}]) \times (\prod_j [L_{init}^j, U_{init}^j])$. Thus $INIT$ captures the cell-to-cell variability in the initial concentration and the rate constant values. In what follows we let $\mathbf{v}$ to range over $\prod_i [L_i^{init}, U_i^{init}]$ and $\mathbf{w}$ to range over $\prod_j [L_{init}^j, U_{init}^j]$.

We will represent our system of ODEs in vector form as $d\mathbf{x}/dt = F(\mathbf{x}, \Theta)$ with $F_i(\mathbf{x}, \Theta) := f_i$. Recall that a function $f_i : \mathbf{V} \to \mathbf{V}$ is a C^1 function if f_i', the derivative of f_i, exists at all $\mathbf{v} \in \mathbf{V}$, and is a continuous function. In the setting of biochemical networks, the expressions in f_i will model kinetic laws such as mass law and Michaelis-Menten [4]. Thus it is reasonable to assume that $f_i \in C^1$ for each i and hence $F : \mathbf{V} \to \mathbf{V}$ is also a C^1 function. As a result, for each $(\mathbf{v}, \mathbf{w}) \in INIT$ the system of ODEs will have a unique solution $\mathbf{X}_{\mathbf{v}, \mathbf{w}}(t)$ [27]. Further, it will satisfy: $\mathbf{X}_{\mathbf{v}, \mathbf{w}}(0) = \mathbf{v}$ and $\mathbf{X}_{\mathbf{v}, \mathbf{w}}'(t) = F(\mathbf{X}_{\mathbf{v}, \mathbf{w}}(t))$. We are also guaranteed that $\mathbf{X}_{\mathbf{v}, \mathbf{w}}(t)$ is a C^0-function (i.e. continuous function) [27], and hence measurable. This fact will be crucial for our SMC procedure.

It will be convenient to define the flow $\Phi_{\mathbf{w}} : \mathbb{R}_+ \times \mathbf{V} \to \mathbf{V}$ for arbitrary initial vectors $\mathbf{v}$. Intuitively, $\Phi_{\mathbf{w}}(t, \mathbf{v})$ is the state reached under the ODE dynamics if the system starts at $\mathbf{v}$ at time 0. The flow will be the C^0-function given by: $\Phi_{\mathbf{w}}(t, \mathbf{v}) = \mathbf{X}_{\mathbf{v}, \mathbf{w}}(t)$. Thus $\Phi_{\mathbf{w}}(0, \mathbf{v}) = \mathbf{X}_{\mathbf{v}, \mathbf{w}(0)=\mathbf{v}}$ and $\partial(\Phi_{\mathbf{w}}(t, \mathbf{v}))/\partial t = F(\Phi_{\mathbf{w}}(t, \mathbf{v}))$ for all t. We will, in fact, work with $\Phi_{\mathbf{w}, t} : \mathbf{V} \to \mathbf{V}$ where $\Phi_{\mathbf{w}, t}(\mathbf{v}) = \Phi_{\mathbf{w}}(t, \mathbf{v})$ for every t and every $\mathbf{v} \in \mathbf{V}$. again, $\Phi_{\mathbf{w}, t}$ is guaranteed to be a C^0 function.

In our application, the dynamics will be of interest only up to a maximal time point T. Fixing such a T, a *trajectory* starting from $\mathbf{v} \in \mathbf{V}$ at time 0 and with $\mathbf{w}$ as the parameter values is denoted $\sigma_{v,w}$. It is the (continuous) function $\sigma_{v,w} : [0, T] \to \mathbf{V}$ satisfying: $\sigma_{v,w}(t) = X_{v,w}(t)$. The behavior of our dynamical system is the set of trajectories given by $BEH = \{\sigma_{v,w} \mid (\mathbf{v}, \mathbf{w}) \in INIT\}$. Our goal is to develop an SMC procedure to verify the dynamical properties of BEH.

3 Statistical Model Checking of ODE Dynamics

3.1 Bounded Linear Time Temporal Logic

To formally express dynamical properties of BEH, we use formulas in a specification logic. We will use bounded linear time temporal logic (BLTL) since our trajectories will be of finite duration. An atomic proposition in our logic will be of the form (i, ℓ, u) with $L_i \leq \ell < u \leq U_i$. Such a proposition will be interpreted as "the current concentration level of x_i is in the interval $[\ell, u]$", and we fix a finite set of such atomic propositions.

We first introduce the syntax and then the semantics of BLTL formulas. The formulas of BLTL are defined as: (i) Every atomic proposition as well as the constants *true*, *false* are BLTL formulas. (ii) If ψ, ψ' are BLTL formulas then $\neg\psi$ and $\psi \vee \psi'$ are BLTL formulas. (iii) If ψ, ψ' are BLTL formulas and $t \leq T$ is a *positive integer* then $\psi\mathbf{U}^{\leq t}\psi'$ and $\psi\mathbf{U}^t\psi'$ are BLTL formulas. We have mildly strengthened BLTL to be able to express that a certain property will hold exactly at t time units from now. This will enable us to encode experimental data in the specification. The derived propositional operators such as $\wedge$, $\supset$, $\equiv$, and the temporal operators $\mathbf{G}^{\leq t}$, $\mathbf{F}^{\leq t}$ are defined in the usual way.

We will interpret the formulas of our logic at the finite set of time points $\mathcal{T} = \{0, 1, \ldots, T\}$. Such a discretization is reasonable since experimental data will be available only at a finite number of discrete time points. Further, qualitative properties of interest are expressible in discrete time. We assume that $\mathcal{T}$ has been chosen appropriately and it includes all the relevant time points with respect to the specified properties.

The semantics of the logic is defined in terms of the relation $\sigma, t \models \varphi$, where σ is a trajectory in BEH and $t \in \mathcal{T}$.

- $\sigma, t \models (i, \ell, u)$ iff $\ell \leq \sigma(t)(i) \leq u$ where $\sigma(t)(i)$ is the ith component of the n-dimensional vector $\sigma(t) \in \mathbf{V}$.
- $\neg$ and $\vee$ are interpreted in the usual way.
- $\sigma, t \models \psi\mathbf{U}^{\leq k}\psi'$ iff there exists k' such that $k' \leq k$, $t + k' \leq T$ and $\sigma, t + k' \models \psi'$. Further, $\sigma, t + k'' \models \psi$ for every $0 \leq k'' < k'$.
- $\sigma, t \models \psi\mathbf{U}^k\psi'$ iff $t + k \leq T$ and $\sigma, t + k \models \psi'$. Further, $\sigma, t + k' \models \psi$ for every $0 \leq k' < k$.

We now define $models(\psi) = \{\sigma \mid \sigma, 0 \models \psi, \sigma \in BEH\}$.

Next, we wish to make statements of the form $P_{\geq r}(\psi)$, where the intended meaning is that the probability that a trajectory in BEH belongs to $models(\psi)$ is at least r. To assign meaning to such statements, we need to define a probability measure over sets of trajectories. Note, however, that the trajectory $\sigma \in BEH$ is completely determined by $\sigma(0)$, the (vector) value it assumes at $t = 0$. Hence we will identify BEH with $INIT$, the set of initial states. To make this explicit, we define the set $Models(\psi) \subseteq INIT$ as:

$(\mathbf{v}, \mathbf{w}) \in Models(\psi)$ iff $\sigma_{\mathbf{v}, \mathbf{w}} \in models(\psi)$. We define the formulas of PBLTL as $P_{\geq r}\psi$ and $P_{\leq r'}\psi$ provided $r \in [0, 1)$, $r' \in (0, 1]$ and ψ is a BLTL formula. We shall say that S, the system of ODEs, meets the specification $P_{\geq r}\psi$ – and this is denoted $S \models P_{\geq r}\psi$ – iff $P(Models(\psi)) \geq r$, while $S \models P_{\leq r'}\psi$ iff $P(Models(\psi)) \leq r'$. Here, and in what follows, P is the standard probability measure assigned to members of the σ-algebra generated by the open intervals contained in $INIT$. It is easy to show that $Models(\psi)$ is a member of this σ-algebra for every ψ. The only case that requires an argument is the one for atomic propositions, and here the measurability of the solution functions $\mathbf{X}_{\mathbf{v}, \mathbf{w}}(t)$ is crucial. The details can be found in the supplementary material [26].

3.2 Statistical Model Checking of PBLTL Formulas

We now introduce a statistical framework for deciding approximately, but with statistical guarantees, whether the model satisfies a property of the form $P_{\geq r}\psi$. Instead of directly approximating the probability of ψ being satisfied [28], we formulate whether $S \models P_{\geq r}\psi$, as a hypothesis test. According to [29], the test is posed between the null hypothesis H0 : $p \geq r + \delta$ and the alternative hypothesis H1 : $p \leq r - \delta$, where $p = P(Models(\psi))$. Here, δ is supplied by the user and signifies an indifference region in which one cannot decide on either H0 or H1. The *strength* of the statistical test is decided by parameters α and β which bound the probability of verifying the property as false when it is in fact true (Type-I error) and verifying it as true when it is in fact false (Type-II error) respectively. Thus the verification is carried out approximately but with guaranteed confidence levels and error bounds. The test proceeds by generating a sequence of sample trajectories $\sigma_1, \sigma_2, \ldots$ by randomly sampling an initial state from $INIT$. One assumes a corresponding sequence of Bernoulli random variables $y_1, y_2 \ldots$, where each y_k is assigned the value 1 if $\sigma_k, 0 \models \psi$; otherwise, y_k is assigned the value 0. We next construct a sequential test that lets us decide if the number of samples taken are sufficient, or whether more samples need to be taken to guarantee the chosen test strength. For each $m \geq 1$, after drawing m samples, we compute a quantity q_m as:

$$q_m = \frac{[r - \delta]^{(\sum_{i=1}^{m} y_i)}[1 - [r - \delta]]^{(m - \sum_{i=1}^{m} y_i)}}{[r + \delta]^{(\sum_{i=1}^{m} y_i)}[1 - [r + \delta]]^{(m - \sum_{i=1}^{m} y_i)}} . \tag{1}$$

The ratio q_m serves as a stopping criterion for the sampling process. Hypothesis H0 is accepted if $q_m \geq \widehat{A}$, and hypothesis H1 is accepted if $q_m \leq \widehat{B}$. If neither is the case then another sample is drawn. The constants $\widehat{A}$ and $\widehat{B}$ are chosen such that it results in a test of strength (α, β). In practice, a good approximation is $\widehat{A} = \frac{1-\beta}{\alpha}$ and $\widehat{B} = \frac{\beta}{1-\alpha}$. A detailed account of our *on-line* model checking algorithm (used to verify each trajectory) can be found in the supplementary material [26].

4 Analysis Methods

Here we present our parameter estimation and sensitivity analysis methods. In doing so, we assume the terminology and notations developed in the previous sections. As a first step, we describe how experimental data can be encoded as a BLTL formula.

Assume, without loss of generality, that $O \subseteq \{x_1, x_2, \ldots, x_k\}$ is the set of variables for which experimental data is available, and which has been allotted as training data to be used for parameter estimation. Assume $\mathcal{T}_i = \{\tau_1^i, \tau_2^i, \ldots, \tau_{T_i}^i\}$ are the time points at which the concentration level of x_i has been measured and reported as $[\ell_t^i, u_t^i]$ for each $t \in \mathcal{T}_i$. The interval $[\ell_t^i, u_t^i]$ is chosen to reflect the noisiness, the limited precision and the cell-population based nature of the experimental data. For each $t \in \mathcal{T}_i$, we define the formula $\psi_i^t = \mathbf{F}^t(i, \ell_t^i, u_t^i)$. Then $\psi_{exp}^i = \bigwedge_{t \in \mathcal{T}_i} \psi_i^t$. We then set $\psi_{exp} = \bigwedge_{i \in O} \psi_{exp}^i$. In case the species x_i has been measured under multiple experimental conditions, the above encoding scheme is extended in the obvious way.

Often qualitative dynamic trends will be available–typically from the literature–for some of the molecular species in the pathway. For instance, we may know that a species shows transient activation, in which its level rises in the early time points, and later falls back to initial levels. Similarly, a species may be known to show oscillatory behavior with certain characteristics. Such information can be described as BLTL formulas that we term to be *trend* formulas. Examples of such formulas can be found in Table1. We let ψ_{qlty} to be the conjunction of all the trend formulas.

Finally, we fix the PBLTL formula $P_{\geq r}(\psi_{exp} \land \psi_{qlty})$, where r will capture the confidence level with which we wish to assess the goodness of the fit of the current set of parameters to experimental data and qualitative trends. We also fix an indifference region δ and the strength of the test (α, β). The constants r, δ, α and β are to be fixed by the user. In our application, it will be useful to exploit the fact that both ψ_{exp} and ψ_{qlty} are conjunctions, and hence can be evaluated separately. As shown in [29], one can choose the strength of each of these tests to be $(\frac{\alpha}{J}, \beta)$, where J is the total number of conjuncts in the specification. This will ensure that the overall strength of the test is (α, β). Further, the results of individual statistical tests can be used to compute the objective function associated with the global search strategy to be described below.

4.1 Parameter Estimation Based on PBLTL Specification

We assume $\Theta_u = \{\theta_1, \theta_2, \ldots, \theta_K\}$ is the set of unknown parameters. For convenience we will assume that the other parameter values are known and that their nominal values do not fluctuate across the cell population. We will also assume nominal values for the initial concentrations and the range of their fluctuations of the form $[L_i^{init}, U_i^{init}]$ for each variable x_i. Again, for convenience, we fix a constant δ'' so that if the current estimate of the values of the unknown parameters is $\mathbf{w} \in \prod_{1 \leq j \leq K}[L^j, U^j]$ then this value will fluctuate in the range $[\mathbf{w}(j) - \delta'', \mathbf{w}(j) + \delta'']$. Setting $L_{init,\mathbf{w}}^j = \mathbf{w}(j) - \delta''$ and $U_{init,\mathbf{w}}^j = \mathbf{w}(j) + \delta''$ we define $INIT_{\mathbf{w}} = (\prod_i[L_i^{init}, U_i^{init}]) \times (\prod_j[L_{init,\mathbf{w}}^j, U_{init,\mathbf{w}}^j])$. The set of trajectories $BEH_{\mathbf{w}}$ is defined accordingly.

To estimate the quality of $\mathbf{w}$, we run our SMC procedure–using $INIT_{\mathbf{w}}$ instead of $INIT$–to verify $P_{\geq r}(\psi_{exp} \land \psi_{qlty})$. Depending on the outcome of this test for the various conjuncts in the specification, we assign a score to $\mathbf{w}$ using an objective function detailed below. We then iterate this scheme for various values of $\mathbf{w}$ generated using a suitable search strategy. The objective function consists of two components, evaluating the contribution from the qualitative properties and the experimental data respectively. It evaluates how many statistical tests carried out with $\mathbf{w}$ resulted in acceptance of the null hypothesis (desired outcome). For the second component, the tests are evaluated species-wise. The corresponding objective value is then composed as a summation of normalized contribution from each species.

Let J_{exp}^i $(= T_i)$ be the number of conjuncts in ψ_{exp}^i, and J_{qlty} the number of conjuncts in ψ_{qlty}. Let $J_{exp}^{i,+}(\mathbf{w})$ be the number of formulas of the form ψ_i^t (a conjunct in ψ_{exp}^i) such that the statistical test for $P_{\geq r}(\psi_i^t)$ accepts the null hypothesis (that is, $P_{\geq r}(\psi_i^t)$ holds) with the strength $(\frac{\alpha}{J}, \beta)$, where $J = \sum_{i \in O} J_{exp}^i + J_{qlty}$. Similarly, let $J_{qlty}^+(\mathbf{w})$ be the number of conjuncts in ψ_{qlty} of the form $\psi_{\ell,qlty}$ that pass the statistical test $P_{\geq r}(\psi_{\ell,qlty})$ with the strength $(\frac{\alpha}{J}, \beta)$. Then $\mathcal{G}(\mathbf{w})$ is computed via:

$$\mathcal{G}(\mathbf{w}) = J_{qlty}^+(\mathbf{w}) + \sum_{i \in O} \frac{J_{exp}^{i,+}}{J_{exp}^i} \tag{2}$$

Thus the goodness to fit of $\mathbf{w}$ is measured by how well it agrees with the qualitative properties as well as the number of experimental data points with which there is acceptable agreement. To avoid over-training the model, we do not insist that every qualitative property and every data point must fit well with the dynamics predicted by $\mathbf{w}$.

The search strategy to evolve candidate parameters will use the values $\mathcal{G}(\mathbf{w})$ to traverse the parameter value space. Global search methods such as Genetic Algorithms (GA) [30], and Stochastic Ranking Evolutionary Strategy (SRES) [10] are computationally more intensive than local methods, but are much better at avoiding local minima. The overall structure of our parameter estimation procedure is presented in Algorithm 1. In practice, one usually maintains a *population* of parameter value vectors in each round, and a round is usually called a *generation*. For convenience, we have assumed that each population is a singleton in the description of Algorithm 1. We use the SRES strategy in our work since it is known to perform well in the context of pathway models [9]. The particular choice of search algorithm, however, is orthogonal to our proposed method.

4.2 Sensitivity Analysis Based on PBLTL Specification

As another application of our SMC procedure, we have constructed a property based sensitivity analysis method by coupling our SMC routine with the global sensitivity analysis technique called multi-parametric sensitivity analysis (MPSA) [12]. We assume we have specified a set of properties (encoded as PBLTL formulas), and are interested in knowing which parameters, when changed, affect these properties significantly. The MPSA procedure involves sampling a large number of parameter combinations from their valid ranges. For each sampled combination, one calculates the objective value with respect to the PBLTL properties according to Equation 2. The objective values allow us to assess the extent to which each parameter affects the model's behavior to the given formulas. Intuitively, if the objective value shows strong dependence on the value of a parameter (over its range) then the output is sensitive to that parameter. The MPSA method employs statistical tests to quantify this dependence, which can be directly interpreted as a measure of sensitivity. The sensitivity is based on computing the Kolmogorov-Smirnov (KS) test to compare the two profiles consisting of (a) the cumulative appearance of *good* intervals along the value space of the parameter and

(b) the same for the *bad* intervals. If these profiles differ significantly then the system is more sensitive to this parameter, and the KS test will assign a higher score to this parameter. Our procedure is outlined in Algorithm 2.

```
input  : ODE model; PBLTL formulas; SMC
         parameters; Number of generations k;
         Initial parameter guess w₀;
output: The best parameter found w_max

initialization: ℓ = 0; 𝒢_max = 0;

while ℓ < k do
   Run SMC on the trajectories defined by
   BEH_{w_ℓ} with respect to the PBLTL formulas;
   Compute 𝒢(w_ℓ);
   if 𝒢(w_ℓ) ≥ 𝒢_max then
   |   w_max = w_ℓ;
   |   𝒢_max = 𝒢(w_ℓ)
   end
   w_{ℓ+1} = Picked by SRES / GA search
   procedure based on w_ℓ;
   ℓ = ℓ + 1;
end
```

Algorithm 1. Parameter estimation

```
input  : ODE model; PBLTL formulas; SMC parameters; Number of
         discretization intervals N_d; Objective function 𝒢; threshold
output: Sensitivity[1 . . . K]

Discretize each parameter into N_d intervals to get (N_d)^K hypercubes;

for i ← 0 to N_d do
   w_i = Sample one hypercube out of the (N_d)^K using LHS;
   Run SMC on BEH_{w_i}; Calculate 𝒢(w_i);
   if 𝒢(w_i) > threshold then
   |   Add w_i to good set;
   else
   |   Add w_i to bad set;
   end
end
for j ← 0 to K do
   Construct cumulative distribution of good and bad intervals in the
   range of parameter j;
   Sensitivity[j] = KS statistic of difference of the two distributions;
end
```

Algorithm 2. Sensitivity analysis

5   Results

We applied our SMC based analysis framework to pathway models taken from the BioModels database [13]. These models have nominal point values for all the rate constants and initial concentrations. We first verified a few properties of the two pathways using SMC. Then, for parameter estimation, we formulated qualitative trends for some species, and generated synthetic experimental data for some other species as follows. We set a ±5% range around the nominal value for the initial concentration of each species and assumed a uniform distribution over the resulting set of initial states. To mimic western blot data, which is cell population based, we averaged 10^4 random trajectories generated by sampling these initial concentration intervals. We then added noise to the data and used a major portion of it for training, and reserved the rest as test data. Finally, we fixed a subset of rate constants to be unknown, and ran our parameter estimation procedure. We let the variability in parameters (δ'') to be 0.5% of the proposed value.

We implemented our method using MATLAB and C++ on a PC with a 3.4Ghz Intel Core i7 processor with 8GB RAM. ODE systems were numerically solved using the SUNDIALS CVODE package [31, 32]. The source code is available at [26]. The code has been optimized to take advantage of the multi-core architecture; all experimental results were run on 8 threads. The parameters used for the statistical model checking algorithm were $r = 0.9$, $\alpha = \beta = \delta = 0.05$ for all our experiments. The choice of these parameters were made so that the probability of satisfaction of the formulas was sufficiently high, and the errors were sufficiently low. The dependence of the performance of the statistical test on the parameters of SMC is well established, we refer the interested reader to [29] for more details. To show the goodness of our estimated parameters (taking into account the variability concerning the initial states and reaction rates),

Table 1. Statistical model checking based verification - The PBLTL Formulas

Pathway	Property	Formula	Result
Thrombin-MLC	sustained activation	$P_{\geq 0.9}(([\text{Phospho MLC} \leq 1]) \wedge (F^{\leq 20}(G^{\leq 20}([\text{Phospho MLC} \geq 3]))))$	false
Thrombin-MLC	transient activation	$P_{\geq 0.9}(([\text{Phospho MLC} \leq 1]) \wedge F^{\leq 20}((([\text{Phospho MLC} \geq 3]) \wedge F^{\leq 20}(G^{\leq 20}((\text{Phospho MLC} \leq 1)))))$	true
Segmentation clock	oscillations	$P_{\geq 0.9}(([\text{Lunatic fringe mRNA} \leq 0.4]) \wedge (F^{\leq 40}([\text{Lunatic fringe mRNA} \geq 2.2] \wedge F^{\leq 40}([\text{Lunatic fringe mRNA} \leq 0.4] \wedge F^{\leq 40}([\text{Lunatic fringe mRNA} \geq 2.2] \wedge F^{\leq 40}([\text{Lunatic fringe mRNA} \leq 0.4]))))))$	true

we generated 1000 trajectories and plotted these to show that the estimated parameters result in a good fit to the data. In each case, experimental data is plotted along with the tolerance interval used in constructing the specification.

For the experiments reported in this section, we used an SRES based global strategy to guide the search. Here we present only the highlights of our experimental results. Many further details including the results obtained using a Genetic Algorithm based search can be found in the supplementary material [26].

5.1   The Case Studies

The segmentation clock network   An oscillating segmentation clock governs the segmentation pattern of the spine in developing vertebrate embryos. It couples signaling pathways of FGF, Notch and Wnt, whose periodic behaviors are produced by negative feedback loops. The ODE model consists of 16 differential equations and 75 kinetic rate parameters. Simulation time (T) was fixed at 200 minutes assumed to be observable at 40 equally spaced time points.

The thrombin-dependent MLC phosphorylation pathway   Endothelial cells form a dynamic barrier between blood/lymph and the underlying connective tissue, and their contraction is crucial to physiological and pathological processes. Agonists such as thrombin play an important role in the contraction function through phosphorylation of MLC, while Rho-kinase is crucial for the sustained contraction of endothelial cells. The pathway model with 105 differential equations and 197 kinetic parameters is considerably large. Simulation time was fixed at 1000 seconds assumed to be observable at 20 equally spaced time points.

5.2   Statistical Model Checking Based Verification

First, we used our SMC framework to verify pathway properties expressed in PBLTL. We used the nominal models (all rate parameter values known, taken from the BioModels database) to verify if they conformed to properties expressed in our logic with high probability. We describe a few such properties along with their BLTL formulas and the result of verification in Table1. For instance, for the MLC phosphorylation pathway, it is known experimentally that the concentration of phosphorylated MLC starts at a low level, and then reaches a high steady state value. Our SMC method shows that the nominal model does not satisfy the property, instead, phosphorylated MLC exhibits a transient profile. This discrepancy has been studied in [33], and attributed to missing components and interactions in the proposed model.

5.3 Parameter Estimation

For the segmentation clock pathway, we assumed 39 of the rate parameters as unknown. We used a combination of dynamic trends and quantitative experimental data. Specifically, we synthesized population based experimental time series data for Axin2 mRNA measured at 14 time points up to 165 minutes. For 5 other species {Notch protein, nuclear NicD, Lunatic fringe mRNA, active ERK and Dusp6 mRNA}, we encoded the dynamic trends as properties in our logic. The dynamic trend of 2 species (cytosolic NicD and Dusp6 protein) were used as test data. Parameter estimation was done with a population of 200 per generation and for 300 generations. The time taken by SRES based search was 2.3 hours. Figure 1 shows simulation profiles with the estimated parameters. Figure 1(a) shows that the model fits training data consisting of the experimental data of Axin2 mRNA and qualitative trends for 3 other species. Figure 1(b) shows dynamic trends of cytosolic NicD used for testing. The simulated time profiles fit the specified test properties (see [26]).

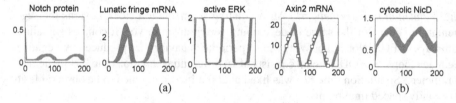

Fig. 1. Parameter estimation results of the segmentation clock pathway. (a) Training data including the experimental data for Axin2 mRNA and the dynamic trends for 3 species), and (b) the test data for one of the species.

To illustrate the scalability of our approach, for the thrombin pathway, we assumed 100 of the kinetic parameters to be unknown. We synthesized population based experimental time series data for 10 species including RGS_2, Rho.GTP, PKC.DAG, MLC_2,

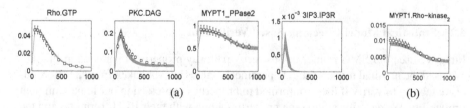

Fig. 2. Parameter estimation results of Thrombin-dependent MLC phosphorylation pathway. (a) Training data, including experimental data of 3 species and dynamic trends of one species, and (b) the test data for one of the species.

CPI-17, Ca-super-2-plus, p115RhoGEF-GTP-alpha, MYPT1-PPase, Rho-kinase.MLC, MYPT1.Rho-kinase$_2$. For thrombinR-active and 3IP3.IP3R we assumed that only the dynamic trend is known. The data of Rho-kinase.MLC and MYPT1.Rho-kinase$_2$ were reserved as test data to evaluate the quality of our parameter estimates, while the data of all other species was used to calibrate the model. Parameter estimation was done with a population of 100 per generation and for 1000 generations. The time taken by SRES based search was 48.8 hours. Figure 2 shows the fit to data of the simulation profiles with the best predicted parameter values for both the training data (Figure 2(a)) and the test data (Figure 2(b)).

5.4 Property Based Sensitivity Analysis

Here we report results just for the segmentation clock pathway (due to the space constraints). We evaluated the sensitivity of parameters against all properties used for parameter estimation. The results are shown in Figure 3(a). It can be seen that the most sensitive parameters are *ksDusp*, *kcDusp*, *VMsMDusp*, *VMdMDusp*, *VMaX*, *VMdX*. This also indicates that the reactions involving Dusp6 degradation and transcription affect the overall dynamics most. Since all these parameters belong to the FGF pathway, we hypothesize that FGF pathway is the most crucial component that drives the behavior of the system.

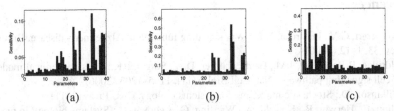

(a)   (b)   (c)

Fig. 3. Sensitivity analysis results. (a-c) Parameter sensitivities of the segmentation clock pathway with respect to (a) all properties, (b) Dusp6mRNA profile, and (c) nuclear nicD profile.

We next searched for parameters affecting the oscillatory property of *Dusp6 mRNA* alone. We found that the same set of parameters as above are the most crucial (see Figure 3(b)). However, when evaluating the oscillatory property of nuclear NicD (Figure 3(c)), we find that the parameters *vsN*, *kt1*, *VdNan* are the most significant. This suggests that although the Notch synthesis (*vsN*), and nuclear NicD transportation (*kt1*) and degradation (*VdNan*) do not significantly affect the overall dynamics, they play a dominant role in segmentation patterning.

6 Conclusion

We have proposed an SMC based approach for studying ODE based bio-pathway models. We have used the temporal logic BLTL to encode both quantitative experimental data and qualitative properties of pathway dynamics. To cater for variability among cells, we assume a uniform distribution over a set of initial states and kinetic rate constants–and impose a reasonable continuity restriction–and show how the probability of the property being met by the behavior of the model can be assessed using an SMC procedure. By combining this method with a global search strategy, we arrive at a parameter estimation procedure as well as a sensitivity analysis technique.

We have demonstrated the applicability of our method with the help of two ODE based bio-pathway models: the segmentation clock network and the thrombin-dependent MLC phosphorylation pathway. Our method successfully obtained good parameter estimates using noisy cell-population data and qualitative knowledge. The results show that our method scales well and can cope with large biological networks. We also show results for performing property based sensitivity analysis, and thereby gain interesting insights about the pathway dynamics that would be difficult to obtain using conventional approaches.

Our parameter estimation method is a generic one and has the potential to be applied to model classes such as continuous time Markov chain (CTMC) models and stochastic differential equation (SDE) models [3]. We plan to explore this in our future work. Another interesting direction will be to develop a GPU-based implementation of our method to exploit the inherent massive parallelism in generating trajectories through numerical integration. In this connection, the platform-aware implementation of a related systems biology application presented in [17] promises to offer helpful pointers.

References

1. De Ferrari, G.V., Inestrosa, N.C.: Wnt signaling function in Alzheimer's disease. Brain Res. Rev. 33, 1–12 (2000)
2. Aldridge, B.B., Burke, J.M., Lauffenburger, D.A., Sorger, P.K.: Physicochemical modelling of cell signalling pathways. Nat. Cell Biol. 8(11), 1195–1203 (2006)
3. Wilkinson, D.: Stochastic modelling for systems biology. CRC Press (2011)
4. Klipp, E., Herwig, R., Kowald, A., Wierling, C., Lehrach, H.: Systems biology in practice: concepts, implementation and application. Wiley-VCH, Weinheim (2005)
5. Spencer, S., Gaudet, S., Albeck, J., Burke, J., Sorger, P.: Non-genetic origins of cell-to-cell variability in TRAIL-induced apoptosis. Nature 459(7245), 428–432 (2009)
6. Snijder, B., Pelkmans, L.: Origins of regulated cell-to-cell variability. Nature Reviews Molecular Cell Biology 12(2), 119–125 (2011)
7. Weiße, A., Middleton, R., Huisinga, W.: Quantifying uncertainty, variability and likelihood for ordinary differential equation models. BMC Systems Biology 4(1), 144 (2010)
8. Younes, H.L.S., Kwiatkowska, M., Norman, G., Parker, D.: Numerical vs. statistical probabilistic model checking. International Journal on Software Tools for Technology Transfer 8, 216–228 (2006)
9. Moles, C.G., Mendes, P., Banga, J.R.: Parameter estimation in biochemical pathways: A comparison of global optimization methods. Genome Res. 13(11), 2467–2474 (2003)
10. Runarsson, T., Yao, X.: Stochastic ranking for constrained evolutionary optimization. IEEE T. Evolut. Comput. 4, 284–294 (2000)
11. Saltelli, A., Ratto, M., Andres, T., Campolongo, F., Cariboni, J., Gatelli, D., Saisana, M., Tarantola, S.: Global sensitivity analysis: the primer. Wiley-Interscience (2008)

12. Cho, K.H., Shin, S.Y., Kolch, W., Wolkenhauer, O.: Experimental design in systems biology, based on parameter sensitivity analysis using a Monte Carlo method: A case study for the TNFα-mediated NF-κB signal transduction pathway. Simulation 79(12), 726–739 (2003)
13. Le Novere, N., Bornstein, B., Broicher, A., Courtot, M., Donizelli, M., Dharuri, H., Li, L., Sauro, H., Schilstra, M., Shapiro, B., Snoep, J., Hucka, M.: BioModels Database: A free, centralized database of curated, published, quantitative kinetic models of biochemical and cellular systems. Nucleic Acids Res. 34, D689–D691 (2006)
14. Jha, S.K., Clarke, E.M., Langmead, C.J., Legay, A., Platzer, A., Zuliani, P.: A bayesian approach to model checking biological systems. In: Degano, P., Gorrieri, R. (eds.) CMSB 2009. LNCS, vol. 5688, pp. 218–234. Springer, Heidelberg (2009)
15. Heath, J., Kwiatkowska, M., Norman, G., Parker, D., Tymchyshyn, O.: Probabilistic model checking of complex biological pathways. Theor. Comput. Sci. 391(3), 239–257 (2008)
16. Li, C., Nagasaki, M., Koh, C.H., Miyano, S.: Online model checking approach based parameter estimation to a neuronal fate decision simulation model in Caenorhabditis elegans with hybrid functional Petri net with extension. Mol. Biosyst. 7(5), 1576–1592 (2011)
17. Liu, B., Hagiescu, A., Palaniappan, S.K., Chattopadhyay, B., Cui, Z., Wong, W., Thiagarajan, P.S.: Approximate probabilistic analysis of biopathway dynamics. Bioinformatics 28(11), 1508–1516 (2012)
18. Donaldson, R., Gilbert, D.: A monte carlo model checker for probabilistic ltl with numerical constraints. University of Glasgow, Dep. of CS, Tech. Rep. (2008)
19. Donaldson, R., Gilbert, D.: A model checking approach to the parameter estimation of biochemical pathways. In: Heiner, M., Uhrmacher, A.M. (eds.) CMSB 2008. LNCS (LNBI), vol. 5307, pp. 269–287. Springer, Heidelberg (2008)
20. Clarke, E.M., Faeder, J.R., Langmead, C.J., Harris, L.A., Jha, S.K., Legay, A.: Statistical model checking in *BioLab*: Applications to the automated analysis of T-cell receptor signaling pathway. In: Heiner, M., Uhrmacher, A.M. (eds.) CMSB 2008. LNCS (LNBI), vol. 5307, pp. 231–250. Springer, Heidelberg (2008)
21. Maler, O.: On under-determined dynamical systems. In: Proceedings of the Ninth ACM International Conference on Embedded Software, pp. 89–96. ACM (2011)
22. Calzone, L., Chabrier-Rivier, N., Fages, F., Soliman, S.: Machine learning biochemical networks from temporal logic properties. T. Comput. Syst. Biol. VI, 68–94 (2006)
23. Rizk, A., Batt, G., Fages, F., Soliman, S.: On a continuous degree of satisfaction of temporal logic formulae with applications to systems biology. In: Heiner, M., Uhrmacher, A.M. (eds.) CMSB 2008. LNCS (LNBI), vol. 5307, pp. 251–268. Springer, Heidelberg (2008)
24. Batt, G., Page, M., Cantone, I., Goessler, G., Monteiro, P., de Jong, H.: Efficient parameter search for qualitative models of regulatory networks using symbolic model checking. Bioinformatics 26(18), i603–i610 (2010)
25. Barnat, J., Brim, L., Krejci, A., Streck, A., Safranek, D., Vejnar, M., Vejpustek, T.: On parameter synthesis by parallel model checking. IEEE/ACM T. Comput. Bi. 9(3), 693–705 (2012)
26. Supplementary information and source code, http://www.comp.nus.edu.sg/~rpsysbio/SMC/
27. Hirsch, M., Smale, S., Devaney, R.: Differential equations, dynamical systems, and an introduction to chaos. Academic Press (2012)
28. Hérault, T., Lassaigne, R., Magniette, F., Peyronnet, S.: Approximate probabilistic model checking. In: Steffen, B., Levi, G. (eds.) VMCAI 2004. LNCS, vol. 2937, pp. 73–84. Springer, Heidelberg (2004)
29. Younes, H.L.S., Simmons, R.G.: Statistical probabilistic model checking with a focus on time-bounded properties. Inform. Comput. 204, 1368–1409 (2006)
30. Goldberg, D.: Genetic algorithms in search, optimization, and machine learning. Addison-Wesley (1989)

31. Hindmarsh, A., Brown, P., Grant, K., Lee, S., Serban, R., Shumaker, D., Woodward, C.: SUNDIALS: Suite of nonlinear and differential/algebraic equation solvers. ACM T. Math. Software 31(3), 363–396 (2005)
32. Vanlier, J., Tiemann, C., Hilbers, P., van Riel, N.: An integrated strategy for prediction uncertainty analysis. Bioinformatics 28(8), 1130–1135 (2012)
33. Maedo, A., Ozaki, Y., Sivakumaran, S., Akiyama, T., Urakubo, H., Usami, A., Sato, M., Kaibuchi, K., Kuroda, S.: Ca^{2+}-independent phospholipase A2-dependent sustained Rho-kinase activation exhibits all-or-none response. Genes Cells 11, 1071–1083 (2006)

Constraint Programming in Community-Based Gene Regulatory Network Inference

Ferdinando Fioretto and Enrico Pontelli

Dept. Computer Science
New Mexico State University
{ffiorett,epontell}@cs.nmsu.edu

Abstract. Gene Regulatory Network (GRN) inference is a major objective of Systems Biology. The complexity of biological systems and the lack of adequate data have posed many challenges to the inference problem. *Community networks* integrate predictions from individual methods in a "meta predictor", in order to compose the advantages of different methods and soften individual limitations. This paper proposes a novel methodology to integrate prediction ensembles using Constraint Programming, a declarative modeling paradigm, which allows the formulation of dependencies among components of the problem, enabling the integration of diverse forms of knowledge. The paper experimentally shows the potential of this method: the addition of biological constraints can offer improvements in the prediction accuracy, and the method shows promising results in assessing biological hypothesis using constraints.

1 Introduction

Within a cellular context, genes interact to orchestrate a multitude of important tasks. These interactions are regulated by different gene products, as proteins called *Transcription Factors (TFs)* and RNA, and they constitute an intricate machinery of regulation referred to as *Gene Regulatory Networks (GRNs)*. In turn *GRN inference* describes the process of inferring the topology of a particular GRN. GRN inference from high-throughput data is of central importance in computational system biology. Its use is crucial in understanding important genetic diseases, such as cancer, and to devise effective medical interventions.

The availability of a wealth of genomic data has encouraged the development of diverse methods for GRN inference. However, data sets are quite heterogeneous in nature, containing information which is limited and difficult to analyze [20]. This reverberates on performance of GRN inference methods, which tend to be biased toward the type of data and experiments. For instance, methods based on linear models perform poorly on highly non-linear data, such as the one produced in presence of severe perturbations like gene knock-outs [11]. To alleviate these difficulties several alternatives have been proposed, such as integrating heterogeneous data into the inference model [17], or integrating a collection of predictions across different inference methods in *Community Networks (CNs)* [13,14]. The former is a promising research direction but it has to

A. Gupta and T.A. Henzinger (Eds.): CMSB 2013, LNBI 8130, pp. 135–149, 2013.

face several challenges which span from how to relate different types of data to data sets normalization processes. The latter has the advantage of promoting the benefits of individual methods while smoothing out their drawbacks. Moreover it does not exclude the use of the former solution within the initial prediction set. The CN integration process poses many challenges, raising questions like: (i) how to take into account strengths and weaknesses of individual inference methods—e.g., the difficulty for Mutual Information (MI) or correlation based methods to discriminate TFs; and (ii) how to leverage additional information which cannot be taken into account by the individual methods.

In this paper, we propose a novel methodology based on *Constraint Programming (CP)* to integrate community predictions. CP is a declarative problem solving paradigm, where logical rules are used to model problem properties and to guide the construction of solutions. CP offers a natural environment where heterogeneous information can be actively handled. The use of constraint expressions allows the incremental refinements of a model. This is particularly suitable to take care of biological knowledge integration, when such knowledge cannot be directly handled by individual prediction methods.

We test our method on a set of 110 benchmarks proposed by the DREAM3 [14] and DREAM4 [16] challenges. We show increases in prediction accuracy with respect to a CN prediction based on the Borda count election method [13]. In addition, we show promising results in assessing biological hypotheses that could be used to guide the biological experimental design process.

2   Background

2.1   Basic Definitions

Gene Regulatory Networks. A GRN can be described by a weighted directed graph $G = (V, E)$, where V is the set of regulatory elements of the network and $E \subseteq V \times V \times [0, 1]$ is the set of regulatory interactions. The presence of an edge $\langle s, t, w \rangle \in E$ indicates that an interaction between the regulatory elements s and t is present with *confidence* value w. The number $|V|$ of regulatory elements of the GRN is referred to as its *size*. If the GRN has no uncertainty, then each edge in E has weight 1. In the problem of *GRN inference*, we are given the set of vertices V and a set of experiments describing the behavior of the regulatory elements. The goal is to accurately detect the set of regulatory interactions E.

Constraint Programming. CP is a declarative programming methodology commonly used to address combinatorial search problems. It focuses on capturing properties of the problem in the form of *constraints*, which are satisfied exclusively by solutions of the problem. CP models are fully declarative and elaboration tolerant, enabling the incremental integration of new knowledge.

A *Constraint Satisfaction Problem (CSP)* is formalized as a triple $\langle \mathcal{X}, \mathcal{D}, \mathcal{C} \rangle$, where $\mathcal{X} = \langle x_1, \ldots, x_n \rangle$ is an n-tuple of variables, $\mathcal{D} = \langle D_1, \ldots, D_n \rangle$ is a corresponding n-tuple of domains (and each D_i is a set of possible values for the variable x_i), and $\mathcal{C} = \langle C_1, \ldots, C_k \rangle$ is a k-tuple of constraints. A constraint C_j

over a set of variables $S_j \subseteq \mathcal{X}$ is a subset of the Cartesian product of the domains of the variables in S_j. A constraint represents the set of joint assignments that can be given to the tuple of variables in S_j. Given an n-tuple $A = \langle a_1, \ldots, a_n \rangle$, we denote with $A|_{S_j}$ the restriction of the tuple to the variables in S_j.

A *solution* of a CSP $\langle \mathcal{X}, \mathcal{D}, \mathcal{C} \rangle$ is an n-tuple $A = \langle a_1, \ldots, a_n \rangle$ where $a_i \in D_i$ (for $1 \leq i \leq n$) and $A|_{S_j} \in C_j$ (for $1 \leq j \leq k$)—i.e., the projection of A onto the set of variables involved in C_j satisfies the relation C_j. Typical resolution algorithms for CSP rely on efficient procedures to explore the search space of possible solutions and on *consistency methods*, where constraints are used to remove infeasible elements from the domains of the variables.

Related Work. A wide variety of GRN inference methods from expression data have been proposed [17]. These include: **(1)** Discrete models based on Boolean networks and Bayesian networks [11]; **(2)** Regression methods like *TIGRESS*— which imposes a regression problem to each gene; **(3)** Methods based on mutual information (MI) theory, such as *ARACNE* [15] and CLR [5], based on statistical likelihood of MI values. *Ensemble learning* has been explored for example by *GENIE3*, which uses a Random Forest approach [10]. Meta approaches have also been explored, such as *INFLEATOR*, based on re-sampling combining *median-corrected z-scores(MCZ)*, *time-lagged CLR (tlCLR)*, and linear ODE models [8]. *Community Networks (CNs)* integrate multiple inference methods to obtain a common consensus prediction. They have been shown to achieve better average confidence across different datasets and produce more robust results with respect to the individual methods being composed [13]. A simple scheme for combining predictions in a community network has been proposed in [13], where each interaction is re-scored by averaging the ranks it obtained within each of all the employed predictions. In the rest of the paper we will refer to it with CN_{rank}.
Constraint Technologies have been recently successfully applied in the field of System Biology [19]. For example, Answer Set Programming has been adopted to address problems in network inconsistencies detection [7] and in metabolic network analysis [18]. CP has been investigated to reason over discrete network models, where GRNs are modeled using multi-valued variables and transition rules [4]. In particular, CP is exploited to represent GRNs' possible dynamics [6].

3 Methods

The CN approach adopted in this work is built by combining four GRN inference procedures and creating an *inference ensemble*. Three of them are top-ranking methods that have been presented in the past DREAM competitions [13]: *(i) TIGRESS* [9], *(ii) INFLEATOR* [8], and *(iii) GENIE3* [10]. The fourth is an "off-the-shelf" widely adopted MI-based method *(CLR)* [5]. We use the *GP-DREAM* web platform (http://dream.broadinstitute.org) to develop the predictions from each of these methods. These methods have been selected to provide robustness and diversity, avoiding method redundancies that could potentially bias the inference ensemble.

3.1   Problem Formalization

Given a set of n genes, a GRN inference problem is formalized as a CSP $\langle \mathcal{X}, \mathcal{D}, \mathcal{C} \rangle$, with $\mathcal{X} = \langle x_1, \ldots, x_{n^2-n} \rangle$; each x_k describes a regulatory relation (without self regulations), and each $D_k = \{0, \ldots, 100\}$ is the set of possible confidence values associated with such relation. A variable x_i is said to be *assigned* if its associated domain D_i has been reduced to a singleton set. We adopt the notation $d(x_i)$ to indicate the value of an assigned variable x_i. For the sake of presentation, we denote with $x_{\langle s,t \rangle}$ the variable associated with the regulatory relation "*s regulates t*" and $D_{\langle s,t \rangle}$ its domain. A solution to the above CSP defines a GRN prediction $G = (V, E)$, with $V = \{1, \ldots, n\}$ and $E = \{\langle s, t, w \rangle \mid d(x_{\langle s,t \rangle}) > 0\}$, where $w = d(x_{\langle s,t \rangle})/100$.

Variables and Domains. The proposed CSP solution leverages the collection of GRN predictions obtained employing all the methods described in Sec. 3 by: **(1)** considerably reducing the size of the solution search space[1] and **(2)** taking into account the discrepancies among the community predictions. These objectives are achieved by mapping the edge confidence levels of each prediction to the corresponding CSP variable domain. The greater the agreement in the inference ensemble, the smaller is the set of values in the domain of the variable representing the relation being considered. The size of each domain captures the degree of uncertainty expressed by an edge prediction within the inference ensemble.

Let us consider a set of predictions $\mathcal{G}$ of a GRN $G = (V, E)$. We denote with G_j each prediction in the inference ensemble, and we denote with E_j the edges of E that have been identified by G_j. We also assume that each prediction has been normalized with respect to the ensemble itself. Furthermore, let θ_d ($0 \le \theta_d \le 1$) be a given disagreement threshold. The procedure described in Alg. 1 reduces the content of the domains in $\mathcal{D}$ to at most three values. For each edge (s, t) we calculate the average confidence value (w_rank)—according to the *Borda* count election method, as presented in [13], which averages the ranked edge confidence values assigned by each prediction—and the discrepancy value (w_d) within $\mathcal{G}$ (line 4). The latter captures the ensemble prediction disagreement for a given edge, averaging the pairwise differences of the edge ranks associated to each prediction of the ensemble. If the discrepancy value exceeds the discrepancy threshold θ_d and the average confidence value is not strongly informative (line 6), we force the domain $D_{\langle s,t \rangle}$ to take account of the prediction disagreement by adding a variation of w_d/2 to the average confidence value. fd is the nearest integer function which converts a prediction confidence value into an integer domain encoding, and it is defined as: $\mathtt{fd}(x) = \lfloor 100\, x + 0.5 \rfloor$. Line 5 ensures the presence of the value w_rank in $D_{\langle s,t \rangle}$. For a given prediction G_j, $w_j^{\#}(s, t)$ is the function ranking the prediction confidence for the edge (s, t) within the confidence values in E_j.

---

[1]   An upper bound for the search space of a GRN inference problem of size n is 101^{n^2}.

Algorithm 1. Domain Variable Population

Require: normalized $G_j \in \mathcal{G}, \theta_d, G = (V, E)$

1: $J \leftarrow |\mathcal{G}|$
2: **for all** $(s, t) \in E$ **do**
3:     $B \leftarrow \emptyset$
4:     $(\texttt{w\_rank, w\_d}) \leftarrow \left(\dfrac{1}{J} \displaystyle\sum_{j=1}^{J} \omega_j^{\#}(s, t), \quad \dfrac{1}{\binom{J}{2}} \displaystyle\sum_{j=1}^{J} \displaystyle\sum_{i=j+1}^{J} |\omega_j^{\#}(s, t) - \omega_i^{\#}(s, t)| \right)$
5:     $B \leftarrow B \cup \{\texttt{fd(w\_rank)}\}$
6:     **if** $\texttt{w\_d} \geq \theta_d \wedge 0.1 < \texttt{w\_rank} < 0.9$ **then**
7:         $B \leftarrow B \cup \left\{ \max\left(0, \texttt{fd}\left(\texttt{w\_rank} - \dfrac{\texttt{w\_d}}{2}\right)\right), \min\left(100, \texttt{fd}\left(\texttt{w\_rank} + \dfrac{\texttt{w\_d}}{2}\right)\right) \right\}$
8:     **end if**
9:     $D_{\langle s, t \rangle} \leftarrow D_{\langle s, t \rangle} \cap B$
10: **end for**

Constraint Modeling. Let us analyze the constraints that can be exploited to enforce the satisfaction of GRNs' specific properties and to take into account collective strengths and individual weaknesses of the CN predictions. Furthermore, we will discuss the propagation rules associated with the various constraints used to reduce the domain size of the variables ensuring constraint consistency.

Sparsity Constraints. It is widely accepted that the GRN machinery is controlled by a relatively small number of genes. Several state-of-the-art methods for reverse engineering GRN encourage sparsity in the inferred networks [13]. Nevertheless, when combining predictions in a community based approach, no guarantees on the sparsity of the resulting prediction can be provided. To address this issue we introduce a sparsity constraint, which is built from two more general constraints: atleast_k_ge and atmost_k_ge. They both enforce a relation among a set of variables and ensure that among the variables involved at least (resp. at most) k of them have values greater or equal than a threshold. Formally, the constraint:

$$\texttt{atleast\_k\_ge}(k, X, \theta) : \quad \left|\{x_i \in X \mid d(x_i) > \theta\}\right| \geq k \tag{1}$$

enforces a lower bound (k) on the number of variables in X whose confidence value is greater than θ; the constraint:

$$\texttt{atmost\_k\_ge}(k, X, \theta) : \quad \left|\{x_i \in X \mid d(x_i) > \theta\}\right| \leq k \tag{2}$$

limits to at most k the variables in X with confidence value greater than θ.

The propagation of the atmost_k_ge constraint is exploited during the solution search to enforce the property (2) by the following:

$$\texttt{atmost\_k\_ge}(k, X, \theta) : \dfrac{S = \{x_i \in X \mid d(x_i) > \theta\}, |S| = k}{\displaystyle\bigwedge_{x_j \in X \setminus S} D_{\langle x_j \rangle} = D_{\langle x_j \rangle} \cap \{0, \ldots, \theta\}} \tag{3}$$

The `atleast_k_ge` cannot benefit from a powerful propagation rule, but early failures can be detected during the solution search by checking the upper bound on the number of variables not yet instantiated which satisfy property (1).

The sparsity constraint `g-sparsity` is a global constraint over the variables in X. It enforces lower and upper bounds on the number of edges whose confidence value is outside a given threshold. Formally, given $k_l, k_m, \theta_l, \theta_m$:

$$\texttt{atleast\_k\_ge}(k_l, X, \theta_l) \cap \texttt{atmost\_k\_ge}(k_m, X, \theta_m) \tag{4}$$

Redundant Edge Constraints. Several state-of-the-art inference methods rely on MI or correlation techniques; the community approach adopted for this work employs *CLR* and *INFLEATOR*, which are both MI-based methods (see Sec. 3). One of the disadvantages of such methods is the difficulty in speculating on the directionality of a given prediction. We define a constraint that has been effective in our experiments in detecting the edge directionality based on the collective decision of the CN predictions, among the non MI- or correlation-based methods.

Let us consider a collection of predictions $\mathcal{G} = \{G_1, \ldots, G_n\}$ for a GRN $G = (V, E)$, and a non-empty set of MI- or correlation-based methods $\mathcal{H} \subseteq \mathcal{G}$. An edge (t, s) is said to be *redundant* if:

$$\forall G_i \in \mathcal{G} \setminus \mathcal{H}. \quad \omega_i(t, s) < \omega_i(s, t) \wedge (\omega_i(s, t) - \omega_i(t, s)) > \beta \tag{5}$$

where $\omega_i(s, t) : V \times V \to [0, 1] \subseteq \mathbb{R}$ expresses the confidence value of the edge (s, t) in the prediction G_i. Given a redundant edge (t, s) we call the edge (s, t) the *required* edge. The `redundant_edge` constraint enforces a relation between two variables $x_{\langle s,t \rangle}$ and $x_{\langle t,s \rangle}$. Let X_R be the set of all the redundant and required variables.[2] For a pair of variables $x_{\langle s,t \rangle}, x_{\langle t,s \rangle} \in X_R$ the constraint:

$$\texttt{redundant\_edge}(x_{\langle s,t \rangle}, x_{\langle t,s \rangle}, \theta_e, L) : \quad x_{\langle s,t \rangle} > \theta_e \wedge \max(D_{\langle t,s \rangle}) < L \tag{6}$$

ensures that the confidence value assigned to the required variable $x_{\langle s,t \rangle}$ is greater than a given threshold value $\theta_e \in \mathbb{N}$, with $0 \le \theta_e \le 100$, and that the domain of the redundant edge variable $x_{\langle t,s \rangle}$ contains no values greater than L. The propagation of the `redundant_edge` constraint is exploited during the solution search to enforce property (6):

$$(x_{\langle s,t \rangle}, x_{\langle t,s \rangle}, \theta_e, L) : \frac{\min(D_{\langle s,t \rangle}) > \theta_e, \ \max(D_{\langle t,s \rangle}) \ge L}{D_{\langle t,s \rangle} = D_{\langle t,s \rangle} \cap \{0, \ldots, L-1\}} \tag{7}$$

Transcriptor Factor Constraints. Often, GRN specific information, such as sequence DNA-binding TFs or functional activity of a set of genes, is available from public sources (e.g., DBD [12]). Moreover, several studies show that similar mRNA expression profiles are likely to be regulated via the same mechanisms [1]. Not every method may be designed to handle such information, or this information can become available in an incremental fashion, and hence not

---

[2] $x_{\langle s,t \rangle}$ is redundant/required if the corresponding edge (s, t) is redundant/required.

suitably usable by prediction methods. We propose constraints that can directly incorporate such information in the CN model.

A regulatory element is a *Transcription-factor (TF)* if it regulates the production of other genes. This property is described through a relation on the out-degree of the involved gene for those edges with an adequate confidence value. The `transc-factor` constraint over a gene s is enforced by an `atleast_k_ge`(k, X_s, θ) constraint with $X_s = \{x_{\langle s,u \rangle} \in \mathcal{X} \mid u \in V\}$, and k representing the co-expressing degree, i.e., the number of genes targeted by the TF.

Multiple TFs can cooperate to regulate the transcription of specific genes; these are referred to as *Co-regulators*. When this information is available it can be expressed by a `coregulator` constraint. The latter involves two TFs, s' and s''; it enforces a relation over a set of variables X, to guarantee the existence of at least k elements that are co-regulated by both s' and s'' for which an interaction is predicted with confidence value greater than θ $(0 < \theta \leq 1)$. Formally:

$$\texttt{coregulator}(k, X, \theta): \qquad \forall x_{\langle s',t' \rangle}, x_{\langle s'',t'' \rangle} \in X$$
$$\left| \{(s', s'', t') \mid s' \neq s'' \wedge t' = t'' \wedge d(x_{\langle s',t' \rangle}) > \theta \wedge d(x_{\langle s'',t'' \rangle}) > \theta\} \right| \geq k \qquad (8)$$

Search Strategy. The proposed modeling of GRN prediction allows a great degree of flexibility in exploring the solution space. We implement two search strategies: **(1)** a classical prop-labeling tree exploration (DFS), where constraint propagation phases are interleaved with non-deterministic branching phases used to explore different value assignments to variables [2], and **(2)** a Monte Carlo (MC)-based prop-labeling tree exploration, which performs a random value assignment to each variable. We set a trial limit for the MC-based solution and a solution number limit for both strategies.

GRN Consensus. A challenge in GRN inference is the absence of a widely accepted objective function to drive the solution search. We decided to generate an ensemble of m solutions and propose three criteria to compute the final GRN prediction. Given a set of m solutions $S = \{S_1, \ldots, S_m\}$, where each $S_i = \langle a_1^i, \ldots, a_{n^2-n}^i \rangle$, let $S|_{x_k} = \bigcup_{i=1}^m \{a_k^i\}$ be the set of values assigned to the variable x_k in the different solutions, and $\texttt{freq}(a, k)$ be the function counting the occurrences of the value a among the assignments to x_k in the solution set. The consensus value a_k^* associated with the variable x_k is computed by:

- *Max Frequency:* $a_k^* = \arg\max_{a \in S|_{x_k}} (\texttt{freq}(a, k))$. This estimator rewards the edge confidence value appearing with the highest frequency in the solution set. The intuition is that edge-specific confidence values appearing in many solutions may be important for the satisfaction of the constraints.
- *Average:* $a_k^* = \frac{1}{m} \sum_{i=1}^m a_k^i$. It computes the average edge consensus among all solution in order to capture recurring predictive trends.
- *Weighted average:* $a_k^* = \frac{1}{\sum_{a \in S|_{x_k}} \texttt{freq}(a,k)^2} \sum_{a \in S|_{x_k}} \texttt{freq}(a, k)^2 a$. This estimator combines the intuitions of the two above by weighting the average edge confidence by the individual quadratic value frequencies.

We also investigated some potential *global* measures—i.e., acting collectively on the prediction values of all edges—in terms of the solution which minimizes the Hamming distance among all edge prediction values. These global measures were always outperformed by the three estimators discussed above.

3.2 A Case Study

We provide an example to illustrate our approach. We adopt the "E.coli2" network from the 10-node DREAM3 subchallenge [14] (Fig. 1). The target network has two co-regulators (G_1 and G_5) which are in turn regulated by gene G_9. The network has 15 interactions.

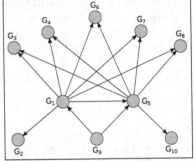

Phase 1: CN Predictions. The inference ensemble was generated by feeding the datasets provided within the DREAM3 challenge to each of the four methods adopted in the community network schema (see Sec. 3). In addition, we generate a CN_{rank} as done in [13], and use it as baseline to build the domain variables (see Alg. 1) and for evaluation.

Phase 2: Modeling the CSP. The execution of Alg. 1 for the prediction disagreements analy-

Fig. 1. An extract of E.coli GRN

sis reduced the initial domain sizes to 1 for 64 cases, and to 3 for the others. The disagreement threshold was set to $\theta_d = 0.20$. As the inference ensemble adopted employs methods that may suffer from the *edge redundancy* problem, we impose a `redundant_edge` constraint for all the edge pairs $(s,t),(t,s)$ that satisfy the definition with $\beta = 0.15$ as:

$$\texttt{redundant\_edge}(x_{\langle s,t\rangle}, x_{\langle t,s\rangle}, 75, 50). \qquad (r)$$

This constraint was able to reduce the value uncertainty for two additional variables—only one element in their domains can possibly satisfy the conditions above for any value choice of the required edge variable.

A sparsity constraint was imposed at a global level as:

$$\texttt{g-sparsity:}\quad \texttt{atleast\_k\_ge}(10, \mathcal{X}, 65) \cap \texttt{atmost\_k\_ge}(25, \mathcal{X}, 65). \qquad (s)$$

Phase 3: Generating the Consensus. We performed 1,000 Monte Carlo samplings and return the first 100 solutions found, which we refer to as *Constrained Community Networks (CCNs)*. To illustrate the effect of constraints integration on the CCNs we consider the best prediction returned by each CSP exhibiting a different combinations of the imposed constraints. We plot it as a graph containing all and only the edges of highest confidence necessary to make such graph weakly connected. These resulting predictions are illustrated in Fig. 2, together with the CN_{rank} (top-right). In each network the green edges (thick with filled arrows) denote the true positive predictions, the red edges (with empty arrows) denote the false positive predictions, and the gray (dotted) edges denote the

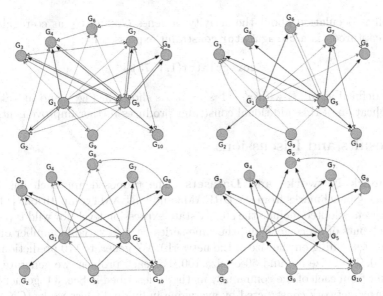

Fig. 2. The CN$_{rank}$ consensus (top-right) and the CCN prediction after the integration of the redundant edge and sparsity constraints (top-left), the TF constraints (bottom-left) and Co-factor constraint (bottom-right).

false negatives. The results are also summarized in Table 1, where we report the AUC scores [3] for the best prediction (CCN$_{best}$) generated and for each CCN generated by the evaluation criteria presented in in Sec. 3.1.

Phase 4: Employing network specific information. Let us now model some specific information about the target network. The target network includes three TFs: G_1, G_5, G_9, which can be modeled via three `transc_factor` constraints as:

$$\texttt{atleast\_k\_ge}(2, N_1, 85), \texttt{atleast\_k\_ge}(2, N_5, 85), \texttt{atleast\_k\_ge}(2, N_9, 85) \qquad (t)$$

with $N_i = \{x_{(i,s)} \mid (\forall G_j \in \mathcal{G})\, \omega_j(i,s) > 0.10\}$. Note that $\omega_j(i,s)$ is the prediction confidence assigned to edge (i,s) by the inference method J in the prediction G_j. Fig. 2 and Table 1 show the improvements using the latter formalization.

Table 1. The effects of constraint integration on the AUC scores for the "Ecoli2" CCNs

Constr.	CN$_{rank}$	CCN$_{best}$	CCN$_{max-f}$	CCN$_{avg}$	CCN$_{w-avg}$
r	0.7271	0.8036	0.7556	0.7644	0.7751
s	0.7271	0.8044	0.7529	0.7164	0.7591
r, s	0.7271	0.8453	0.7778	0.7609	0.7760
r, s, t	0.7271	0.9209	0.7458	0.8489	0.8587
r, s, t, c	0.7271	0.9378	0.7929	0.8622	0.8729

Finally, speculation about the activity of genes G_1 and G_2 as co-regulators can be captured via a `coregulator` constraint expressed by:

$$\texttt{coregulator}(1, V, 75) \hspace{3cm} (c)$$

with V defined as in (8) with $s' = 1, s'' = 5$. As shown in Fig. 2 and in Table 1, the application of this additional constraint produces further improvements.

4   Results and Discussions

Benchmark Networks and Datasets. The proposed approach has been tested using benchmarks from the DREAM3 and DREAM4 competitions [14,16]. The datasets adopted include the steady state expression levels for wild type and for knock-outs of every gene and the time-series data (a variable number of trajectories, depending on the size of the network). We generate 110 predictions: 50 of size 10, 25 of size 50, and 50 of size 100. For each problem we generate four consensus from each of the community methods described in Sec. 3 together with a consensus network constructed by averaging individual edges ranks (CN_{rank}).

Validation. To measure prediction accuracy against the corresponding reference network we adopted the AUC score [3], which relates the ratio between the *true positive* rate and the *false positive* rate. An AUC value of 0.5 corresponds to a random prediction, whereas a value of 1.0 indicates perfect prediction.

Settings. For each experiment we perform a $1,000$ Monte Carlo samplings and return the first 100 solutions found. We observed that the DFS was always outperformed by the MC search and therefore not reported. To guide the parameter selection for the sparsity constraint, given the thresholds θ_l, θ_m (see Eq. (4)), we identity the bounds k_l and k_m which would make the constraint unsatisfiable and use them to set the sparsity parameters. In this way, k_l and k_m are set so that they are bounded, respectively, above by $|\{x_i | x_i \in \mathcal{X} \wedge \max(D_{x_i}) > \theta_l\}|$, and below by $|\{x_i | x_i \in \mathcal{X} \wedge \min(D_{x_i}) > \theta_m\}|$, provided that $k_l < k_m$. The closer are their values to the respective bounds, the more restrictive is the constraint.

The `g_sparsity` (s) and `redundant_edge` (r) constraints have been enabled for all the experiments, with parameters $k_l = \frac{n^2}{10}$, $k_m = \frac{n^2}{4}$, $\theta_l, \theta_m, \theta_e$ in $\{65, 75\}$, and $L = 50$ (from ref. (4) and (6)). The latter was applied to all the pairs of edges satisfying (6) with $\beta = 0.15$. The disagreement threshold θ_d was set to 0.2 (see Alg. 1). We observed that such settings, for both search and constraints parameters, produced stable results across the whole benchmark set, which in turn was designed to capture a variety of network topologies to asseses GRN inference algorithms. We generate four CN consensus (CCNs), one for each estimator described in Sec. 3.1 ($\text{CCN}_{\text{max-f}}$, CCN_{avg}, $\text{CCN}_{\text{w-avg}}$) and CCN_{best}, as best prediction with respect to the AUC score, and compare them against CN_{rank}. The estimators-based CCNs may outperform the CCN_{best} as they are not elements of the set of solutions returned. We experimentally verified their constraints consistency, which was always satisfied.

Table 2. Average AUC score improvements (in percentage) with respect to CN_{rank}

	Dream3 10	Dream4 10	Dream3 50	Dream3 100	Dream4 100
$CCN^{s\,r}_{best}$	+10.52	+7.01	+3.63	+1.75	−0.17
$CCN^{s\,r}_{w\text{-avg}}$	+3.01	+1.96	+1.49	+0.43	+0.05
$CCN^{s\,r\,t}_{best}$	+15.02	+8.43	+8.49	+4.13	+2.29
$CCN^{s\,r\,t}_{w\text{-avg}}$	+5.42	+2.48	+6.32	+3.21	+4.21

Experiments. We first focused on examining the predicted CCNs using the sparsity and redundant edge constraints to leverage community-method features and networks properties. We categorize the benchmarks by DREAM edition and size, and average their respective AUC scores. Table 2 reports the percentage of the average AUC improvements for the best CCN_{best} and best $CCN_{w\text{-avg}}$ with respect to CN_{rank} across all the benchmarks (first two columns). Our choice of reporting only the weighted average estimator, among all those defined in Sect. 3.1, is driven by the observation that the former offers higher stability to parameter tuning and in general outperforms the other two. The CCNs achieved higher average prediction accuracy with respect to CN_{rank} for small and medium size networks, while performance improvements decreased for bigger networks. This is probably due to the high permissiveness of the CSP model for bigger networks. We show next that the application of additional constraints overcomes such effect.

We extended the set of constraints to include specific knowledge about individual networks. We enabled the transcriptor-factor constraint over a set of randomly selected genes which were verified TFs in the target networks. The TFs set sizes were chosen to be at most 30%, 15% and 10%, respectively, for the networks of size 10, 50 and 100; the co-expressing degree was set as $k = 2$ and $\theta = 85$. We performed 5 repetitions and for each TF t the set of possible regulators X has been chosen among the variables $x_{\langle t,s \rangle}$ such that $\omega^{\#}(t,s) > 0.25$. Moreover to promote such constraint we increased the uncertainty for the regulation $x_{\langle t,s \rangle}$ such that $\max D_{\langle t,s \rangle} \leq 50$. These parameters were chosen in accordance to the study presented in [1]. The integration of additional knowledge produced improvements of the AUC scores for both the *best* and the *weighted average* measures—see the last two columns of Table 2. A complete summary of the results is reported in Table 3.

The CCNs outperformed in general CN_{rank}, and $CCN_{w\text{-avg}}$ offers larger improvement for the bigger networks with respect to the version without the TF constraint. This supports our hypothesis that the addition of biological knowledge can better guide the predictions even if re-adopting the same inference ensemble. From a preliminary analysis of the incorrect predicted regulations supported by the TF constraint we observed that many of the erroneous inferences relate genes located in different regions of the graph. This effect could be attenuated by clustering the consensus graph for different connectivities, and targeting the TF constraint on the same cluster (if no prior knowledge on the specific TF is given). We plan to investigate this direction as future work.

Table 3. AUC scores for the benchmark test set categorized by DREAM challenge and network size. The left part of the Table summarizes the CCNs results computed by enabling the redundant edge and the sparsity constraint. The right part of the Table summarizes the CCNs returned by extending the model to include the transcriptor factor constraints. The table reports the best predictions found by each estimator. The average AUC scores outperforming CN_{rank} are highlighted in bold.

Network	CN_{rank}	CCN^{sr}_{best}	$CCN^{sr}_{max\text{-}f}$	CCN^{sr}_{avg}	$CCN^{sr}_{w\text{-}avg}$	CCN^{srt}_{best}	$CCN^{srt}_{max\text{-}f}$	CCN^{srt}_{avg}	$CCN^{srt}_{w\text{-}avg}$
DREAM3 Size 10									
Ecoli1	0.7192	0.8101	0.7443	0.7422	0.7319	0.8642	0.7756	0.7664	0.7664
Ecoli2	0.7271	0.8453	0.7778	0.7778	0.7778	0.9209	0.8444	0.8676	0.8456
Yeast1	0.7413	0.8550	0.7625	0.7350	0.7637	0.8613	0.7325	0.7688	0.7685
Yeast2	0.6191	0.7145	0.6560	0.6123	0.6640	0.7606	0.6794	0.6646	0.6557
Yeast3	0.5428	0.6507	0.5742	0.5588	0.5622	0.6935	0.6457	0.5822	0.5842
Avg	0.6699	**0.7700**	**0.7023**	**0.6852**	**0.7000**	**0.8201**	**0.7355**	**0.7299**	**0.7241**
DREAM4 Size 10									
Net1	0.7493	0.7858	0.7244	0.7324	0.7324	0.8471	0.7680	0.7600	0.7644
Net2	0.6943	0.7981	0.7618	0.7188	0.7188	0.8218	0.7331	0.7365	0.7348
Net3	0.8018	0.8622	0.8396	0.8329	0.8364	0.8649	0.8062	0.8329	0.8338
Net4	0.8501	0.9171	0.8771	0.8601	0.8601	0.9201	0.8511	0.8701	0.8721
Net5	0.8718	0.9541	0.9006	0.9006	0.9038	0.9348	0.8974	0.8921	0.8857
Avg	0.7934	**0.8635**	**0.8207**	**0.8090**	**0.8103**	**0.8777**	**0.8112**	**0.8183**	**0.8182**
DREAM3 Size 50									
Ecoli1	0.6678	0.7317	0.6801	0.6871	0.6991	0.7919	0.7195	0.7724	0.7740
Ecoli2	0.7010	0.7214	0.7083	0.7075	0.7064	0.8205	0.7481	0.7954	0.7954
Yeast1	0.6539	0.6817	0.6545	0.6496	0.6586	0.7205	0.6520	0.6782	0.6842
Yeast2	0.6273	0.6609	0.6466	0.6429	0.6442	0.6866	0.6780	0.6712	0.6725
Yeast3	0.6181	0.6536	0.6236	0.6236	0.6340	0.6731	0.6360	0.6580	0.6579
Avg	0.6536	**0.6899**	**0.6626**	**0.6621**	**0.6685**	**0.7385**	**0.6867**	**0.7150**	**0.7168**
DREAM3 Size 100									
Ecoli1	0.7704	0.7831	0.7647	0.7716	0.7746	0.8131	0.7921	0.8128	0.8128
Ecoli2	0.7152	0.7353	0.7179	0.7186	0.7226	0.7826	0.7479	0.7693	0.7692
Yeast1	0.6975	0.7141	0.7031	0.6984	0.7014	0.7337	0.7086	0.7181	0.7187
Yeast2	0.6116	0.6371	0.6116	0.6164	0.6134	0.6524	0.6201	0.6394	0.6438
Yeast3	0.5596	0.5723	0.5605	0.5629	0.5642	0.5794	0.5684	0.5633	0.5704
Avg	0.6709	**0.6884**	**0.6716**	**0.6736**	**0.6752**	**0.7122**	**0.6874**	**0.7006**	**0.7030**
DREAM4 Size 100									
Net1	0.7829	0.7797	0.7523	0.7788	0.7785	0.7975	0.7606	0.8273	0.8273
Net2	0.7511	0.7396	0.7228	0.7549	0.7535	0.7773	0.7637	0.8085	0.8085
Net3	0.7158	0.7254	0.6956	0.7191	0.7200	0.7455	0.7288	0.7454	0.7492
Net4	0.7408	0.7380	0.7200	0.7429	0.7429	0.7604	0.7230	0.7652	0.7742
Net5	0.7222	0.7220	0.7054	0.7198	0.7205	0.7466	0.7241	0.7626	0.7643
Avg	0.7426	0.7409	0.7192	**0.7431**	**0.7431**	**0.7655**	0.7400	**0.7818**	**0.7847**

4.1   Other Uses: Validating Biological Hypothesis

The underlying technology adopted in this work allows us to test biological hypotheses, expressed in form of constraints, that may assist the phase of experimental design. The solver verifies the existence of a set of solutions consistent with the given hypotheses and its size can be related to confidence strength of the answer.

Consider a case study based on the "E.coli2" network presented in Sec. 3.2 (Fig. 1) to verify the hypothesis on the presence of a co-regulatory interaction. We perform 90 experiments, one for each pair of vertexes s', s'' of the network, with $s' \neq s''$, involved in a constraint of the type coregulator$(2, V, 75)$ with V defined as in (8) and employ the same settings as the one adopted in Sec. 3.2. Among the entire set of problems only four satisfied the imposed hypothesis returning a non-empty set of solutions. These were the ones associated with the putative co-regulators $(G_1, G_5), (G_5, G_3), (G_5, G_6)$ and (G_1, G_8), generating respectively $115, 151, 32$ and 48 solutions. This notably restricts the number of possible biological tests to be performed, and also assigns higher probability to the first two co-regulators as they were able to generate more consistent solutions. We tried to shrink the set of putative co-regulators even more by imposing a stronger constraint: coregulator$(3, V, 70)$ with the same settings used above. This produced only one consistent set of solutions, associated with the pair (G_1, G_5), containing 151 elements. The result confirms the biological value of the experiment (see Figure 1). We stress that the hypothesis tested leverage the collective knowledge as well as additional network specific constraints (e.g., sparsity, redundant_edge) which are collectively handled in the CP model.

5   Conclusions

In this paper we introduced a novel approach based on CP to infer GRNs by integrating a collection of predictions in a CN. Our approach does not impose any hypothesis on the datasets adopted nor on the type of inference methods. We introduced a class of constraints able to (1) enforce the satisfaction of GRNs' specific properties and (2) take account of the community prediction collective agreements on each edge, and of method-specific limitations. Experiments over a set of 110 benchmarks proposed in past editions of the DREAM challenges show that our approach can consistently outperform the consensus networks constructed by averaging individual edges ranks, as proposed in [13] (up to 15.02% for small networks and 4.13% for big networks). We have shown how knowledge specific about target networks could provide further improvements in the AUC measure. This was possible as our model encourages the modular integration of biological knowledge, in form of logical rules, and proposes a set of candidate solutions satisfying the imposed constraints rather than an arbitrary one chosen among many. We introduced three estimators to compute a consensus from the set of consistent candidates and verified their consistency among the imposed constraints. We also show the potential of the proposed solution to assess biological hypotheses by verifying the consistency of the constrained model. This can be helpful in assisting the biological experimental design.

We plan to investigate new optimization measures by taking into account local and global network properties, e.g., the number of specific network motifs in a target GRN region, or the scale free degree in a given a portion of the graph. This can be achieved by including soft constraints in our model. We also plan to use this information to address method-specific biases towards different connectivity patterns. On the CP side, we will extend existing constraints, for instance by studying the most likely set where a TF constraint could be targeted, and model new constrains and propagators to capture different type of biological knowledge, such us information about cell conditions at the time of the experiments.

References

1. Allocco, D., et al.: Quantifying the relationship between co-expression, co-regulation and gene function. BMC Bioinformatics 5(1), 18 (2004)
2. Apt, K.: Principles of Constraint Programming. Cambridge University Press (2009)
3. Baldi, P., et al.: Assessing the accuracy of prediction algorithms for classification: an overview. Bioinformatics 16(5), 412–424 (2000)
4. Corblin, F., Fanchon, E., Trilling, L.: Applications of a formal approach to decipher discrete genetic networks. BMC Bioinformatics 11, 385 (2010)
5. Faith, J.J., et al.: Large-scale mapping and validation of escherichia coli transcriptional regulation from a compendium of expression profiles. PLoS Biol. 5(1) (2007)
6. Fromentin, J., et al.: Analysing gene regulatory networks by both constraint programming and model-checking. In: Conf. Proc. IEEE Eng. Med. Biol. Soc., pp. 4595–4598 (2007)
7. Gebser, M., et al.: Detecting inconsistencies in large biological networks with answer set programming. CoRR, abs/1007.0134 (2010)
8. Greenfield, A., et al.: Dream4: Combining genetic and dynamic information to identify biological networks and dynamical models. PLoS ONE 5(10), e13397 (2010)
9. Haury, A., et al.: Tigress: Trustful inference of gene regulation using stability selection. BMC Syst. Biol. 6(1), 145 (2012)
10. Huynh-Thu, V.A., et al.: Inferring regulatory networks from expression data using tree-based methods. PLoS ONE 5(9), e12776 (2010)
11. Kim, S., et al.: Dynamic Bayesian network and nonparametric regression for modeling of GRNs from time series gene expression data. Biosystems, 104–113 (2003)
12. Kummerfeld, S.K., Teichmann, S.A.: DBD: A transcription factor prediction database. Nucl. Acids Res. 34(suppl. 1), 74–81 (2006)
13. Marbach, D., et al.: Wisdom of crowds for robust gene network inference. Nat. Meth. 9(8), 796–804 (2012)
14. Marbach, D., et al.: Revealing strengths and weaknesses of methods for gene network inference. Proc. Natl. Acad. Sci. U S A, 6286–6291 (2010)
15. Margolin, A.A., et al.: Aracne: An algorithm for the reconstruction of gene regulatory networks in mammalian cellular context. BMC Bioinformatics 7(S1) (2006)
16. Prill, R.J., et al.: Towards a rigorous assessment of systems biology models: The dream3 challenges. PLoS ONE 5(2), e9202 (2010)
17. Sîrbu, A., et al.: Integrating heterogeneous gene expression data for gene regulatory network modelling. Theory in Biosciences 131(2), 95–102 (2012)

18. Soh, T., Inoue, K.: Identifying necessary reactions in metabolic pathways by minimal model generation. In: ECAI, pp. 277–282. IOS Press (2010)

19. Videla, S., Guziolowski, C., Eduati, F., Thiele, S., Grabe, N., Saez-Rodriguez, J., Siegel, A.: Revisiting the training of logic models of protein signaling networks with ASP. In: Gilbert, D., Heiner, M. (eds.) CMSB 2012. LNCS, vol. 7605, pp. 342–361. Springer, Heidelberg (2012)

20. Zhou, X., et al.: Genomic Networks: Statistical Inference from Microarray Data. Wiley (2006)

ABC–Fun: A Probabilistic Programming Language for Biology

Anastasis Georgoulas[1], Jane Hillston[1,2], and Guido Sanguinetti[1,2]

[1] School of Informatics, University of Edinburgh
[2] SynthSys — Synthetic and Systems Biology, University of Edinburgh

Abstract. Formal methods have long been employed to capture the dynamics of biological systems in terms of Continuous Time Markov Chains. The formal approach enables the use of elegant analysis tools such as model checking, but usually relies on a complete specification of the model of interest and cannot easily accommodate uncertain data. In contrast, data-driven modelling, based on machine learning techniques, can fit models to available data but their reliance on low level mathematical descriptions of systems makes it difficult to readily transfer methods from one problem to the next. Probabilistic programming languages potentially offer a framework in which the strengths of these two approaches can be combined, yet their expressivity is limited at the moment.

We propose a high-level framework for specifying and performing inference on descriptions of models using a probabilistic programming language. We extend the expressivity of an existing probabilistic programming language, Infer.NET Fun, in order to enable inference and simulation of CTMCs. We demonstrate our method on simple test cases, including a more complex model of gene expression. Our results suggest that this is a promising approach with room for future development on the interface between formal methods and machine learning.

1 Introduction

Continuous Time Markov Chains (CTMCs) have long been established as a framework for the description and analysis of dynamical systems, including those encountered in the life sciences. Of particular interest, and also widespread, is their use within high-level formalisms, such as process algebras. Adopting a high-level language rather than working with the CTMC itself offers various advantages: a friendlier language to specify the system, easier modification and some degree of inbuilt error-checking, among others. It also gives access to an array of tools to analyse and reason about the behaviour of the system, such as stochastic simulation and model-checking. The major weakness of this framework is that the models implicitly assume complete mechanistic knowledge of the system. Therefore it does not offer support for integrating experimental data into models or inferring parameters from observations.

In contrast, machine learning techniques applied to models of biological processes are designed to predict a system's behaviour in the presence of uncertainty.

A. Gupta and T.A. Henzinger (Eds.): CMSB 2013, LNBI 8130, pp. 150–163, 2013.

An important category of such methods is concerned with using (possibly noisy) observations from the system in order to refine our understanding of it, a task often referred to as *learning* and which can be broken down into two aspects: learning the structure of the system in question, or learning its parameters (for a given structure). These methods mostly work on mathematical descriptions of systems, often in the form of ordinary, partial or stochastic differential equations. For example, Bayesian Networks, a graphical framework for describing proba-bilistic models, while intuitive and widely used are still essentially a front-end to the underlying equations. Working with the low-level description negates the advantages afforded by high-level languages, as described above, thus limiting the applicability of the inference techniques.

Some common ground between the two approaches may be found in the field of *probabilistic programming*. Probabilistic programming promises to of-fer a high-level language that can be used for both describing and learning non-deterministic systems. Using a programming language and the expressive power it affords makes the the process of specifying a system easier, while at the same time offering a range of other features such as modularity and type sys-tems. Additionally, we also obtain a unified, general framework for automatically performing inference on a given model, eliminating the need to write bespoke solutions and learning algorithms for every system of interest.

While the field of probabilistic programming has generated considerable inter-est in recent years, current probabilistic programming languages are limited in the types of models they can describe, as well as in the inference methodologies they implement. In particular, to our knowledge, continuous time, non-parametric mod-els such as CTMCs cannot be handled by current probabilistic programs. Several techniques have been proposed, including approximations based on finite dimen-sional projections [1], sampling methodologies (e.g.[15,3]), and variational infer-ence approaches [12]. However, these methods all require a low level mathematical description of the system (usually a way of approximately solving the chemical master equation).

In this paper, we explore the potential for a framework which encompasses the strengths of both high-level formalisms and machine learning by extending an existing probabilistic programming language, Infer.NET Fun [2], in order to enable stochastic simulation and approximate Bayesian inference for CTMCs. We call the resulting approach ABC–Fun and illustrate it on two examples of biological significance, showing the potential of probabilistic programming as an effective tool for modelling in systems biology. Our focus, however, is not on presenting a fully-formed solution but rather on exploring the applicability of a novel approach, with the ultimate aim of facilitating interfacing of models and experiments.

The rest of the paper is structured as follows: we give an overview of proba-bilistic programming and the platform we are using, Infer.NET Fun; we briefly describe our implementation of Gillespie's stochastic simulation algorithm, the basis of our approach, before presenting the inference process; finally, we describe our experiments and discuss their results.

2   Background

2.1   Bayesian Inference

In this work, we focus on the *Bayesian* approach to learning, which uses probability distributions to model and quantify uncertainty about all aspects of the system under study, including its structure or parameters. Assume the system to be characterised by a set of parameter values Θ (e.g. transition rates of a certain CTMC). We are also given a set of (partial) observations of the system **y**. The principal ingredients of the Bayesian approach are two: the *prior distribution* $p(\Theta)$ encodes any initial beliefs about the values of the parameters. The *likelihood* $p(\mathbf{y}|\Theta)$ (sometimes called observation model) gives the probability of the observations given the values of the parameters. Since the observations are fixed, this is a function of the parameter values. Bayes' rule combines these two ingredients to provide a mathematically sound way of estimating the impact of the observations on our beliefs over the parameters,

$$p(\Theta|\mathbf{y}) = \frac{1}{Z}p(\mathbf{y}|\Theta)\,p(\Theta)\,. \tag{1}$$

$p(\Theta|\mathbf{y})$ is the *posterior distribution* over the parameters, which quantifies the uncertainty over the parameters implied by the data and the prior beliefs.

A major computational hurdle in applying Bayes' rule is the estimation of the proportionality constant Z in equation (1). This term, the *marginal likelihood* or *evidence*, represents the probability of the data under all possible settings of the parameters; its value is obtained by performing (usually analytically intractable) integrals over the parameter space, which become prohibitive in even moderate dimensions. In the case of CTMCs, this problem is further compounded by the fact that (in general) even the likelihood cannot be computed analytically: the probability of the state of a CTMC taking a particular value at a certain time can only be obtained by solving the chemical master equation, which is impossible in most cases. In general, Bayesian inference in CTMCs remains a challenging problem: current methods either resort to approximations to the chemical master equations [12,1] or sampling approximations [15,3]. In all of these approaches, inference relies on a low level mathematical description of the system as a Markov transition system, and often specific characteristics of the system (e.g. functional form of the transition rates) are hard-wired in any accompanying code, greatly reducing the ease of portability and applicability of the approach.

2.2   Probabilistic Programming

Probabilistic programs can be thought of as an extension of conventional, deterministic programs, in which expressions describe stochastic experiments. Rather than having a concrete value, then, an expression corresponds to a whole distribution over values and evaluating it means performing the experiment and recording its outcome [13]. Constraining some variables within an expression to have a specified value is equivalent to performing inference, with the observations representing constraints.

Historically, probabilistic programming languages have been primarily targeted at graphical models, a popular class of models in machine learning. Briefly, a graphical model is a specification of a finite number of random variables and the (conditional) dependence relationship which define their joint distribution. The name graphical model derives from the fact that such models can be represented as graphs or networks; this graphical representation enables a quick and intuitive formulation of the model, and also encodes several properties which are important for simplifying inference. For a thorough review, we refer the reader to the excellent book [7].

Examples of languages for probabilistic programming include IBAL [13] and Church [5]. Both of these use a functional language to specify probabilistic models, equipped with a way of performing inference on them. The generation of the inference code is automated and tailored to the model at hand, which means the user can focus on describing the model and not on adjusting or rewriting code for every different model. However, the automation of inference comes at a cost, either in terms of the class of models that can be considered (IBAL for example only considers finite graphical models), or of the inference methodology employed (Church only allows the Metropolis-Hastings sampling algorithm, which requires an analytically tractable likelihood function).

Infer.NET [9] is a probabilistic programming framework developed by Microsoft Research for specifying probabilistic models and performing Bayesian inference on them. More specifically, it offers a high-level, programming language interface for the description of graphical models. Further to this, Infer.NET includes an inference engine that can use a number of different algorithms, such as Expectation Propagation and Gibbs Sampling, to obtain estimates of the distribution of the model's parameters, informed by the knowledge of some observed data. Infer.NET also provides bindings for programming languages such as C# and Python and these can be used to describe a model and the desired inference queries, which are then compiled into source code. The resulting code can then itself be compiled and executed, returning the results of the inference queries. An additional component of Infer.NET is Fun ([2,6]), an F# interface that aims to make the process of describing a model even more similar to "conventional" programming. As such, its syntax is very lightweight and consists of simple additions to the standard F# syntax, resulting in a user-friendly framework.

A model expressed in Fun can be used in two different yet related ways. The first is to view it as a generative model, that is, a description of how data points are generated. In this case, every **random** expression produces a sample from the given distribution. The model can therefore be "run forwards", giving rise to a set of values in a style similar to ancestral sampling.

The second way to use a model is to pass it to the Infer.NET inference engine. The model is then "run backwards". We use **observe** expressions to specify observed data and condition the model on them — when such an expression is encountered, if its condition is not met, the execution is marked as failed. The result of this procedure is that we can obtain the posterior distributions on the random variables of the model, i.e. the distribution when only considering those executions which satisfy all the observations.

To illustrate what models probabilistic programming languages can handle, and see why they are not sufficient for CTMCs, we consider the kind of systems that can be modelled in Fun. In the simplest case, a Fun model can describe a static system with a finite number of random variables. In this case, one would describe how the variables depend on one another by specifying their conditional distributions. Some of these distributions may be parametrized, and the unknown parameters are also treated like random variables, in the Bayesian style.

A more complex example which can still be modelled in Fun is a dynamical system with a known number of steps. The state at each time depends only on the previous state, thus the system can be represented as a Markov Chain. Describing this would result in a recursive definition, which is not supported by Infer.NET. However, in the functional programming paradigm, this can be reformulated as a folding operation, to avoid explicit recursion. Folding a function over an array involves applying it to all elements of the array sequentially. The result of every application is used to obtain a new function, which is then applied to the next element, and so on. This technique has been applied to one of the examples in [6], but it should be noted that this is made possible because the number of steps in the system is known *a priori*.

3   Probabilistic Programming for CTMCs

In this section we highlight the limitations of Fun and describe how these can be addressed in an economic way by exploiting Fun's parent language, F#.

Infer.NET, and Fun in particular, is designed to address graphical models, i.e. models with a known, finite number of random variables. In particular, the number of random variables is hard-wired in the definition of a Fun model type. In the case of CTMCs, however, the number of transitions that may occur in a given time is generally unknown, therefore so is the number of random variables (since every transition is associated with two random variables, one each for the choice of transition and delay). This means both that we do not have an array to fold over, and so cannot eliminate the recursion, and that we must use lists instead of arrays, since the latter have a fixed size while the former can grow indefinitely. However, these are not features supported by Fun or Infer.NET, so we must leave Fun for a different language. A major implication of this is that the inbuilt inference engine of Infer.NET cannot be used, so that alternative inference strategies need to be used (detailed in Section 3.2).

As we would like to retain some of the functionality of Fun, such as drawing samples, we turn to its base language, F#, and use Fun as a library for the operations we need. This way, we can retain useful language structures and features such as random number generators for various different probability distributions, although the final code will not be compatible with the Infer.NET inference engine. Moreover the F# code has been packaged as a library for FUN making it available to other users who are not necessarily familiar with the implementation details for CTMCs. This illustrates our aim to lift the data generation and inference techniques into a high-level language, supporting their use by a wide-range of users as transparently as possible.

3.1   Implementing the Stochastic Simulation Algorithm

The ability to simulate CTMCs is central for both modelling and inference. We describe here the ABC-Fun implementation of Gillespie's Stochastic Simulation Algorithm [4]; this is given in considerable detail both because of its importance and in order to provide the reader with a concrete example of ABC-Fun syntax.

We handle models as chemical reactions: a model with N species and K reactions is comprised of a stoichiometry matrix and a list of kinetic laws. The stoichiometry matrix is implemented as a list of lists; each of its K sub-lists has length N, corresponds to a reaction and contains the updates for the population of each species when that reaction occurs. Each reaction also has an associated kinetic law, which is a function that acts on a list of integers (the state of the system) and returns a real value (the rate of the reaction); these are collected in the second component of the model. A model can also be parametrized – such a model is essentially a function that takes a list of parameters and returns a concrete model, as described above.   We define a `Model` constructor, which combines the two elements (list of rates and stoichiometry matrix) and creates a `model` object. Technically, this has the F# type:

```
(int list -> float) list * (int list) list -> model
```

This provides an interface through which one can specify models, without concern for how the simulation is performed. For example, a model of a single species birth-process can be encoded as follows:

```
let r1_1 l = 0.1 * float (List.head l)
let r1_2 l = 20.0     // constant rate
let rateLaws1 = [r1_1; r1_2]
let stoich1 = [[-1]; [1]]
let m1 = Model(rateLaws1, stoich1)
```

A trace (one possible run of the CTMC) can then be obtained simply by calling the function `pathSample`, which accepts a model, an initial state and the stopping time, and returns a trajectory through the state space (a list of states and a list of the corresponding times). Its type is therefore

```
model -> int list -> float -> int list list * float list
```

In order to generate a trace, we must make explicit the sampling steps involved in the SSA. To do so in a probabilistic programming language, we use some of the Fun in-built functions (primarily the random number generators). The SSA can be recast in probabilistic programming terms if we consider that, at every step, the next reaction to occur is a discrete random variable, with each possible value having a probability that can be calculated from the reaction rates at the current state. Similarly, the time to the next state is also a random variable. Both of these probabilities are encoded in the kinetic parameters of the model. In order to simulate the CTMC, we keep the parameter fixed; we will see in the next section how to vary the parameters in an inference scheme. The code below shows how we implement the standard version of the SSA in F# using Fun as a library. The part shown here specifies how the next reaction and delay are chosen. It is straightforward to use this in order to recursively build the trajectory, keeping a list of the states and transition times and stopping when we reach the final time.

```
let nextStateAndTime state rateFunctions stoich =
  let rates = [| for r in rateFunctions -> r state |]
  let sumRates = Seq.sum rates
    if sumRates > 0.0 then
      let delay = ExponentialSample sumRates
      let reaction = random(Discrete (Vector.FromArray rates))
      let newState = updateState state (List.nth stoich reaction)

      (newState, delay)
    else //all rates are 0, we can stop
      (state, infinity)
```

Note that the code above uses the **random** construct from Fun to sample from a distribution. **ExponentialSample** is also defined using **random** to draw from an exponential distribution with the specified rate:

```
let ExponentialDist rate = GammaFromShapeAndRate(1.0, rate)
let ExponentialSample rate = random(ExponentialDist rate)
```

updateState simply calculates the next state given the current state and the row of the stoichiometry matrix corresponding to the chosen reaction. As both these arguments are represented as lists, this can be expressed as the pairwise sum of their elements.

3.2   Choosing an Inference Engine

As explained earlier, the limitations of the Infer.NET engine mean that we must adopt a different inference method. For the purposes of this work, we use Approximate Bayesian Computation (ABC) [19], a parameter inference scheme that constructs an approximation of the posterior distribution by repeated simulations of the system. This enables us to use our implementation of the SSA to also perform inference, as described below.

ABC works by generating samples of parameters. For each such parameter, the behaviour of the system is simulated, in our case producing a path through the state space. If the path obtained this way matches the observed data, under some given metrics and tolerance, the parameter sample is kept, otherwise it is discarded. The process is repeated until a sufficient number of samples has been accepted, and the resulting set of accepted parameter values serves as the approximate representation of the posterior. ABC can therefore be thought of as a way of converting simulation into an inference technique. The choice of tolerance can be significant, as there is a trade-off between accuracy and efficiency of the algorithm. The lower the tolerance, the harder it is to accept a sample, which means more sampling attempts will be required in order to reach the desired number of accepted samples (as the rejection rate will be higher) but the final set will be more representative of the true posterior distribution.

There are different versions of the algorithm (as described, for example, in [17]), depending on how the parameter samples are generated. The simplest approach is to sample independently from a prior distribution, which may however prove to be inefficient, while another is to have each sample depend on the previous one, giving rise to a Markov Chain Monte Carlo (MCMC) scheme.

Choosing a metric to evaluate the distance between a simulated trace and the observed data is an important issue. Assume we have a series of successive observations along with the corresponding measurement times $\{(y_1, t_1), ..., (y_N, t_N)\}$, and a simulated trace $\{(x_1, \tau_1), ..., (x_M, \tau_M)\}$, with $M > N^1$. In this work, the first thing we do is "shift" and thin the trace, keeping the value at every t_i. Formally, we define a new time-series $\hat{x}$ such that $\hat{x}_i = x_m$, where $m = argmax_j(\tau_j \leq t_i)$, for $i = 1, .., N$.

The simplest way to calculate the distance between x and y is to take the absolute difference between $\hat{x}$ and y, averaged over all points:

$$d(x,y) = \frac{1}{N} \sum_{i=1}^{N} |\hat{x}_i - y_i| \qquad (2)$$

In the case of CTMCs, a plausible alternative could be to rescale the difference between $\hat{x}$ and y adaptively according to the value of y; this can be justified by noticing that noise in CTMCs is usually multiplicative. We therefore also consider the following metric

$$\tilde{d}(x,y) = \frac{1}{N} \sum_{i=1}^{N} \frac{|\hat{x}_i - y_i|}{\sqrt{y_i}} \qquad (3)$$

which may be more suitable when the observations span a wide range of values.

In this work, we perform parameter estimation using the MCMC version of the ABC algorithm. This works by constructing a Markov chain in parameter space which asymptotically will converge to an approximation of the posterior distribution. Given data and an initial set of parameter values, we sample a new set of parameter values from a proposal distribution which depends on the current parameters (in our application, a Gaussian centred at the old parameter values). We then simulate the system (by running the model forwards using these values), compare the trace obtained this way with the input data and decide whether to accept it or not, depending on the "distance" between the two traces, our tolerance level and the prior and proposal distributions. We

---

[1] Notice that our measurements are counts at individual times, not transition times; therefore, an analytical expression for the likelihood is in general not available.

repeat this sampling scheme for a previously specified number of steps, then use the samples collected this way as a representation of the posterior distribution of the parameters.

4   Experiments

4.1   Birth-Death Process

We have tested ABC–Fun on a simple model of a birth-death process, in which a single species can be created at a constant rate or degraded according to mass-action kinetics (i.e. at a rate proportional to its amount). We initially fix the death rate and reduce the problem to a one-dimensional one, where we try to estimate the birth rate. We use a uniform prior, reflecting a belief that all parameter values are equally likely in the absence of any data.After 10^5 steps, the distribution of samples is clearly centred around the true parameter value (Figure 1). A common issue with MCMC schemes is the difficulty of assessing whether the process has converged, i.e. whether the samples are truly representative of the posterior distribution and have "overcome" the influence of the initial state. Experimenting with different initial samples indicated that this convergence to the true value was robust, although it required more samples when the initial point was further away from the true value (Figure 2).

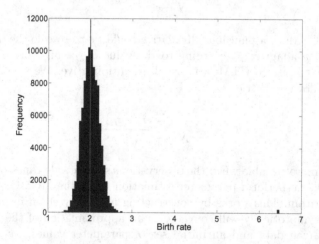

Fig. 1. Histogram showing the number of accepted samples after 100000 steps of the ABC-MCMC algorithm when inferring only the birth rate. The true value of the parameter is 2, shown by the vertical line.

We then considered the problem of inferring the full model, i.e. estimating both rates. The parameters chosen for the simulation were a birth rate of 2 and a death rate of 0.02, leading to the steady state of 100 being reached after about 250 time units. The data that was used as input for the inference came

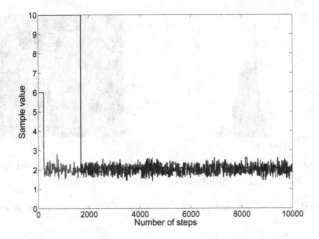

Fig. 2. Accepted samples when inferring only the birth rate, showing convergence to the true value. The algorithm was run for 10000 steps, starting from an initial value of 6 (red) or 10 (blue). Similar results occur for all tested initializations.

from a single stochastic simulation, taking 20 samples from the resulting trace (approximately 15 during the transient phase and 5 at the steady state).

This time, the results are not as clear-cut: the heat map (Figure 3b) indicates there are multiple value pairs that match the observed data. This reflects an identifiability issue with the system, which can be easily explained if we consider that the probability of choosing one reaction over the other depends only on the ratio between the two rates. Therefore, there exist multiple parameterizations which would give the same relative probabilities, and the unknown parameters can only be estimated up to a multiplicative constant. It is possible to reduce this uncertainty by considering information about the timing of the reactions, as the duration of the reactions does depend on the concrete values of the parameters, rather than just their ratio — intuitively, higher rates will result in faster reactions.

The results show that the highest number of accepted samples is concentrated around the true values of the parameters. Additionally, the other areas with a significant number of samples lie on a diagonal line, indicating a constant ratio between the two parameters, matching our expectation. If we plot the ratio of the two parameters for each of the accepted samples (Figure 3a), we can see that this quantity is most often close to 100, its true value. In short, the inference procedure manages to distinguish the true parameter pair from the others that give a similar behaviour.

4.2 Regulation of Gene Expression in Single Cells

As a more biologically meaningful example, we consider the classic on/ off model of gene expression (e.g. [14]). Here, the rate of mRNA production is assumed to depend solely on the state of the gene promoter: thus, mRNA can be produced at

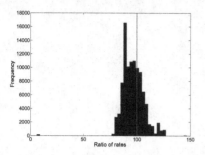
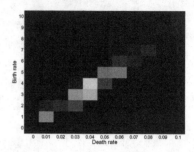

(a) histogram of birth to death rate ratio for the accepted samples (true value: 100)   (b) heat map of accepted samples (true values: 2 and 0.02)

Fig. 3. Accepted samples when inferring both kinetic rates in the new experimental configuration

a high rate (promoter ON) or a low rate (promoter OFF). The inference task is to reconstruct both mRNA production/ decay rates and the promoter occupancy state from gene expression time series. This model was recently used in [18] to tease apart bursting kinetics in mRNA production; there the parameters were estimated by maximum likelihood. Bayesian inference methodologies for this model have been recently proposed assuming mRNA concentrations to be continuous variables [16,11]; here, we consider the Bayesian inference problem when mRNA counts are discrete, and are thus governed by a birth/ death process whose birth rate depends on an unobserved binary process[2]. The importance of this model lies not only in its fundamental role as a mechanism for gene expression, but also in the possibility of using it as a building block for modelling complex gene regulatory networks [10].

To slightly simplify the task, we assumed that the promoter state only performed two transitions within the time frame under consideration (i.e., it starts in the OFF condition, turns ON at a random time, and then turns OFF again). We then tested the ABC–Fun approach on simulated data under ten different configurations of the model parameters/ switching times. For these experiments we used the modified distance metric $\tilde{d}$ (Equation 3). Figure 4 shows the posterior probabilities of the ON and OFF times in a particular run, with the true values indicated by a vertical line. As we can see, the posterior distribution is approximately centred around the true value. The inferred posterior distributions for the promoter activity (difference between birth rate in the two states) and the decay rate are shown in Figure 5. We can see that the posterior distribution has substantial mass concentrated around the true value, but is quite wide due to the identifiability problems already mentioned in Section 4.1. Results for other configurations of the parameters gave qualitatively similar results.

---

[2] This can be seen as a special case of Bayesian inference for Markov Jump Processes [12,20], albeit employing a different inference methodology.

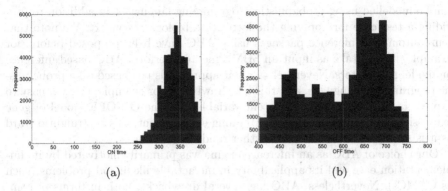

Fig. 4. Accepted samples when inferring the (a) switch-on and (b) switch-off time (real value shown by red vertical line)

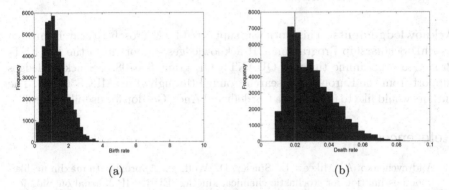

Fig. 5. Accepted samples when inferring the (a) promoter activity A) and (b) degradation rate λ (real value shown by red vertical line)

5   Discussion and Conclusions

We have presented ABC–Fun, a probabilistic programming language which handles biological models expressed as CTMCs with uncertain rates. Our approach uses features of an existing probabilistic programming language, Infer.NET-Fun; however, the non-parametric nature of CTMCs cannot directly be handled by Fun, so that we have to extend it by defining new types in F#, and use a different inference engine to perform approximate Bayesian inference. Our initial results on two simple but biologically relevant models show that this approach can be a valuable addition to the systems biology toolkit: in particular, the two different models only required minimal coding changes. We expect this high portability to be an increasingly important feature as systems biology matures to handle ever more complex models.

Our method extends the range of systems and inference methodologies that can be modelled using probabilistic programming languages, in addition to providing a test case for applying the latter in a biological context. We note that semi-automated inference packages using ABC have been proposed before: for example, [8] will take as input an SBML file and perform ABC-based inference on model parameters. Nevertheless, their approach is not based on a probabilistic programming language, and this has drawbacks: for example, it is not easy to express in SBML models with latent variables like the ON-OFF model of gene expression. In a probabilistic programming environment, this is straightforward as it is merely the addition of a further random variable.

Our choice of ABC as an inference engine was primarily motivated by its implementation ease and its applicability in intractable likelihood problems (such as CTMCs). Nevertheless, ABC has several drawbacks, both in terms of computational efficiency, and in terms of relying on a tolerance parameter which is difficult to tune in a principled way. Exploring alternative inference approaches which can ameliorate these problems will be key to extending our methodology to larger, more relevant models.

Acknowledgements. This work was supported by Microsoft Research through its PhD Scholarship Programme. JH acknowledges support from the EU FET-Proactive programme through QUANTICOL grant 600708. GS acknowledges support from the European Research Council through grant MLCS306999. The authors would like to thank Luca Cardelli and Andy Gordon for useful discussion.

References

1. Andreychenko, A., Mikeev, L., Spieler, D., Wolf, V.: Approximate maximum likelihood estimation for stochastic chemical kinetics. EURASIP Journal on Bioinformatics and Systems Biology 2012(1), 9 (2012)
2. Borgström, J., Gordon, A.D., Greenberg, M., Margetson, J., Van Gael, J.: Measure transformer semantics for Bayesian machine learning. In: Barthe, G. (ed.) ESOP 2011. LNCS, vol. 6602, pp. 77–96. Springer, Heidelberg (2011)
3. Boys, R., Wilkinson, D., Kirkwood, T.: Bayesian inference for a discretely observed stochastic kinetic model. Statistics and Computing 18, 125–135 (2008)
4. Gillespie, D.T.: Exact stochastic simulation of coupled chemical reactions. Journal of Physical Chemistry 81(25), 2340–2361 (1977)
5. Goodman, N.D., Mansinghka, V.K., Roy, D.M., Bonawitz, K., Tenenbaum, J.B.: Church: a language for generative models. In: UAI, pp. 220–229 (2008)
6. Gordon, A., Aizatulin, M., Borgström, J., Claret, G., Graepel, T., Nori, A., Rajamani, S., Russo, C.: A model-learner pattern for Bayesian reasoning. In: Proceedings of the ACM SIGPLAN Conference on Principles of Programming Languages, POPL (2013)
7. Koller, D., Friedman, N.: Probabilistic Graphical Models. MIT Press, Cambridge (2010)
8. Liepe, J., Barnes, C., Cule, E., Erguler, K., Kirk, P., Toni, T., Stumpf, M.P.: ABC-SysBio—approximate Bayesian computation in Python with GPU support. Bioinformatics 26(14), 1797–1799 (2010)

9. Minka, T., Winn, J., Guiver, J., Knowles, D.: Infer.NET 2.5, Microsoft Research Cambridge (2012), http://research.microsoft.com/infernet
10. Ocone, A., Millar, A.J., Sanguinetti, G.: Hybrid Regulatory Models: a statistically tractable approach to model regulatory network dynamics. Bioinformatics 29(7), 910–916 (2013)
11. Opper, M., Ruttor, A., Sanguinetti, G.: Approximate inference for Gaussian-jump processes. In: Advances in Neural Information Processing Systems 24 (2010)
12. Opper, M., Sanguinetti, G.: Variational inference for Markov jump processes. In: Platt, J., Koller, D., Singer, Y., Roweis, S. (eds.) Advances in Neural Information Processing Systems 20, pp. 1105–1112. MIT Press, Cambridge (2008)
13. Pfeffer, A.: The Design and Implementation of IBAL: A General-Purpose Probabilistic Language. In: Getoor, L., Taskar, B. (eds.) Introduction to Statistical Relational Learning. The MIT Press (2007)
14. Ptashne, M., Gann, A.: Genes and signals. Cold Harbor Spring Laboratory Press, New York (2002)
15. Rao, V., Teh, Y.W.: Fast MCMC sampling for Markov jump processes and continuous time Bayesian networks. In: UAI (2011)
16. Sanguinetti, G., Ruttor, A., Opper, M., Archambeau, C.: Switching regulatory models of cellular stress response. Bioinformatics 25(10), 1280–1286 (2009)
17. Sisson, S.A., Fan, Y., Tanaka, M.M.: Sequential Monte Carlo without likelihoods. Proceedings of the National Academy of Sciences 104(6), 1760–1765 (2007)
18. Suter, D.M., Molina, N., Gatfield, D., Schneider, K., Schibler, U., Naef, F.: Mammalian genes are transcribed with widely different bursting kinetics. Science 332(6028), 472–474 (2011)
19. Toni, T., Welch, D., Strelkowa, N., Ipsen, A., Stumpf, M.P.: Approximate Bayesian computation scheme for parameter inference and model selection in dynamical systems. Journal of the Royal Society Interface 6(31), 187–202 (2009)
20. Zechner, C., Pelet, S., Peter, M., Koeppl, H.: Recursive Bayesian estimation of stochastic rate constants from heterogeneous cell populations. In: IEEE CDC-ECE, pp. 5837–5843 (2011)

A Temporal Logic Approach to Modular Design of Synthetic Biological Circuits

Ezio Bartocci[1], Luca Bortolussi[2,3], and Laura Nenzi[4]

[1] Vienna University of Technology, Austria
[2] DMG, University of Trieste, Italy
[3] CNR/ISTI, Pisa, Italy
[4] IMT, Lucca, Italy

Abstract. We present a new approach for the design of a synthetic biological circuit whose behaviour is specified in terms of signal temporal logic (STL) formulae. We first show how to characterise with STL formulae the input/output behaviour of biological modules miming the classical logical gates (AND, NOT, OR). Hence, we provide the regions of the parameter space for which these specifications are satisfied. Given a STL specification of the target circuit to be designed and the networks of its constituent components, we propose a methodology to constrain the behaviour of each module, then identifying the subset of the parameter space in which those constraints are satisfied, providing also a measure of the robustness for the target circuit design. This approach, which leverages recent results on the quantitative semantics of Signal Temporal Logic, is illustrated by synthesising a biological implementation of an half-adder.

Keywords: Synthetic Biology, Parameter Synthesis, Temporal Logic.

1 Introduction

Synthetic Biology [15, 28] is an emerging discipline that aims at the rational design of *artificial* living systems with a predictable behaviour, either by creating new biological entities that do not exist in nature or by redesigning the existing ones. Even though important technological developments have been achieved in this field, the *de-novo* design of biological circuits implementing a desired behaviour results to be a very hard task, especially for large scale networks. Biological systems are complex to understand and to be engineered: the non-linear nature of interactions reflects in the emergence of systemic behavioural properties, not directly derivable from the knowledge of the individual parts. To model and control such systems we need to understand the relationships between the emergent behaviour and the topology of such complex interactions. A possible approach is to divide the whole system in "subunits" and to look at the structure of the interactions between them. This subdivision is often suggested by the way we describe (the components of) those systems. The idea is that compositionality at the specification level, to a certain extent, has to be reflected into compositionality at the behavioural level. This should depend on the properties satisfied by a single "subunit" and on the wiring between them. This way to approach the study of a system is called *modularity* and the "subunits" of the system are called *modules*. Modularity

A. Gupta and T.A. Henzinger (Eds.): CMSB 2013, LNBI 8130, pp. 164–177, 2013.
© Springer-Verlag Berlin Heidelberg 2013

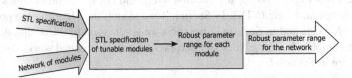

Fig. 1. Overview of the proposed approach

can be effectively achieved in *Synthetic Biology*, combining a *bottom-up* [32] and a *top-down* [31] methodology. The former consists in the assembling of a set of well-characterised modules [32] together to build sophisticated biological circuits and devices. The latter [31] aims to identify and characterise the possible "subunits" and this is also helpful to understand real biological systems, for example to discover unknown structures or behaviours or to better understand and test current knowledge.

To unveil the system dynamics, it is important to correlate the denotation of a module with some of its specific behaviours, and understand how the global properties emerge from these local ones. This can be performed better if the emergent behaviours are specified in a formal language. We consider here a *logical* characterisation in terms of (linear) temporal logic formulae. In particular, we focus our attention on genetic regulatory circuits, seen as networks of interacting genetic modules (each representing, for instance, a logic gate). Each module has a set of inputs and outputs (usually transcription factors), and its local behaviour is specified by temporal logic properties.

In particular, we characterise the behaviour of *logic gates* with the addition of constraints on the response time. Logic gates are physical devices implementing a boolean function and they are the fundamental bricks upon which all the other logic circuits, including multiplexers, arithmetic logic units, memories and microprocessors, are built. They are primarily implemented using electronic transistors acting as electronic switches. In the last decade, genetic circuits acting as logic gates have been successfully identified and synthesised [23]. This lead researchers to hope to engineer cells to turn them into miniature computers.

The main idea of this paper, sketched in Figure 1, is to translate the structural compositionality of networks of modules into compositionality of local behaviours, exploting it to enforce a set of global behaviours to the network. This is realised by identifying a subset of parameters for which the truth of local properties implies the truth of the global specification, exploiting the modular structure of the network. We thus interpret the network of modules as a composition of their local properties, connecting the emergent behaviours with the topology of interaction of those local properties. The technical core of our approach is the quantitative semantics of Signal Temporal Logic [21], which can be seen as a measure of robustness of the satisfaction of a certain formula, and which comes with simulation-based methods to compute the robustness score and to identify a region of the parameter space in which the formula holds true.

The contributions of this paper are thus twofold: a design methodology for biological circuits based on a high level logical specification of behaviours and an algorithmic procedure exploiting compositionality to make parameter synthesis more effective, which gives as a byproduct a measure of robustness of the implementation.

The paper is structured as follows: in Section 2 we introduce the background material. In Section 3 we discuss the logical characterisation of the basic modules in terms of Signal Temporal Logic (STL). In Section 4 we sketch the algorithmic approach to parameter synthesis and in Section 5 we show an application to the design of an half-adder, a fundamental building block of microprocessors. The related works and the final discussion are in Section 6.

2   Background Material

Modelling of Gene-Regulatory Networks. In this paper we consider deterministic models of gene regulatory networks, given by a set of non-linear Ordinary Differential Equation (ODE) [17]. For simplicity, we consider lumped models of gene expression, in which mRNA is not explicitly represented (cf. Remark 2 for a further discussion on this point). We assume to have n genes and proteins. Concentration of protein i at time t, $i = 1, \ldots, n$, is denoted by the variable $x_i[t]$, while $x = (x_1, \ldots, x_n)$ denotes the vector of concentration variables. The ODE for $x_i[t]$ will then be of the form

$$\frac{dx_i}{dt} = f_i(x) = f_i^+(x) - f_i^-(x),$$

where f_i^+ is a function giving the net production rate of x_i, while f_i^- is its degradation rate, which is usually a linear function of the form $\mu_i x_i$, for some $\mu_i > 0$. The function f_i^+, instead, encodes the regulatory mechanism of gene i, and is a combination of Michaelis-Menten or Hill functions [28].

Signal Temporal Logic. Temporal logic [24] provides a very elegant framework to specify in a compact and formal way an emergent behaviour in terms of *time-dependent* events. Among the myriads of temporal logic extensions available, Signal Temporal Logic [21] (STL) is very suitable to characterise behavioural patterns in time series of real values generated during the simulation of a dynamical system. STL extends the dense-time semantics of Metric Interval Temporal Logic [1] (MITL), with a set of parametrised numerical predicates playing the role of atomic propositions. STL provides two different semantics: a boolean semantics that returns yes/no depending if the observed trace satisfies or not the STL specification, and a quantitative semantics that in addition returns a measure of robustness of the specification. Recently, Donze et. al [12] proposed a very efficient monitoring algorithm for STL robustness, now implemented in the Breach [9] tool. The combination of robustness and sensitivity-based analysis of STL formulae have been successfully applied in several domains, ranging from analog circuits [16] to systems biology [10, 11], to study the parameter space and also to refine the uncertainty of the parameter sets. In the following we recall [13] the syntax and the quantitative semantics of STL that will be used in the rest of the paper. The boolean semantics can be inferred using the sign of the quantitative result (positive for true and negative for false).

Definition 1 (STL syntax). *The syntax of the STL is given by*

$$\varphi := \top \mid \mu \mid \neg\varphi \mid \varphi_1 \wedge \varphi_2 \mid \varphi_1 \, \mathcal{U}_{[a,b]} \, \varphi_2$$

where $\top$ is a true formula, conjunction and negation are the standard boolean connectives, $[a, b]$ is a dense-time interval with $a < b$ and $\mathcal{U}_{[a,b]}$ is the until operator.

The atomic predicate $\mu : \mathbb{R}^n \to \mathbb{B}$ is defined as $\mu(\mathbf{x}) := (y(\mathbf{x}) \geqslant 0)$, with $\mathbf{x}[t] = (x_1[t], ..., x_n[t])$, $t \in \mathbb{R}_{\geqslant 0}$, $x_i \in \mathbb{R}$, and $y : \mathbb{R}^n \to \mathbb{R}$ a real-valued function.

The (bounded) *until* operator $\varphi_1 \, \mathcal{U}_{[a,b]} \, \varphi_2$ requires φ_1 to hold from now until, in a time between a and b time units, φ_2 becomes true. The *eventually* operator $F_{[a,b]}$ and the *always* operator $G_{[a,b]}$ can be defined as usual: $F_{[a,b]}\varphi := \top \mathcal{U}_{[a,b]}\varphi$, $G_{[a,b]}\varphi := \neg F_{[a,b]} \neg \varphi$.

Definition 2 (STL Quantitative Semantics).

$$
\begin{aligned}
\rho(\mu, \mathbf{x}, t) &= y(\mathbf{x}[t]) \qquad \text{where } \mu \equiv (y(\mathbf{x}[t]) \geqslant 0) \\
\rho(\neg\varphi, \mathbf{x}, t) &= -\rho(\varphi, \mathbf{x}, t) \\
\rho(\varphi_1 \wedge \varphi_2, \mathbf{x}, t) &= \min(\rho(\varphi_1, \mathbf{x}, t), \rho(\varphi_2, \mathbf{x}, t)) \\
\rho(\varphi_1 \, \mathcal{U}_{[a,b]}\varphi_2, \mathbf{x}, t) &= \max_{t' \in t+[a,b]} (\min(\rho(\varphi_2, \mathbf{x}, t'), \min_{t'' \in [t,t']} (\rho(\varphi_1, \mathbf{x}, t''))))
\end{aligned}
$$

where ρ is the quantitative satisfaction function, returning a real number $\rho(\varphi, \mathbf{x}, t)$ quantifying the degree of satisfaction of the property φ by the signal $\mathbf{x}$ at time t. Moreover, $\rho(\varphi, \mathbf{x}) := \rho(\varphi, \mathbf{x}, 0)$.

3 Logical Characterisation of Modules

The approach for the synthesis of biological circuits is based on the idea of combining simple genetic networks according to a specific design. These basic building blocks, or modules, are usually composed of a single or few genes, and express a specific transcription factor (or signal) in response to an input signal, generally the presence or absence of activators or repressors influencing the module behaviour. In most of the proposed approaches [28, 29], such modules are the biological equivalent of the logic gates of electronics, and as such they encode simple boolean functions, like AND, OR, or NOT, that can be combined together to build more complex circuits. Logic gates are usually described by their truth table. However, when moving from electronics to biology, the temporal dimension becomes much more relevant, and it cannot be neglected. Furthermore, biological modules considered in literature often produce more complex input/output (I/O) responses than a boolean I/O relationship, like pulses and oscillations [28]. For this reason, we find more convenient to describe the I/O behaviour of a module by a set of temporal logic properties.

More precisely, we define a module $\mathcal{M}$ to be a *genetic network* containing n genes, that produce proteins whose concentration is indicated by $\boldsymbol{x} = (x_1, \ldots, x_n)$. The genes of $\mathcal{M}$ are also regulated by additional n_I external transcription factors, which are the *inputs* of the module. A subset of n_O of the produced proteins constitutes the output of the module. The behaviour of such a module is characterised by a set of STL formulae of the form $\varphi_I \to \varphi_O$, expressing an I/O relationship, which can be arbitrarily complex. Here φ_I depends only on the concentration of the input signals $\mathbf{x_I} = (x_{I_1}, ..., x_{I_{n_I}})$ and φ_O only on the concentration of the output signals $\mathbf{x_O} = (x_{O_1}, ..., x_{O_{n_O}})$. Modules can be easily connected into a network, by using one output of a module as the input of another module (see Figure 2). Such networks can still have external inputs, while a subset of outputs of their modules will be identified as the output of the network. Furthermore, the network behaviour can also be characterised in terms of a temporal I/O relationship given by STL formulae of the form $\varphi_I \to \varphi_O$. In this sense, a network is nothing but a more complex module, which can then be used as a building block itself, resulting in a hierarchical compositional approach to circuit design.

Example: Logic Gates. As an example, in this paper we consider modules corresponding to AND, OR, and NOT logic gates. For instance, a simple biological implementation of an AND gate can be obtained by a module in which a single gene, producing the output protein, is activated by two input signals, both required to start the gene expression. This requirement can be enforced directly at the level of the gene promoter [23] or by letting the complex formed by two input proteins activate the gene [20]. We stick to the first formulation. The truth table of the gate is shown in Table 1. To each input and output protein, we associate two thresholds, θ_+ and θ_-. The value *true* in the truth table corresponds to a concentration of the corresponding protein above θ_+, while the value *false* corresponds to the concentration being below θ_-. In the truth table we also provide a high level specification of the temporal behaviour of the gate, in terms of the *maximum response time* δ and the *minimum duration* λ of the output signal. The former is an upper bound on the time needed by the gate to stabilise. The latter, instead, specifies for how long the output remains up or down. This in turn implies a constraint on the duration of the input signal: if we want the output to remain up for λ units of time, then both inputs have to remain up for at least $\lambda + \delta$ units of time. We can easily turn such a truth table into a set of STL formulae, a formula for each row. For instance, the row four of Table 2 gives:

$$G_{[0,\lambda+\delta]}(x_A \geq \theta_{A+} \wedge x_B \geq \theta_{B+}) \to F_{[0,\delta]}G_{[0,\lambda]}(x_C \geq \theta_{C+}), \qquad (1)$$

where x_A and x_B are the input signals and x_C is the output. The mathematical model associated with this gate will be given by the non-linear ODE:

$$\dot{x}_C = H_{AND}(x_A, x_B, x_C, \mathbf{k}) = k_{AB} \frac{x_A^n}{K_A^n + x_A^n} \frac{x_B^n}{K_B^n + x_B^n} - k_C x_C, \qquad (2)$$

where $\mathbf{k} = (k_{AB}, k_C, K_A, K_B, n)$ is a tuple of 5 parameters: k_{AB}, the maximum production rate (here we assume a zero basal expression rate), k_C, the degradation rate, K_A and K_B, governing the Hill activation function, and n, governing the steepness of the Hill function.

The other basic logic gates can be modelled in a similar fashion [23]: the OR gate can be obtained from the AND gate by a non-collaborative activation of gene expression (e.g., replacing in the ODE model the product of Hill functions by a single Hill function depending on the sum of the two concentrations), while the NOT gate can be modeled by a gene whose production is repressed by the input protein. For actual biological implementations, see for instance the discussion in [23, 28].

Example: XOR Gate. Figure 2 shows how to build a XOR gate using AND, OR, and NOT gates. We stress here that the circuit architecture, seen as an implementation of a boolean function, can be obtained by classical techniques (e.g. by Karnaugh maps [18]). To fully specify the *extended truth table* of the XOR gate, like for the AND gate (cf. Table 1), we need to specify additional information about the maximum response time and the minimal duration of the output signal for the network. These two quantities obviously depend on the corresponding ones of the constituent modules. Here we will specify a target temporal behaviour for the network and we will consequently constrain the temporal behaviour of modules.

Suppose we fix a maximum response time δ and a minimum duration λ of the output signal for the XOR gate. Looking at Figure 2, we clearly see that the input signal to the XOR gate has to go through no more than three gates before influencing the output. Hence, if each gate has a maximum response time of $\delta/3$, we obviously obtain a response time for the XOR bounded by δ. To enforce the constraint on the minimum duration of the output signal, we just need to make the output signals of internal gates last sufficiently long to trigger an output signal of the network of the target duration. This can be done by simply taking into account the maximum response delay of each gate. In the XOR example, we obtain that the AND gates need to have a minimal duration of $\lambda + \delta/3$, while the NOT gates of $\lambda + 2\delta/3$. Clearly, the input signal of the network needs to stay on for $\lambda + \delta$ units of time.

Constraints for Arbitrary *Acyclic* Networks of Logic Gates. This simple compatibility analysis is easily generalised to arbitrary *acyclic* networks of logic gates, to which we restrict ourselves for the moment. Dealing with feedback loops is more complicated and is left to future investigation.

Consider a generic module/logic gate in an acyclic network, with target maximum delay δ and target output signal duration λ. For each module $\mathcal{M}$ (with a single output) of such a network, let $\ell_f(\mathcal{M})$ be the length of the *longest* path from $\mathcal{M}$ to an output module (i.e. a module producing one output of the network) and $\ell_b(\mathcal{M})$ be the length of the *longest* path from $\mathcal{M}$ to an input module (i.e. a module with an external input). Due to the acyclic nature of the network, both such quantities are finite and can be easily computed by a visit of the graph. Then the processing of an input signal passing from $\mathcal{M}$ has to go through at most $\ell_f(\mathcal{M}) + \ell_b(\mathcal{M}) + 1$ modules, so that a maximum delay of $\delta(\mathcal{M}) = \delta/(\ell_f(\mathcal{M}) + \ell_b(\mathcal{M}) + 1)$ guarantees the response time bound on the network. As for the minimum duration of the output for module $\mathcal{M}$, we can obtain it by the recursive relation $\lambda(\mathcal{M}) = \delta(\mathcal{M}) + \max\{\lambda(\mathcal{M}')\}$, where $(\mathcal{M}, \mathcal{M}')$ is an edge of the network, i.e. $\mathcal{M}'$ is a module receiving as input an output of $\mathcal{M}$. These relationships are easily extended to modules with more than one output, defining a max response time constraint for each output.

We observe here that this compatibility analysis between delays and durations has a counterpart in the STL characterisation of module behaviours. The main idea is that we can express the consistency of the output-input links by the STL formulae like:

$$F_{[\nu_1, \nu_1 + \gamma_1]} G_{[0, \mu_1]}(x \geq \theta_+) \rightarrow G_{[\nu_2, \nu_2 + \mu_2]}(x \geq \theta_+), \qquad (3)$$

Table 1. Extended truth table for the AND gate

Inputs		Output	Input\Output
max delay=δ		min. duration=λ	
pA	pB	pC	STL Formula
low	low	low	$G_{[0,\lambda+\delta]}(x_A \leq \theta_{A-} \wedge x_B \leq \theta_{B-}) \to F_{[0,\delta]}G_{[0,\lambda]}(x_C \leq \theta_{C-})$
low	high	low	$G_{[0,\lambda+\delta]}(x_A \leq \theta_{A-} \wedge x_B \geq \theta_{B+}) \to F_{[0,\delta]}G_{[0,\lambda]}(x_C \leq \theta_{C-})$
high	low	low	$G_{[0,\lambda+\delta]}(x_A \geq \theta_{A+} \wedge x_B \leq \theta_{B-}) \to F_{[0,\delta]}G_{[0,\lambda]}(x_C \leq \theta_{C-})$
high	high	high	$G_{[0,\lambda+\delta]}(x_A \geq \theta_{A+} \wedge x_B \geq \theta_{B+}) \to F_{[0,\delta]}G_{[0,\lambda]}(x_C \geq \theta_{C+})$

This formula states that if a variable is eventually expressed for μ_1 units of time, starting between time ν_1 and $\nu_1 + \gamma_1$, it is for sure expressed for μ_2 units of time, starting at time ν_2. If we set $\mu_1 = \lambda + \delta$, $\mu_2 = \lambda$, $\gamma_1 = \delta$, and $\nu_2 = \nu_1 + \delta$, with $\nu_1 \geq 0, \lambda, \delta > 0$ arbitrary, we obtain that the formula (3) is valid. According to the previous discussion, we need to choose $\lambda = \lambda(\mathcal{M})$ and $\delta = \delta(\mathcal{M})$.

Remark 1. In principle, we can consider more complex building blocks than logic gates, for instance modules acting as switches or oscillators. To this end, we need to generalise the technique for combining modules. More specifically, effective connection of modules is enforced by requiring the validity of formula (3), which is of the form $\varphi_O \to \varphi_I$. Such a formulation in terms of validity of STL formulae can be extended to more general output properties (or proper subformulae thereof). For instance, we can describe oscillations as signals being eventually above a high threshold for some time, and then falling below a low threshold for a subsequent period of time (this property holding globally). The subformulae describing these two behaviours can then be matched with input formulae of the kind considered in this paper.

4   Parameter Synthesis

Consider a network composed by modules representing logic gates, fix a network specification in terms of an extended truth table/ STL formulae, and consider an ODE model of the network, depending on a tuple of parameters **k**. We now tackle the problem of identifying parameters **k** such that the network satisfies the specifications. According to the previous section, in order to satisfy the temporal constraints at the network level, we can simply enforce local constraints at the module level. The key intuition of our approach is that modularity can be further exploited, doing parameter synthesis for each module, with a guarantee that the so obtained parametrisation will satisfy the global specification at the network level. Furthermore, we will identify a *set* of compatible parameter values rather than a single point. Within the set, furthermore, we can identify an *optimal* parametrisation, by maximising the satisfaction level of the properties, according to STL quantitative semantics. We can also search a biological database, like BioBricks, to find genes with the synthesised kinetic constraints.

At the heart of the proposed approach resides the STL characterization of (the biological implementation of) logic gates. Essentially, we will restrict to a single gate, fixing the temporal constraints to those implied by the network requirements and by its structure, and find a subset of the parameter space in which the STL formulae characterising the gate behaviour hold true. This can be done algorithmically, using the simulation approach to parameter synthesis of [10], based on sensitivity analysis and STL quantitative semantics and implemented in Breach [9]. For the simple class of logic gates considered here, we can also do this analytically. Modularity is the key to the efficiency of our approach: as we treat independently each gate, we just need to explore a low dimensional parameter space, which makes the (computational) procedure feasible.

Modularity of Parameter Synthesis for Logic Gates. The main difficulty we have to solve is related to the fact that modules are connected in the network, hence they are not independent. Indeed, the expression of a gene is driven by the dynamical behaviour of its input transcription factors. The idea to get around this problem is to do a *worst case analysis*, showing that a specific parameter combination satisfies the properties for the "worst possible input signal", and that this implies the satisfaction for all possible input signals compatible with the input constraints. This will result in a conservative, but computationally efficient, estimate. We can define the notion of "worst case input signal" in terms of the STL characterisation of module behaviour. Given an input signal $x_I[t]$ of a module $\mathcal{M}$, $t \in [0, T]$, we denote with $x_{x_I, k}[t]$ the trajectory of the module, with input $x_I[t]$ and parameters k.

Definition 3. *An input signal $\hat{x}_I[t]$, $t \in [0, T]$ is a worst-case input signal for the STL specification $\varphi_{Input} \to \varphi_{Output}$ of the behaviour of a module $\mathcal{M}$ if and only if, for each parameter configuration k such that $\rho(\varphi_{Input}, \hat{x}_I) \geq 0$ (and φ_{Input} true) and $\rho(\varphi_{Output}, x_{\hat{x}_I, k}) > 0$, the following property holds:*

- *for each other input signal x_I satisfying $\rho(\varphi_{Input}, x_I) \geq 0$ (and φ_{Input} true), it holds that $\rho(\varphi_{Output}, x_{x_I, k}) \geq \rho(\varphi_{Output}, x_{\hat{x}_I, k})$.*

The characterisation of such a "worst possible input signal" depends on the structure of the target STL formula and on the system of ODE describing a particular module. We provide now such a characterisation for the basic logic gate models considered in this paper and for the STL formulae associated with their extended truth tables. Consider the property $G_{[0, \lambda + \delta]}(x_A \geq \theta_{A+} \wedge x_B \geq \theta_{B+}) \to F_{[0, \delta]} G_{[0, \lambda]}(x_C \geq \theta_{C+})$, which describes a row of the extended truth table of an AND gate. This property is of the desired form $\varphi_{Input} \to \varphi_{Output}$. Now, φ_{Input} identifies a subset of trajectories of the space of functions from $[0, \lambda + \delta]$ to $\mathbb{R}^2$, i.e. those that satisfy the inequality $x_A \geq \theta_{A+} \wedge x_B \geq \theta_{B+}$ for all $t \in [0, \lambda + \delta]$. Among those functions, we consider $\hat{x}_A[t] \equiv \theta_{A+}$ and $\hat{x}_B[t] \equiv \theta_{B+}$, which satisfy φ_{Input} but have quantitative satisfaction score equal to zero. Furthermore, for any other trajectory $x_A[t], x_B[t]$ that satisfies φ_{Input}, we have $x_A[t] \geq \hat{x}_A[t]$ for each $t \in [0, \lambda + \delta]$, and similarly for x_B. By monotonicity of Hill functions, this implies that the vector field of the AND gate satisfies $f_{AND}(x_A[t], x_B[t], x_C, k) \geq f_{AND}(\hat{x}_A[t], \hat{x}_B[t], x_C, k)$ for any $x_C \geq 0$. It then follows, by integrating the vector field, that $x_C[t] \geq \hat{x}_C[t]$ for $t \in [0, \lambda + \delta]$. Looking at the satisfaction function of φ_{Output}, defined by

$$\rho(\varphi_{Output}, x_C) = \max_{\hat{t} \in [0, \delta]} (\min_{t \in [\hat{t}, \hat{t} + \lambda]} (x_C[t] - \theta_{C+})),$$

it is easy to see that $x_C[t] \geq \hat{x}_C[t]$ for $t \in [0, \lambda + \delta]$ implies $\rho(\varphi_{Output}, x_C) \geq \rho(\varphi_{Output}, \hat{x}_C)$. Hence, any configuration of parameters such that $\rho(\varphi_{Output}, \hat{x}_C) > 0$

will imply the truth of φ_{Output} for any input signal satisfying φ_{Input}, and therefore the truth of $\varphi_{Input} \rightarrow \varphi_{Output}$. It follows that $\hat{x}_A, \hat{x}_B$ is a worst-case input signal.

For the AND gate, a similar approach allows us to deal with the other three STL properties associated with the other rows of the truth table. In these cases, we need to find an upper bound for $x_C[t]$, as we need to satisfy the output property $F_{[0,\delta]}G_{[0,\lambda]}(x_C \leq \theta_{C-})$. To achieve this, we just need to set $x_J[t]$ to θ_{J-}, if the input J is false, and to γ_J if the input J is true, where γ_J is the maximum concentration level for the input x_J, obtained by dividing maximum production rate by the degradation rate (here $J = A, B$). In fact, in this way we maximise the production rate. All this analysis is easily extended to OR and NOT gates, and is captured in the following proposition.

Proposition 1. *Let x_O be the output of a AND or OR logic gate and let x_J be a generic input. Fix the attention on a row of the extended truth table.*

- *If x_O is high, and x_J high, then $\hat{x}_J \equiv \theta_{J+}$.*
- *If x_O is high, and x_J low, then $\hat{x}_J \equiv 0$.*
- *If x_O is low, and x_J high, then $\hat{x}_J \equiv \gamma_J$.*
- *If x_O is low, and x_J low, then $\hat{x}_J \equiv \theta_{J-}$.*

Similarly, let x_O be the output of a NOT logic gate[1] and let x_J be its input. Then

- *If x_O is high, then x_J is low and $\hat{x}_J \equiv \theta_{J-}$.*
- *If x_O is low, then x_J is high and $\hat{x}_J \equiv \theta_{J+}$.*

We stress that this proposition not only allows us to do parameter synthesis modularly, but also to *find a lower bound on the robustness score* of each parameterization.

Remark 2. The worst case analysis presented in this section relies on the monotonicity of the robustness score with respect to the input signal. This follows from the monotone dependence of the output on the input (in fact, $\frac{\partial f}{\partial x_J} > 0$), and of the robustness score on the output. The construction of the worst case input is easily generalised to more complex scenarios satisfying a generalised monotonic property of the robustness score, following [27]. As an example, consider a model of the gene expression in which the gene produces the mRNA, and mRNA is in turn translated into the protein. In this case, for an AND gate, we have an ODE for mRNA similar to the one above, namely $\frac{dm_C}{dt} = f_{AND}(x_A, x_B, m_C, \mathbf{k})$, while the ODE for the protein becomes $\frac{dx_C}{dt} = f_C(m_C, x_C, \mathbf{k}) = k_t m_C - k_d x_C$, with k_t the translation constant and k_d the protein degradation constant. The monotonic dependence of the robustness score (when both inputs are on) from inputs essentially follows because a larger input concentration will produce more mRNA, which in turn will result in a higher expression of the protein, giving a larger robustness degree (input/ output properties are the same). If such a monotonic dependence fails, determining the worst case input can be more challenging. We will tackle this issue in our future work.

---

[1] The difference between AND/ OR and NOT gates is in the fact that the input is an activator in the first two cases and a repressor in the last one.

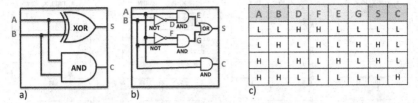

A	B	D	F	E	G	S	C
L	L	H	H	L	L	L	L
L	H	L	H	L	H	H	L
H	L	H	L	H	L	H	L
H	H	L	L	L	L	L	H

Fig. 2. a) Half Adder implemented using two logic gates (XOR, AND), b) Half Adder implemented combining six logic gates, c) truth table for the Half Adder

Sketch of the Algorithm. Assuming the temporal constraints on the extended truth tables of modules have been derived from those of the network, the algorithm for parameter synthesis then work as follows: for any module/gate of the network, and any row in the extended truth table, fix the values of input signals to the worst case ones, and then do STL parameter synthesis to identify a subset of the parameter space in which the STL formula associated with the row is true. Take the intersection of these sets for each row in the truth table of each module[2].

The STL parameter synthesis can be performed applying the sensitivity-based algorithm [10] implemented in the Matlab toolbox Breach [9]. This is a general approach, applicable to any module for which a worst-case input signal has been identified. However, for logic gates AND, OR, and NOT, we can further exploit their simplicity and characterise analytically a subset of parameters for which the STL specification is satisfied. This is due to the fact that, once the input signals are fixed, the non-linear model of the gate reduces to a linear set of ODEs, for which we can compute the solution in closed form. The details of the computation are reported in [2].

5   Example: Half-Adder

The half-adder is a digital component that perfoms the sum of two bits A and B and provides two outputs, the sum (S) and the carry (C) signal representing an overflow into the next digit of a multi-digit addition. The value of the sum is $2C + S$. Figure 2 a) shows the simplest half-adder design and it incorporates a XOR gate for S and an AND gate for C. Figure 2 b) shows an alternative design using two NOT gates, two AND gates and one OR gate instead of a XOR gate. This is the design of the half-adder we intend to use, thus exploiting the characterisation of worst-case inputs for AND, OR, and NOT gates given in Proposition 1. Figure 2 c) shows the output of each component gate of the half-adder, for each pair of inputs.

We applied the algorithm discussed in the previous section to such a network layout, fixing the maximum total delay of the half-adder to 12 time units. Applying the method to enforce time constraints to each module, we obtain that all the gates that are part of the XOR gate must have a maximum time delay of 4 time units, while the AND gate whose output is C can have a maximum response time bounded by 12 time units. Before doing parameter synthesis, we also rescaled the concentration

---

[2] We use the convention that parameters not influencing a gate are set to their whole domain by the STL procedure.

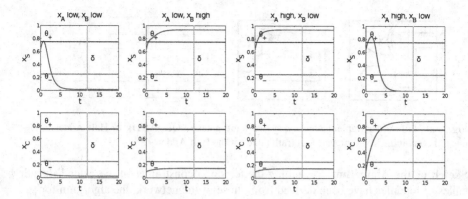

Fig. 3. The red curves represent the output signals of the Half-Adder gate, S and C, in the four different combination of the inputs A and B, one for each column; the horizontal lines are the threshold concentrations (θ_+ in blue and θ_- in green); the yellow vertical line represents the time bound δ.

of each protein to the interval [0,1]. In this way, activation and deactivation thresholds are relative to the maximum steady state expression level of each protein. For this example, we then arbitrarily fixed all the activation thresholds to $\theta_+ = 0.75$ and the deactivation thresholds to $\theta_- = 0.25$, and then synthesised set of parameters consistent with the STL network specification and with such thresholds. We obtained the following bounds for parameters, with indices in the n and α parameters referring to the output variable and indices in the K parameters referring to the input and output protein, as from Figure 2 b). AND gate: $n_C, n_E, n_G \geq 3.2129, 0.3406 \leq K_{AC}, K_{BC}, K_{AE}, K_{DE}, K_{BG}, K_{FG} \leq 0.4228, \alpha_C \geq 0.3074, \alpha_E, \alpha_G \geq 0.9222$. OR gate: $n_S \geq 3.1681, 0.4050 \leq K_{ES}, K_{GS} \leq 0.5090, \alpha_S \geq 0.9222$. NOT gates: $n_D, n_F \geq 2.5372, 0.4192 \leq K_{AF}, K_{BD} \leq 0.4966, \alpha_D, \alpha_F \geq 0.9222$. Constraints are similar for all gates of a given class (e.g. all AND gates) as a consequence of the rescaling of variables in [0,1]. Obviously, in a further step matching actual biological components to the circuit design, this rescaling has to be properly accounted for (for instance, by rescaling also the parameters of the biological components). Picking a value for each parameter consistent with the previous constraints, we can observe in Figure 3 that the dynamics of the network indeed satisfies the specifications of a half-adder.

We remark that, even if in this example we fixed the activation and deactivation thresholds and did parameter synthesis for the other parameters of the model, in the formal derivation we considered such threshold as parameters themselves.

6   Discussion

In this paper we focused on the design techniques for synthetic biological systems. We developed an approach based on two ideas: the specification of system properties in terms of signal temporal logic, and the exploitation of modularity to obtain an efficient procedure to identify a set of parameters for which the network satisfies its STL specification. In particular, we considered the parameter synthesis problem for *networks of logic gates*, implemented as simple genetic networks. For *acyclic networks*, we are able to identify efficiently a set of parameters satisfying STL formulae encoding not only the desired boolean behaviour of the network, but also constraints on its response time.

Modularity allows us to synthesise parameters efficiently, processing each gate component independently. This is possible by isolating each module from the network assuming the worst possible input, which we formally characterised for the basic logic gates considered. We then showed the approach at work with a network implementing an half-adder.

The approach of this paper can be complemented by looking at databases of biological components, like BioBricks [19], for actual combinations of gene and promoters that satisfy the constraints on parameters. A delicate point for this plan is that we are implicitly requiring each module to produce different, non-interfering, output proteins, a not necessarily biologically realistic hypothesis. We will look at possible ways of relaxing this constraint, as in [33]. Other directions for future work include the generalisation of Proposition 1 to deal with more complex modules, for instance feed-forward networks implementing pulse generation or a low-pass filter. Moreover, we will consider the problem of dealing with more complex network topologies, having feedback loops. We expect to make some progress in this direction by suitably rephrasing parameter synthesis as the computation of a fixed point. Finally, we will also take into account the effects of stochasticity, for instance by exploiting moment closure techniques [30].

Related Work. De novo design of a synthetic biological circuit [8] implementing a desired behaviour is a very computational intensive task. The majority of the existing approaches relies on brute-force techniques running sophisticated optimization (i.e. evolutionary algorithms [14], simulating annealing [7]) algorithms to tune the kinetic parameters [6, 25, 29] values in order to match the desired beahaviour.

These methodologies, lacking of compositionality, do not scale well and they are very computationally expensive for large networks. A more rational approach for automatic design was proposed by Marchisio and Stelling in [4, 22] where they show a workflow design taking as input a truth table and generating as output several possible circuit schemes, ranking them in the order of complexity. The choice of a truth table as a input specification for the target circuit design may be not enough when we need to guarantee that the result is produced after a proper delay. Additionally, the design needs to take in consideration the signal compatibility among the "wired" devices (a problem treated in [33]): the output signal of one device must match (in terms of low/high thresholds) with the input signal the other design. The novelty of our contribution is using signal temporal logic as specification language both for the target circuit and for the available components, adding also time constraints in the design process. Furthermore, the device compatibility is rephrased in terms of a STL formula, of the form $\varphi_O \rightarrow \varphi_I$, and the correct matching is elegantly obtained by requiring this formula to be valid.

Another related approach, is the one proposed by Batt et al. in [3], where the authors approximate the behaviour of genetic regulatory networks with piecewise multi-affine systems. In this class of models, the state-space is partitioned in hyper-rectangles exhibiting useful convexity properties [5] that allows to compute an over-approximation of the reachable sets. The authors exploit this characteristic to guide the parameter space partitioning in search of the intervals for which the gene networks is enforced to satisfy a particular behaviour expressed in a linear temporal logic formula. However, their approach is not modular, and only the rates of production and degradation of the proteins can be chosen as possible parameters. Furthermore, by using an over-approximation, the property usually expresses invariants and the parameter ranges found are very coarse, without discriminating trajectories with different time-constraints.

Finally, among the vast literature on combinatorial circuit design, we mention [26], where authors study the timing behaviour of a acyclic circuits by means of timed automata. Our approach is simpler and motivated by the inherent precision of delays in ODE models. However, the techniques of [26] could be helpful to relax the timing constraints we impose and to deal with intrinsic variability of biochemical systems.

Acknowledgements. Work partially funded by the EU-FET project QUANTICOL (nr. 600708) and by FRA-UniTS.

References

1. Alur, R., Feder, T., Henzinger, T.A.: The benefits of relaxing punctuality. J. ACM (1996)
2. Bartocci, E., Bortolussi, L., Nenzi, L.: Supplementary material for "A temporal logic approach to modular design of synthetic biological circuits". CoRR (2013), http://arxiv.org/abs/1306.4493v1
3. Batt, G., Yordanov, B., Weiss, R., Belta, C.: Robustness analysis and tuning of synthetic gene networks. Bioinformatics 23(18), 2415–2422 (2007)
4. Beal, J., Weiss, R., Densmore, D., Adler, A., Appleton, E., Babb, J., Bhatia, S., Davidsohn, N., Haddock, T., Loyall, J., Schantz, R., Vasilev, V., Yaman, F.: An End-to-End Workflow for Engineering of Biological Networks from High-Level Specifications. ACS Synth. Biol. 1(8), 317–331 (2012)
5. Belta, C., Habets, L.C.G.J.M.: Controlling a class of nonlinear systems on rectangles. IEEE Trans. of Automatic Control 51(11), 1749–1759 (2006)
6. Chen, B.S., Hsu, C.Y., Liou, J.J.: Robust design of biological circuits: evolutionary systems biology approach. J. Biomed. Biotechnol. 2011, 304236 (2011)
7. Dasika, M.S., Maranas, C.D.: OptCircuit: an optimization based method for computational design of genetic circuits. BMC Syst. Biol. 2, 24 (2008)
8. Densmore, D., Anderson, J.C.: Combinational logic design in synthetic biology. In: IEEE International Symposium on Circuits and Systems, ISCAS 2009., pp. 301–304 (2009)
9. Donzé, A.: Breach, A toolbox for verification and parameter synthesis of hybrid systems. In: Touili, T., Cook, B., Jackson, P. (eds.) CAV 2010. LNCS, vol. 6174, pp. 167–170. Springer, Heidelberg (2010)
10. Donzé, A., Clermont, G., Langmead, C.J.: Parameter synthesis in nonlinear dynamical systems: Application to systems biology. Journal of Computational Biology 17(3), 325–336 (2010)
11. Donzé, A., Fanchon, E., Gattepaille, L.M., Maler, O., Tracqui, P.: Robustness analysis and behavior discrimination in enzymatic reaction networks. PLoS One 6(9), e24246 (2011)
12. Donzé, A., Ferrère, T., Maler, O.: Efficient robust monitoring for STL. In: Sharygina, N., Veith, H. (eds.) CAV 2013. LNCS, vol. 8044, pp. 264–279. Springer, Heidelberg (2013)
13. Donzé, A., Maler, O.: Robust satisfaction of temporal logic over real-valued signals. In: Chatterjee, K., Henzinger, T.A. (eds.) FORMATS 2010. LNCS, vol. 6246, pp. 92–106. Springer, Heidelberg (2010)

14. Francois, P., Hakim, V.: Design of genetic networks with specified functions by evolution in silico. PNAS 101(2), 580–585 (2004)
15. Fu, P., Panke, S.: Systems Biology and Synthetic Biology. John Wiley & Sons (2009)
16. Jones, K.D., Konrad, V., Nickovic, D.: Analog property checkers: A ddr2 case study. Formal Methods in System Design 36(2), 114–130 (2010)
17. De Jong, H.: Modeling and simulation of genetic regulatory systems: A literature review. Journal of Computational Biology 9, 67–103 (2002)
18. Karnaugh, M.: The map method for synthesis of combinational logic circuits. Trans. American Institute of Electrical Engineers 72(2) (1953)
19. Knight, T.: Idempotent vector design for standard assembly of biobricks. Technical Report MIT Synthetic Biology Working Group, MIT (2003)
20. Madec, M., Lallement, C., Gendrault, Y., Haiech, J.: Design methodology for synthetic biosystems. In: Proc. of MIXDES, pp. 621–626 (2010)
21. Maler, O., Nickovic, D.: Monitoring temporal properties of continuous signals. In: Lakhnech, Y., Yovine, S. (eds.) FORMATS/FTRTFT 2004. LNCS, vol. 3253, pp. 152–166. Springer, Heidelberg (2004)
22. Marchisio, M.A., Stelling, J.: Automatic design of digital synthetic gene circuits. PLoS Comput. Biol. 7(2), e1001083 (2011)
23. Myers, C.J.: Engineering Genetic Circuits. Chapman & Hall/CRC (2009)
24. Pnueli, A.: The temporal logic of programs. In: IEEE Annual Symposium on Foundations of Computer Science, pp. 46–57 (1977)
25. Rodrigo, G., Jaramillo, A.: AutoBioCAD: Full Biodesign Automation of Genetic Circuits. ACS Synth. Biol. (November 2012)
26. Salah, R.B., Bozga, M., Maler, O.: On timing analysis of combinational circuits. In: Larsen, K.G., Niebert, P. (eds.) FORMATS 2003. LNCS, vol. 2791, pp. 204–219. Springer, Heidelberg (2004)
27. Smith, H.L.: Systems of odes which generate an order preserving flow. A survey of results. SIAM Review 30(1), 87–113 (1988)
28. Szallasi, Z., Stelling, J., Periwal, V.: System Modelling in Cellular Biology: from concepts to nuts and bolts. The Mit Press (2006)
29. Terzer, M., Jovanovic, M., Choutko, A., Nikolayeva, O., Korn, A., Brockhoff, D., Zurcher, F., Friedmann, M., Schutz, R., Zitzler, E., Stelling, J., Panke, S.: Design of a biological half adder. IET Synthetic Biology 1(1-2), 53–58 (2007)
30. Van Kampen, N.G.: Stochastic Processes in Physics and Chemistry. Elsevier (1992)
31. von Dassow, G., Meir, E., Munro, E.M., Odell, G.M.: The segment polarity network is a robust developmental module. Nature (2000)
32. Voy, B.H., Scharff, J.A., Perkins, A.D., Saxton, A.M., Borate, B., Chesler, E.J., Branstetter, L.K., Langston, M.A.: Extracting gene networks for low-dose radiation using graph theoretical algorithms. PLoS Comput. Biol. (2006)
33. Yaman, F., Bhatia, S., Adler, A., Densmore, D., Beal, J.: Automated selection of synthetic biology parts for genetic regulatory networks. ACS Synth. Biol. 1(8), 332–344 (2012)

A Lattice-Theoretic Framework
for Metabolic Pathway Analysis

Yaron A.B. Goldstein and Alexander Bockmayr

DFG Research Center MATHEON, Freie Universität Berlin,
Arnimallee 6, 14195 Berlin, Germany
{yaron.goldstein,alexander.bockmayr}@fu-berlin.de

Abstract. Constraint-based analysis of metabolic networks has become
a widely used approach in computational systems biology. In the simplest
form, a metabolic network is represented by a stoichiometric matrix and
thermodynamic information on the irreversibility of certain reactions.
Then one studies the set of all steady-state flux vectors satisfying these
stoichiometric and thermodynamic constraints.

We introduce a new lattice-theoretic framework for the computational
analysis of metabolic networks, which focuses on the support of the flux
vectors, i.e., we consider only the qualitative information whether or not
a certain reaction is active, but not its specific flux rate. Our lattice-
theoretic view includes classical metabolic pathway analysis as a special
case, but turns out to be much more flexible and general, with a wide
range of possible applications.

We show how important concepts from metabolic pathway analysis,
such as blocked reactions, flux coupling, or elementary modes, can be
generalized to arbitrary lattice-based models. We develop corresponding
general algorithms and present a number of computational results.

Keywords: metabolic networks, constraint-based analysis, lattices.

1 Introduction

Constraint-based modeling has become a very successful approach for the analysis of genome-scale reconstructions of metabolic networks [1–4]. Given a set of
metabolites M and a set of reactions R, the network is represented by its stoi-
chiometric matrix $S \in \mathbb{R}^{M \times R}$, and a subset of irreversible reactions $\text{Irrev} \subseteq R$.
The steady-state flux cone $C = \{v \in \mathbb{R}^R \mid Sv = 0, v_{\text{Irrev}} \geq 0\}$ contains all
steady-state flux vectors satisfying the stoichiometric and thermodynamic constraints. Based on this cone, many analysis methods have been introduced over
the years, among them *Flux Balance Analysis* (FBA) [5,6], *Elementary Mode
Analysis* (EMA) [7–9], and *Flux Coupling Analysis* (FCA) [10,11].

While these methods are now well-established, various ideas have been explored on how to modify or extend the underlying modelling framework. A
lot of research concerns the question of how to include regulatory information

A. Gupta and T.A. Henzinger (Eds.): CMSB 2013, LNBI 8130, pp. 178–191, 2013.
© Springer-Verlag Berlin Heidelberg 2013

into the metabolic model (e.g. [12]). This has lead to diverse FBA strategies like rFBA [13] or SR-FBA [14]. Elementary mode computation has been extended to include transcriptional regulatory networks in [15]. Further, there has been a discussion on whether stronger thermodynamic constraints should be applied [16,17]. Others combine the idea of FBA to analyse optimal-growth steady-states with the insight that this condition alone does not constrain the system to a single possible state, but to a mathematical space of different (biologically) optimal states [18]. Still other approaches give up the steady-state assumption and use completely different modelling approaches, e.g. hyperpaths that are constructed by ordering the reactions of a network based on their (graph-theoretical) distance to nutrients [19]. So far each modification of the basic modelling approach required a specific reformulation and adaptation of the algorithms and analysis tools.

In this paper, we introduce the algebraic framework of lattices as a unifying approach to metabolic pathway analysis. We will present the necessary concepts that will allow us to adopt a broad range of modelling ideas within a unique generic framework. We have already tested ways to include optimal-growth or thermodynamic constraints as an option into our analysis tools. As a next step, we intend to create a formalism for regulatory constraints, which can be added to lattice-based models. Once implemented and tested, we will be able to perform EMA and especially FCA with regulatory or thermodynamic constraints.

Finite lattices [20, 21] are some of the simplest algebraic structures, but they have proven to be useful in many applications, such as abstract interpretation [22], knowledge representation [23], or distributed computing [24]. As we will see, they can be employed naturally to describe qualitative, pathway-based metabolic models, including the steady-state flux cone and related constraint-based methods. Regarding qualitative modelling, our work is related to [25], who use the concept of abstract interpretation to give knockout predictions in reaction networks.

Here we will introduce lattice-based EMA and a very fast FCA method. Our implementation L4FC (Lattices for Flux Coupling) can be used for traditional, flux-cone-based FCA. But it also allows applying other lattice-based modelling approaches, by simply changing one particular method that looks for pathways through a given reaction in the model.

Lattice-based models are independent from the steady-state assumption. In our models, we can use the flux cone, but we do not have to. The only algebraic requirement a lattice-based model has to fulfill is one that is easily proven for most approaches: any two pathways or states a and b can be combined to a new one that uses together all the reactions of a and b. This already defines a semi-lattice, which in our setting will automatically be a lattice.

Our approach allows for more flexibility in choosing the model constraints and provides general analysis tools that we can immediately use without spending much time on adapting them to our needs. As we will see, lattice-based modelling is fully compatible with the traditional steady-state flux cone and many of its extensions. But, it is also open for completely new ideas.

2   Lattice Theory in Metabolic Pathway Analysis

Many important questions in metabolic pathway analysis involve only qualitative information: Which reactions participate in a pathway? Which are the minimal sets of reactions needed to realize certain biological functions? Which reactions are coupled to each other? To answer these and other questions, we do not need the quantitative information of reaction rates. Instead we can consider a pathway to be simply a subset a of the reaction set R, $a \subseteq$ R, satisfying certain properties. This idea has appeared before in the literature, e.g. as *activity sets* [26] or *flux patterns* [27]. As a unifying framework for various modelling approaches in metabolic pathway analysis, we propose in this paper the algebraic concept of (semi-)lattices.

A *semi-lattice* [21] is an algebraic structure $(L, \circ)$ consisting of a set L and a binary operation $\circ$ which satisfy the following axioms:

- L is $\circ$-closed, i.e., if $a, b \in L$ then $a \circ b \in L$.
- $\circ$ is associative and commutative, i.e., $a \circ (b \circ c) = (a \circ b) \circ c$ and $a \circ b = b \circ a$.
- $\circ$ is idempotent, i.e., $a \circ a = a$.

A *lattice* can be defined as an algebraic structure $(L, \vee, \wedge)$ such that $(L, \vee)$ and $(L, \wedge)$ are semi-lattices and in addition for any $a, b \in L$, we have $a \wedge (a \vee b) = a$, and $a \vee (a \wedge b) = a$. An example is the lattice $(2^X, \cup, \cap)$ of all subsets of a set X, together with the usual set operations of union and intersection.

In the context of metabolic pathway analysis, we will look at semi-lattices $(L, \cup)$, where $L \subseteq 2^R$ and R is the finite set of reactions in the metabolic network. As we will see, many metabolic models are indeed union-closed, which simply means that the union of two pathways is a pathway again. As noted in [28], such a finite semi-lattice is already a lattice if there exists a neutral element $0 \in L$, with $0 \cup a = a$, for all $a \in L$. This holds if $\emptyset \in L$. Thus for any $L \subseteq 2^R$, we can obtain a *lattice* $(L, \cup, \wedge)$ if the following two axioms are satisfied:

- L is $\cup$-closed, i.e., if $a, b \in L$ then $a \cup b \in L$.
- There is an element $0 \in L$ such that $0 \cup a = a$, for all $a \in L$.

With these two axioms, we can define a second operation $\wedge$ on L, so that $(L, \cup, \wedge)$ becomes a lattice:

$$a \wedge b := \bigcup_{c \subseteq a, c \subseteq b} c . \tag{1}$$

The operation $\wedge$ is well-defined because $0 \subseteq a$, for all $a \in L$.

Similarly to this construction, we can prove that every finite lattice L has a unique maximum 1_L:

$$1_L = \bigcup_{a \in L} a . \tag{2}$$

Since $a \subseteq 1_L$, for all $a \in L$, we call 1_L the *maximum* of L. In Sect. 4, we will use the maximum to reformulate the concept of blocked reactions and flux coupling in metabolic network analysis.

Additionally, there are ways to describe finite lattices based on special sets of elements, the so-called *minimal* and *irreducible elements*, discussed e.g. in [21]. As we will see, these correspond exactly to the concept of elementary modes in the steady-state flux cone.

Lattices are sometimes also introduced as specially ordered sets. A *partial ordering* on pathways can naturally be defined by $a \leq b \Leftrightarrow a \subseteq b$. This reflects the idea that a pathway that is contained in another should be considered smaller in some sense. Because of their order-theoretical roots, many concepts in lattice theory should be understood in this context, e.g. the minimal elements, or the maximum.

The order-theoretical point of view also provides an interesting way of visualizing the relationship of different pathways via the so-called *Hasse diagram*. A Hasse diagram represents a finite, partially ordered set in a compact way. It can be seen as a directed graph with the elements of the set as nodes, and certain pairs of elements as edges. An element a_1 is connected to another element a_2 by an edge iff a_1 *is covered by* a_2, i.e., if $a_1 < a_2$ and there is no other element a with $a_1 < a < a_2$. All edges are implicitly oriented from bottom to top. In lattice-based metabolic models, we can draw a Hasse diagram where the elements are the reaction sets in our model. Two sets a_1, a_2 are connected if $a_1 \subset a_2$ and there is no other set from the model in between. The Hasse diagram of a lattice provides a lot of useful information. An element is irreducible iff it covers only one other element, i.e., there is only one edge going downwards. A reducible reaction set always covers at least two different reaction sets. Since our lattices are $\cup$-closed, it is easy to see that each reaction set that covers three or more other sets can always be written as the union of any two of those reaction sets that it covers. This allows us to identify how pathways can be decomposed into smaller reaction sets. An example is given in Fig. 2.

3   Steady-State Flux Spaces Can Be Modeled as Lattices

Constraint-based analysis of metabolic networks is based on the steady-state flux cone $C = \{v \in \mathbb{R}^{\mathsf{R}} \mid Sv = 0, v_{\mathrm{Irrev}} \geq 0\}$, where $S \in \mathbb{R}^{\mathsf{M} \times \mathsf{R}}$ is the stoichiometric matrix over the set of metabolites M and $\mathrm{Irrev} \subseteq \mathsf{R}$ is the set of irreversible reactions. Constraint-based methods include Flux Balance Analysis (FBA), Elementary Mode Analysis (EMA), or Flux Coupling Analysis (FCA), which allow for growth prediction, structural understanding, or target prediction in metabolic engineering [5–11].

We will show here how two of these approaches, namely EMA and FCA, may be reformulated in lattice-theoretic terms. Proving that we can work on a lattice L^C induced by the flux cone C, will allow us to use the general framework of lattice theory, which simplifies the development of optimized and unified algorithms. As a first step, we prove that any polyhedron $P \subseteq \mathbb{R}^{\mathsf{R}}$ induces a lattice. For this we look at the support of the vectors.

Proposition 1. *Given $P = \{x \in \mathbb{R}^n \mid Ax \leq b\}$, with $A \in \mathbb{R}^{m \times n}, b \in \mathbb{R}^m$, let*

$$L^P := \{\operatorname{supp} x \mid x \in P\}$$

with supp $x = \{r \in \mathbb{R} \mid x_r \neq 0\}$. *Then* $(L^P, \cup)$ *is a finite lattice.*

Proof. Let $a_1, a_2 \in L^P$ with $a_i = \mathrm{supp}\left(x^{(i)}\right)$. Define $x^{(\lambda)} = \lambda x^{(1)} + (1 - \lambda) x^{(2)}$ for $\lambda \in [0,1]$. P is a polyhedron, thus $x^{(\lambda)} \in P$ and supp $\left(x^{(\lambda)}\right) \subseteq a_1 \cup a_2$. Now we only have to show that there is $\lambda^* \in [0,1]$ with supp $\left(x^{(\lambda^*)}\right) = a_1 \cup a_2$. So let us look at the cases where this equality does not hold. We have $x_i^{(\lambda)} = 0$ if and only if $\lambda x_i^{(1)} + (1 - \lambda) x_i^{(2)} = 0$. So for each $i \in a_1 \cup a_2$ there is at most one λ such that $i \notin \mathrm{supp}\left(x^{(\lambda)}\right)$. Because there are less than $|\mathbb{R}| + 1$ values for λ with supp $\left(x^{(\lambda)}\right) \subsetneq a_1 \cup a_2$, we know that the desired $\lambda^* \in [0,1]$ must exist.   □

So we know that the flux cone C induces a lattice:

$$L^C := \{\mathrm{supp}\, v \mid Sv = 0, v_{\mathrm{Irrev}} \geq 0\} . \tag{3}$$

But we can also work on bounded flux vectors, where we assume minimal and maximal reaction rates $l, u \in \mathbb{R}^{\mathsf{R}}$:

$$L^C_{l \leq v \leq u} := \{\mathrm{supp}\, v \mid Sv = 0, l \leq v \leq u\} . \tag{4}$$

A special case of a bounded flux space is the space of all optimal-growth flux vectors, used in FBA and studied e.g. in [18]:

$$L^C_{\mathrm{opt}} := \{\mathrm{supp}\, v \mid Sv = 0, l \leq v \leq u, v_{\mathrm{Biomass}} = \max\} . \tag{5}$$

Fig. 1 shows an example network for this case. As we will see in Sect. 4, lattice theory allows us to define concepts equivalent to EFMs and FCA on these bounded flux spaces, too.

Finally, given a lattice $L \subseteq 2^{\mathsf{R}}$ and a subset $Q \subseteq \mathsf{R}$, we define

$$L_{\perp Q} := \{a \in L \mid a \cap Q = \emptyset\}, \tag{6}$$
$$L_Q := \{a \cap Q \mid a \in L\} . \tag{7}$$

Clearly, $(L_{\perp Q}, \cup)$ resp. $(L_Q, \cup)$ satisfy the two lattice axioms from Sect. 2. Therefore, we get two new lattices, which we call L *without* Q resp. L *projected on* Q.

4   Methods

4.1   Elementary Modes in Lattices

An *elementary mode* [7] is a steady-state flux vector $v \in C$ that is *irreducible* in the sense that it cannot be written in the form $v = v^1 + v^2$, with $v^1, v^2 \in C, \mathrm{supp}\, v^1, \mathrm{supp}\, v^2 \subsetneq \mathrm{supp}\, v$. As proven in [8], a flux vector $v \in C \setminus \{0\}$ is irreducible if and only if supp v is *minimal* (w.r.t. $\subseteq$). In the context of this paper, it is interesting to note that an elementary mode is uniquely determined by its support, i.e., given two elementary modes $v, v' \in C$ with supp $v = \mathrm{supp}\, v'$, there exists $\lambda \neq 0$ such that $v = \lambda v'$ [8].

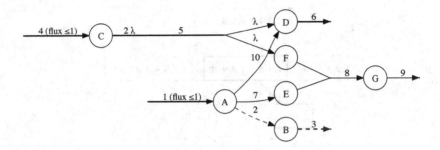

Fig. 1. Example network with metabolites $A, \dots, G$ and reactions $1, \dots, 10$. For example, reaction 5 corresponds to the chemical reaction $2C \to D + F$. Without constraints on the input reactions 1 and 4, none of the reactions is blocked and flux through reaction 6 is unbounded. However, if we include bounds on the input fluxes $v_1, v_4 \le 1$, then we obtain a maximal flux of $v_6 = 1$. The corresponding optimal solution space is given by $v_1 = v_6 = 1, v_2 = v_3 = 0, v_4 = 2\lambda, v_5 = v_7 = v_8 = v_9 = \lambda, v_{10} = 1 - \lambda$ with $\lambda \in [0, 0.5]$. In particular, reactions 2 and 3 become blocked.

In general lattices, minimal and irreducible elements have to be distinguished. [21] defines two sets of lattice elements, which we write as $\mathcal{M}(L)$ and $\mathcal{I}(L)$:

$$\mathcal{M}(L) := \{e \in L \mid \forall a \in L : a \subsetneq e \Rightarrow a = 0\} \ ,$$

$$\mathcal{I}(L) := \{b \in L \mid \forall A \subseteq L : b = \bigcup_{a \in A} a \Rightarrow b \in A\} \ .$$

We call $\mathcal{M}(L)$ the set of (non-trivial) *minimal elements* of L and $\mathcal{I}(L)$ the set of *irreducible elements* of L. The irreducible elements generate the lattice, i.e., for all $a \in L$ there exist $b_1, \dots, b_t \in \mathcal{I}(L)$ such that $a = \bigcup_{i=1}^{t} b_i$. Clearly, all minimal elements are irreducible, i.e., $\mathcal{M}(L) \subseteq \mathcal{I}(L)$. Lattices where both sets are the same are called *atomic*. While the lattice L^C is atomic, this does not hold for the lattice L^C_{opt} of all optimal-growth pathways, cf. Fig. 2. Therefore, for general lattices, the two concepts are different.

In [27] the notion of *elementary flux patterns* was introduced to describe the generating pathways through subsystems $Q \subseteq \mathsf{R}$ of a metabolic network. These may be interpreted as the set of irreducible, but not as the set of minimal elements, in a suitably defined lattice $L_Q := \{a \cap Q \mid a \in L\}$ (cf. (6)).

4.2 Lattice Maxima Give a New View on FCA

Flux coupling analysis (FCA) [10, 11, 29] studies blocked and coupled reactions in the steady-state flux cone C. It has been used for exploring a wide range of biological questions such as network evolution, gene essentiality, or gene regulation [30–35]. Here we offer an extended lattice-theoretic view of FCA, which

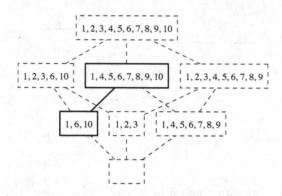

Fig. 2. Hasse diagram for the lattice L^C corresponding to the network in Fig. 1. Each possible support of a flux vector is represented by a box, the empty reaction set (zero flux) as an empty box. Reaction sets that do not represent optimal-growth flux vectors are contained in dashed boxes. In the space of optimal-growth flux vectors, there is only one minimal element: $\mathcal{M}\left(L_{\mathrm{opt}}^C\right) = \{\{1,6,10\}\}$. To describe the whole lattice L_{opt}^C, we need another (non-minimal) irreducible element: $\mathcal{I}\left(L_{\mathrm{opt}}^C\right) = \{\{1,6,10\},\{1,4,5,6,7,8,9,10\}\}$.

allows us to apply this tool not only on the classical flux lattice L^C (cf. (3)), but also on many other structures, such as the lattices defined in (4)-(7).

A reaction $r \in \mathsf{R}$ is *blocked*, if $v_r = 0$, for all $v \in C$. Two unblocked reactions r, s are *directionally coupled* ($r \overset{=0}{\rightarrow} s$) if $v_r = 0$ implies $v_s = 0$, for all $v \in C$, and *partially coupled* ($r \overset{=0}{\leftrightarrow} s$) if both $r \overset{=0}{\rightarrow} s$ and $s \overset{=0}{\rightarrow} r$ [10,11]. If neither $r \overset{=0}{\rightarrow} s$ nor $s \overset{=0}{\rightarrow} r$, then r, s are *uncoupled*. There is also the special case of *fully coupled* reactions, which correspond to *enzyme subsets* [36]. In the case of the flux cone, we can find those pairs using the kernel matrix [29].

Blocked and coupled reactions can be naturally defined in the more general lattice-theoretic framework. A reaction $r \in \mathsf{R}$ is *blocked* in a lattice $L \subseteq 2^{\mathsf{R}}$ if and only if $r \notin a$, for all $a \in L$. For unblocked reactions $r, s \in \mathsf{R}$, we define the *coupling* relations in L:

$$r \to s :\Leftrightarrow \forall a \in L \; : \; (r \notin a \Rightarrow s \notin a) \,,$$
$$r \leftrightarrow s :\Leftrightarrow \forall a \in L \; : \; (r \in a \Leftrightarrow s \in a) \,.$$

Now we come back to the unique maximum 1_L in a lattice L. From (2), we know that a reaction $r \in \mathsf{R}$ is blocked in L if and only if $r \notin 1_L$. Next we look at the lattice $L_{\perp\{r\}} := \{a \in L \mid r \notin a\}$, cf. (6). Using again (2), we see that two unblocked reactions $r, s \in 1_L$ are directionally coupled if and only if s is blocked in $1_{\perp\{r\}} := 1_{L_{\perp\{r\}}}$. Therefore, we get:

Proposition 2. *Given a lattice $L \subseteq 2^{\mathsf{R}}$ and a reaction $r \in \mathsf{R}$, we have:*

$$r \text{ is blocked in } L \Leftrightarrow r \notin 1_L \,. \tag{8}$$

For two unblocked reactions $r, s \in R$, we have:

$$r \to s \Leftrightarrow s \notin 1_{\perp\{r\}} \ . \tag{9}$$

In Sect. 5, we will give a fast algorithm for determining 1_L and $1_{\perp\{r\}}$, which will allow us to perform FCA in a simplified way.

5   Algorithms and Implementation

5.1   Finding Maxima in General Lattices

We first present an algorithm that can be used to perform FCA in any lattice-based model. It is designed in a way that it is easily adaptable to all kinds of models and still very fast. We achieve this by re-using intermediate results $a \in L_{\perp\{r\}}$, which we call collect in a set of *witnesses* $\mathcal{W}$. Using those witnesses, we search a maximum via nested intervals.

At the beginning, we do not know anything, so we assume $lb = \emptyset \subseteq 1_{\perp\{r\}} \subseteq ub = R$ with lower and upper bounds lb and ub. Each element $a \in L_{\perp\{r\}}$ that we obtain improves the lower bound. Every time we find that there is no $a \in L_{\perp\{r\}}$ with $s \in a$, we can decrease ub by removing s. Finally, we get $lb = ub$, which is then our maximum $1_{\perp\{r\}}$.

Algorithm 1 *FCA*
> $\mathcal{W} = \emptyset$
> **for** $r \in R$ **do**
> > $1_{\perp\{r\}} = R \setminus \{r\}$
>
> **for** $r \in R$ **do**
> > $\mathcal{W}_{\perp r} = \{a \in \mathcal{W} \mid r \notin a\}$
> > $lb = \bigcup_{a \in \mathcal{W}_{\perp r}} a, \ ub = R$
> > **for** $s \in R$ **do**
> > > **if** $s \in ub \setminus lb$ **then**
> > > > $a = Test(r, s)$
> > > > **if** $s \in a$ **then**
> > > > > $lb = a \cup lb, \ \mathcal{W} = \{a\} \cup \mathcal{W}$
> > > >
> > > > **else**
> > > > > $ub = ub \cap 1_{\perp\{s\}}$
> >
> > $1_{\perp\{r\}} = ub$

Alg. 1 uses the fact that lattices are $\cup$-closed. Therefore, we can combine each pair of already known pathways to create a new, larger feasible solution. This gives us a lower bound for the maxima $1_{\perp\{r\}}$. By keeping already calculated pathways as witnesses in $\mathcal{W}$, we get a major improvement in running time.

The algorithm does not use any specific properties of the flux cone. It is defined for any lattice-based model. To use it, we include the method Test(r,s) that returns a lattice element $a \in L$ with $r \notin a \ni s$, if such an element exists, and $\emptyset$ otherwise. This method is the only part of the code depending on model-specific information or constraints.

To implement this method for traditional FCA, we can solve in Test the following linear program (LP) (with a trivial objective function):

$$\min \{0 \cdot v \mid Sv = 0, v_{\text{Irrev}} \geq 0, v_r = 0, v_s = \sigma\} \ . \tag{10}$$

For reversible reactions $s \in R \setminus \text{Irrev}$, this linear program has to be solved twice, i.e., $\sigma \in \Omega_s = \{1, -1\}$, for irreversible reactions $s \in \text{Irrev}$, we use $\Omega_s = \{1\}$. If we find a feasible solution $v \in L_{\perp\{r\}}$, the method Test returns $a = \{r \in R \mid v_r \neq 0\}$, otherwise it returns $\emptyset$.

Lemma 1. *The LP* (10) *is infeasible for all* $\sigma \in \Omega_s$ *if and only if* $r \overset{=0}{\to} s$.

Proof. $\Rightarrow$: If r is not directionally coupled to s, there exists $a \in L^C$ s.t. $r \notin a \ni s$. Because of the definition of L^C there exists $v \in C$ with $a = \{i \in R \mid v_i \neq 0\}$. Thus, $v_r = 0 \neq v_s$. Because C is a cone, v is scalable by positive scalars $\lambda > 0$. Thus, there exists a feasible solution of LP (10).

$\Leftarrow$: If LP (10) is feasible, it follows that we have found a flux vector v with support $a = \{i \in R \mid v_i \neq 0\}$. We further know that $s \in a$, but $r \notin a$, thus r is not directionally coupled to s. $\qquad\Box$

Theorem 1. *Let* $L \subseteq 2^R$ *be a lattice and* $\emptyset \subseteq W \subseteq L$ *a list of known lattice elements (witnesses). Then Alg. 1 computes the maxima* $1_{\perp\{r\}}$ *needed for FCA (cf. Prop. 2).*

Proof. Given a reaction $r \in R$, we show that Alg. 1 computes $1_{\perp\{r\}}$. Since $W \subseteq L$ is a set of lattice elements, we have

$$\texttt{lb} = \bigcup_{a \in W_{\perp r}} a \ \subseteq \ \bigcup_{a \in L_{\perp\{r\}}} a = 1_{\perp\{r\}} \ . \tag{11}$$

Therefore, lb is a lower bound for $1_{\perp\{r\}}$ before we enter the inner loop. Since $L_{\perp\{r\}} \subseteq L$, we know $1_{\perp\{r\}} \subseteq 1_L$. Thus, ub is an upper bound before we enter the inner loop. Let $s \in \texttt{ub} \setminus \texttt{lb}$ be minimal. Let a be the result of Test(r, s) in the inner loop. By the definition of Test(r, s), we know that $a \in L_{\perp\{r\}}$ and, if $a \neq \emptyset$, then $s \in a$. Assume $s \in a$. Then the new $\texttt{lb} = a \cup \texttt{lb}$ is an element of $L_{\perp\{r\}}$, with $s \in \texttt{lb}$. Thus, for the next iteration, it holds that $s \notin \texttt{ub} \setminus \texttt{lb}$. Now assume $s \notin a$. This means that $s \notin 1_{\perp\{r\}}$. It follows $1_{\perp\{r\}} \subseteq 1_{\perp\{s\}}$. Since $\texttt{ub} \supseteq 1_{\perp\{r\}}$, it follows $\texttt{ub} \cap 1_{\perp\{s\}} \supseteq 1_{\perp\{r\}}$ is an upper bound. Because of the first loop in Alg. 1, we know $s \notin 1_{\perp\{s\}}$. Thus, in the next iteration, we have $s \notin \texttt{ub} \setminus \texttt{lb}$. $\qquad\Box$

Remark 1. We can accelerate Alg. 1 by replacing loops over R with loops over the set of all unblocked reactions 1_L.

Remark 2. Obviously, we can also modify the algorithm and the LP (10) to calculate this maximum 1_L of the lattice. For that, we have to replace Test(r, s) with a method Test(s) that does not use the constraint $v_r = 0$ in (10).

5.2 FCA in n Steps

Alg. 1 provides a method that can be used for any lattice-based model for which we can implement the method $\text{Test}(r, s)$. The constraints on this method are as simple as they could be: *find a pathway that goes through s but not through r, if possible.* Any $a \in L_{\perp\{r\}}$ with $s \in a$ is suitable. This simplicity is one of the many reasons why this algorithm is so easily adaptable to other lattice-based models. But, there may be cases where we can go even simpler. If there is a direct way to find the lattice maxima $1_{\perp\{r\}}$, we may compare this with Alg. 1. We will do this for classical FCA defined on the flux cone C. According to Prop. 1, the set L^C of all supports of flux vectors is indeed a lattice. That means there is a feasible flux vector $v^* \in C$ with $1_{L^C} = \{r \in \mathsf{R} \mid v_r^* \neq 0\}$. Obviously, the support of this flux vector has maximal cardinality.

Figueiredo et al. [9] introduce a mixed-integer linear program (MILP) that enumerates the (cardinality) shortest elementary modes. To achieve this, they add binary variables $a_i = 1 \Leftrightarrow v_i \neq 0$ to the LP (10). A slight variation of their MILP already provides the solution to find the lattice maximum 1_{L^C} in one single step. Since [9] is interested in finding elements of small cardinality, their objective function is $\min \sum_{i \in \mathsf{R}} a_i$. Here, we want to find an element of maximal cardinality. So we change the function to $\max \sum_{i \in \mathsf{R}} a_i$. Doing that we find the unique $a \in L$ with $a = 1_{L^C}$. For finding the maxima $1_{\perp\{r\}}$, we just have to (re-)add the single constraint $v_s = 0$ or alternatively $a_s = 0$.

5.3 Implementation

We have implemented the algorithm for general lattices in the language C#. Our program L4FC (Lattices for Flux Coupling) accepts files in METATOOL format [36] or separate files for stoichiometric information and irreversibility constraints. The implementation makes full use of the flexibility of lattices: The main program first computes the set of (un-)blocked reactions, before it calculates the FCA-relevant maxima $1_{\perp\{r\}}$. The calculation of those $|\mathsf{R}| + 1$ maxima is encapsulated into a separate calculator class. Our current version uses the idea of nested intervals introduced in Alg. 1. The model-specific method $\text{Test}(r, s)$ is implemented in form of a Gurobi model [37] that solves the LPs (10). This design allows us to include other modelling approaches in an easy and elegant way by implementing new calculator classes. The source code is available at GitHub https://github.com/goldsteiny/L4FC and is licensed under CC BY-NC-SA 3.0. The projects history and future updates will be linked to www.hoverboard.io/L4FC.

6 Discussion

We have run our program on seven widely studied genome-scale metabolic networks from the BiGG database [38] as well as the more recent reconstruction *E. coli* iJO1366 [39]. This selection is comparable to other FCA benchmarks,

Table 1. Runtime behavior of L4FC applied on 7 genome-scale metabolic networks. In addition, we report on the number of LPs solved and the number of pathways found. The computation was done into two steps: First we calculate the set of blocked reactions, then we search for the pairs of unblocked reactions that are coupled.

Model	Step	Solution size	# LPs	$|\mathcal{W}|$	Time (sec)
	Total		11100	4322	242.0
E. coli iJO1366	find unblocked	1718 reactions	1579	469	9.8
2583 reactions	find couples	58613 couples	9521	3853	232.2
	Total		12606	4525	219.5
E. coli iAF1260	find unblocked	1543 reactions	1518	424	8.3
2382 reactions	find couples	39260 couples	11088	4101	211.2
	Total		2485	591	6.4
H. pylori iIT341	find unblocked	436 reactions	190	44	0.3
554 reactions	find couples	62006 couples	2295	547	6.1
	Total		2203	886	8.3
M. barkeri iAF692	find unblocked	483 reactions	340	75	0.6
690 reactions	find couples	76746 couples	1863	811	7.7
	Total		4141	1699	25.3
M. tuberculosis iNJ661	find unblocked	744 reactions	497	158	1.3
1025 reactions	find couples	60750 couples	3644	1541	23.9
	Total		4329	741	9.6
S. aureus iSB619	find unblocked	465 reactions	394	65	0.5
743 reactions	find couples	30160 couples	3935	676	9.0
	Total		5189	1483	31.1
S. cerevisiae iND750	find unblocked	631 reactions	963	129	3.0
1266 reactions	find couples	15511 couples	4226	1354	28.0

e.g. [11, 29]. Table 1 summarizes the results. No calculation took longer than 4 minutes, five of them less than 40 seconds. Given these results we can conclude that the new generic algorithm L4FC has a runtime in the same order of magnitude as F2C2, the fastest dedicated tool currently available [29].

Taking a closer look at the results, we see that the calculation of the blocked reactions takes around 5 − 20% of the total running time. Similar observations can be made about the number of LPs to be solved and the number of feasible reaction sets found during this first step of the program. This is remarkable, because this first phase calculates only 1 maximum, 1_L, whereas the second phase calculates $|1_L| \sim |R|$ maxima. This large disproportion is a direct consequence of our use of nested intervals, where we 1) re-use all elements found in phase 1 to get better lower bounds and 2) directly apply earlier found upper bounds $1_{\perp\{s\}}$ to improve our approximation of $1_{\perp\{r\}}$ for $s < r$. Doing the iteration $\texttt{ub} = \texttt{ub} \cap 1_{\perp\{s\}}$ is an obvious improvement over $\texttt{ub} = \texttt{ub} \setminus \{s\}$, and is quite easy to understand with lattices in mind. Using this, we achieve similar run time improvements as discussed in [29], where transitivity tables are analysed and proven.

We ran our algorithm on a machine with Intel Core i7-2600 (3.4 GHz, 4 cores, hyperthreading) and 4GB RAM. We used Gurobi 5.1 with Windows 7 Professional, Service Pack 1 (64-bit), .NET Framework 4.0.30319. As tolerance values for zero flux, we used $|v_i| \leq 10^{-8} \Rightarrow i \notin \operatorname{supp}(v)$.

7 Summary

We have shown that the concept of EFMs and FCA can be extended to general lattice-based models. Using this algebraic framework, we can now apply these methods to new classes of models. For example, we can run FCA on the space of all optimal-growth flux vectors.

We have introduced a new algorithm for computing the set of unblocked reactions 1_L and performing FCA, using only lattice properties. This allows an easy adaptation to any lattice-based model. We have further implemented the algorithm for traditional FCA of the flux cone and shown on a benchmark set of genome-scale metabolic networks like *E. coli* iJO1366 that our generic tool L4FC is comparable in speed to dedicated FCA algorithms.

Acknowledgement. This work was funded by the Gerhard C. Starck Stiftung in terms of a PhD stipend. We thank Alexandra Grigore and Arne C. Müller for their comments on this paper and many helpful discussions.

References

1. Papin, J.A., Stelling, J., Price, N.D., Klamt, S., Schuster, S., Palsson, B.O.: Comparison of network-based pathway analysis methods. Trends in Biotechnology 22(8), 400–405 (2004)
2. Terzer, M., Maynard, N.D., Covert, M.W., Stelling, J.: Genome-scale metabolic networks. Wiley Interdiscip. Rev. Syst. Biol. Med. 1(3), 285–297 (2009)
3. Schellenberger, J., Que, R., Fleming, R.M.T., Thiele, I., Orth, J.D., Feist, A.M., Zielinski, D.C., Bordbar, A., Lewis, N.E., Rahmanian, S., Kang, J., Hyduke, D.R., Palsson, B.O.: Quantitative prediction of cellular metabolism with constraint-based models: The COBRA Toolbox v2.0. Nature Protocols 6(9), 1290–1307 (2011)
4. Lewis, N.E., Nagarajan, H., Palsson, B.: Constraining the metabolic genotype-phenotype relationship using a phylogeny of in silico methods. Nat. Rev. Microbiol. 10(4), 291–305 (2012)
5. Varma, A., Palsson, B.: Metabolic Flux Balancing: Basic Concepts, Scientific and Practical Use. Nature Biotechnology 12(10), 994–998 (1994)
6. Orth, J.D., Thiele, I., Palsson, B.O.: What is flux balance analysis? Nature Biotechnology 28(3), 245–248 (2010)
7. Schuster, S., Hilgetag, C.: On elementary flux modes in biochemical reaction systems at steady state. Journal of Biological Systems 2(2), 165–182 (1994)
8. Schuster, S., Hilgetag, C., Woods, J.H., Fell, D.A.: Reaction routes in biochemical reaction systems: Algebraic properties, validated calculation procedure and example from nucleotide metabolism. J. Math. Biol. 45, 153–181 (2002)

9. de Figueiredo, L.F., Podhorski, A., Rubio, A., Kaleta, C., Beasley, J.E., Schuster, S., Planes, F.J.: Computing the shortest elementary flux modes in genome-scale metabolic networks. Bioinformatics 25(23), 3158–3165 (2009)
10. Burgard, A.P., Nikolaev, E.V., Schilling, C.H., Maranas, C.D.: Flux Coupling Analysis of Genome-Scale Metabolic Network Reconstructions. Genome Research 14(2), 301–312 (2004)
11. David, L., Marashi, S.A., Larhlimi, A., Mieth, B., Bockmayr, A.: FFCA: a feasibility-based method for flux coupling analysis of metabolic networks. BMC Bioinformatics 12(1), 236 (2011)
12. Jensen, P., Lutz, K., Papin, J.: TIGER: toolbox for integrating genome-scale metabolic models, expression data, and transcriptional regulatory networks. BMC Syst. Biol. 5, 147 (2013)
13. Covert, M.W., Schilling, C.H., Palsson, B.: Regulation of Gene Expression in Flux Balance Models of Metabolism. J. Theoretical Biology 213(1), 73–88 (2001)
14. Shlomi, T., Eisenberg, Y., Sharan, R., Ruppin, E.: A genome-scale computational study of the interplay between transcriptional regulation and metabolism. Molecular Systems Biology 3, 101 (2007)
15. Jungreuthmayer, C., Ruckerbauer, D.E., Zanghellini, J.: regEfmtool: Speeding up elementary flux mode calculation using transcriptional regulatory rules in the form of three-state logic. BioSystems 113(1), 37–39 (2013)
16. Beard, D.A., Babson, E., Curtis, E., Qian, H.: Thermodynamic constraints for biochemical networks. Journal of Theoretical Biology 228, 327–333 (2004)
17. Müller, A.C., Bockmayr, A.: Fast thermodynamically constrained flux variability analysis. Bioinformatics 29(7), 903–909 (2013)
18. Kelk, S.M., Olivier, B.G., Stougie, L., Bruggeman, F.J.: Optimal flux spaces of genome-scale stoichiometric models are determined by a few subnetworks. Scientific Reports 2, 580 (2012)
19. Carbonell, P., Fichera, D., Pandit, S.B., Faulon, J.L.: Enumerating metabolic pathways for the production of heterologous target chemicals in chassis organisms. BMC Systems Biology 6(1), 10 (2012)
20. Birkhoff, G.: Lattices and their applications. Bulletin of the American Mathematical Society 44(12), 793–801 (1938)
21. Davey, B.A., Priestley, H.A.: Introduction to lattices and order. Cambridge University Press (1990)
22. Cousot, P., Cousot, R.: Abstract interpretation: a unified lattice model for static analysis of programs by construction or approximation of fixpoints. In: Fourth Annual ACM Symposium on Principles of Programming Languages, Los Angeles, pp. 238–252. ACM Press (1977)
23. Oles, F.J.: An application of lattice theory to knowledge representation. Theoretical Computer Science 249(1), 163–196 (2000)
24. Garg, V., Mittal, N., Sen, A.: Applications of lattice theory to distributed computing. ACM SIGACT Notes (2003)
25. John, M., Nebut, M., Niehren, J.: Knockout Prediction for Reaction Networks with Partial Kinetic Information. In: Giacobazzi, R., Berdine, J., Mastroeni, I. (eds.) VMCAI 2013. LNCS, vol. 7737, pp. 355–374. Springer, Heidelberg (2013)
26. Nuño, J.C., Sánchez-Valdenebro, I., Pérez-Iratxeta, C., Meléndez-Hevia, E., Montero, F.: Network organization of cell metabolism: monosaccharide interconversion. The Biochemical Journal 324, 103–111 (1997)
27. Kaleta, C., de Figueiredo, L.F., Schuster, S.: Can the whole be less than the sum of its parts? Pathway analysis in genome-scale metabolic networks using elementary flux patterns. Genome Research 19(10), 1872–1883 (2009)

28. Nation, J.B.: Revised Notes on Lattice Theory (2012),
    http://www.math.hawaii.edu/~jb/
29. Larhlimi, A., David, L., Selbig, J., Bockmayr, A.: F2C2: a fast tool for the computation of flux coupling in genome-scale metabolic networks. BMC Bioinformatics 13(75), 57 (2012)
30. Notebaart, R.A., Teusink, B., Siezen, R.J., Papp, B.: Co-regulation of metabolic genes is better explained by flux coupling than by network distance. PLoS Comput. Biol. 4, e26 (2008)
31. Notebaart, R.A., Kensche, P.R., Huynen, M.A., Dutilh, B.E.: Asymmetric relationships between proteins shape genome evolution. Genome Biol. 10, R19 (2009)
32. Pál, C., Papp, B., Lercher, M.J.: Adaptive evolution of bacterial metabolic networks by horizontal gene transfer. Nat. Genet. 37, 1372–1375 (2005)
33. Yizhak, K., Tuller, T., Papp, B., Ruppin, E.: Metabolic modeling of endosymbiont genome reduction on a temporal scale. Mol. Syst. Biol. 7, 479 (2011)
34. Montagud, A., Zelezniak, A., Navarro, E., de Córdoba, P.F., Urchueguía, J.F., Patil, K.R.: Flux coupling and transcriptional regulation within the metabolic network of the photosynthetic bacterium *Synechocystis* sp. PCC6803. Biotechnol. J. 6, 330–342 (2011)
35. Szappanos, B., Kovács, K., Szamecz, B., Honti, F., Costanzo, M., Baryshnikova, A., Gelius-Dietrich, G., Lercher, M., Jelasity, M., Myers, C., Andrews, B., Boone, C., Oliver, S., Pál, C., Papp, B.: An integrated approach to characterize genetic interaction networks in yeast metabolism. Nat. Genet. 43(7), 656–662 (2011)
36. Pfeiffer, T., Sanchez-Valdenebro, I., Nuno, J., Montero, F., Schuster, S.: METATOOL: for studying metabolic networks. Bioinformatics 15(3), 251–257 (1999)
37. Gurobi Optimization Inc: Gurobi 5.1 (2012)
38. Schellenberger, J., Park, J.O., Conrad, T.M., Palsson, B.O.: BiGG: A Biochemical Genetic and Genomic knowledgebase of large scale metabolic reconstructions. BMC Bioinformatics 11(213), 213 (2010)
39. Orth, J.D., Conrad, T.M., Na, J., Lerman, J.A., Nam, H., Feist, A.M., Palsson, B.: A comprehensive genome-scale reconstruction of Escherichia coli metabolism-2011. Molecular Systems Biology 7(535) (2011)

On the Hybrid Composition and Simulation of Heterogeneous Biochemical Models

Katherine Chiang[1,2], François Fages[1], Jie-Hong Jiang[2], and Sylvain Soliman[1]

[1] EPI Contraintes, Inria Paris-Rocquencourt, France
[2] Graduate Institute of Electronics Engineering, National Taiwan University, Taiwan

Abstract. Models of biochemical systems presented as a set of formal reaction rules with kinetic expressions can be interpreted with different semantics: as either deterministic Ordinary Differential Equations, stochastic continuous-time Markov Chains, Petri nets or Boolean transition systems. While the formal composition of reaction models can be syntactically defined as the (multiset) union of the reactions, the hybrid composition of models in different formalisms is a largely open issue. In this paper, we show that the combination of reaction rules with conditional events, as the ones already present in SBML, does provide the expressive power of hybrid automata and can be used in a non standard way to give meaning to the hybrid composition of heterogeneous models of biochemical processes. In particular, we show how hybrid differential-stochastic and hybrid differential-Boolean models can be compiled and simulated in this framework, through the specification of a high-level interface for composing heterogeneous models. This is illustrated by a hybrid stochastic-differential model of bacteriophage T7 infection, and by a reconstruction of the hybrid model of the mammalian cell cycle regulation of Singhania et al. as the composition of a Boolean model of cell cycle phase transitions and a differential model of cyclin activation.

1 Introduction

Systems biology aims at elucidating the high-level functions of the cell from their biochemical basis at the molecular level [24]. A lot of work has been done for collecting genomic and post-genomic data, making them available in databases [5,25], and organizing the knowledge on pathways and interaction networks into models of cell metabolism, signaling, cell cycle, apoptosis, etc. now published in model repositories (e.g. http://biomodels.net/). In particular, the Systems Biology Markup Language (SBML) [23] provides a common exchange format for biochemical *reaction models* and is nowadays supported by a majority of modeling tools.

According to the knowledge available on the system and to the nature of the queries that will be asked to the model, e.g. qualitative or quantitative predictions, these reaction rule-based models can be interpreted (and simulated) under different semantics as either:

- ordinary differential equations (differential semantics),
- continuous-time Markov chains (stochastic semantics),
- Petri nets (discrete semantics),

A. Gupta and T.A. Henzinger (Eds.): CMSB 2013, LNBI 8130, pp. 192–205, 2013.

– Boolean transition systems (Boolean semantics),
– and many variants.

Some modeling tools support several of these different interpretations which can also be related by approximation [16,17,18] or abstraction [11] relationships.

In the perspective of applying engineering methods to the analysis and control of biological systems, the issue of building complex models by composition of elementary models is a central issue. While reaction rule-based models can be formally composed simply by the multiset union of reaction rules, and interpreted by one common semantics, there is also a need to compose models with different semantics. What we call a *hybrid model* is a model obtained by composition of models with heterogeneous semantics (differential, stochastic, Boolean, etc.), and *hybrid simulation* is the topic of simulating such hybrid models.

Hybrid simulation is a classical topic in physics on the one hand, e.g. for numerically solving equations describing stochastic systems using ordinary differential equations whenever possible in place of stochastic equations in order to speed-up simulations [3,31], and on the other hand, in computer science for programming and verifying hybrid systems which have both discrete and continuous dynamics [9,4,21]. Hybrid modeling is also used in systems biology for reducing the complexity of many modeling task, e.g. [29,4,13,6,26,1,33], or for speeding up stochastic simulations [32,19,22].

In this paper, we show that the combination of reaction rules with conditional events, as the one already present in SBML, does provide the expressive power of hybrid automata and can be used in a non standard way to give meaning to the hybrid composition of heterogeneous reaction models. In particular, we show how hybrid differential-stochastic and hybrid differential-Boolean models can be compiled and simulated in this formal framework of reactions plus events, through the specification of a high-level interface for composing reaction models.

This interface for composing models has been implemented as a preprocessor for Biocham [7,10]. This preprocessor transforms stochastic reaction models in events that implement Gillespie's direct method for stochastic simulation and that can be combined with the simulation of differential reaction models. Similarly, it transforms Boolean state transition models in events with extra conditions that express the links to the continuous variables and parameters of the differential reaction model.

This approach is illustrated through the hybrid stochastic-differential composition and simulation of bacteriophage T7 infection [3], and a reconstruction of the hybrid model of the mammalian cell cycle regulation of Singhania et al. [33] as the composition of a Boolean model of cell cycle phase transitions and a differential model of cyclin activation.

2   The Expressive Power of Events with Kinetic Reactions

2.1   Reactions Rules with Kinetics

In the spirit of the Chemical Reaction Network Theory [12], we define our systems of study as sets of reaction rules r_i, however as in SBML [23] any function can be used as reaction rate. In the following this will be represented using Biocham syntax [10] as: v_i for $\sum_j l_{ij} \times S_j \Rightarrow \sum r_{ij} \times S_j$, where v_i is a continuous function[1] of parameters of the

---

[1] It would be possible to admit non-continuous functions as rates (e.g., conditional statements in v_i), and that is actually the case in many tools, however the same result can be obtained with the event mechanism described in the next section.

system and of species concentrations defining the rate of reaction i (mass action kinetics of parameter k are abbreviated as $MA(k)$), l_{ij} and r_{ij} are stoichiometric coefficients, and the S_j are the species of the model.

According to the data available on the system and to the nature of the queries that will be asked to the model, e.g. qualitative or quantitative predictions, these reaction models can be interpreted (and simulated) under different semantics: differential, stochastic, discrete or boolean. We recall here the basics of these semantics. An Ordinary Differential Equation (ODE) system can be defined from a reaction model as follows: $\frac{d[S_j]}{dt} = \sum_i (r_{ij} - l_{ij}) \times v_i$

The differential semantics corresponds to the limit of the Continuous-Time Markov Chain defined using the v_i as propensities, and realizing the solution of the Chemical Master Equation [16]. The differential semantics usually leads to numerical integration, whereas the stochastic semantics is either used for exact or approximate simulation, or for stochastic Model-checking (see for instance [27]).

The discrete semantics forgets about the rates v_i but keeps the stoichiometric information, for instance as weights in a Petri net representation [8,14].

Finally, the Boolean semantics forgets about precise stoichiometry and keeps only information about whether or not a species is active. It can be defined as an abstraction of the previous discrete semantics [11].

2.2   Semantics of Events

In this section, we present a generic notion of *events* compatible with the differential semantics of reaction models and then describe how it relates to existing concepts, most notably the events of SBML and Biocham.

An event is basically twofold, it is built by a *condition*, determining when it fires, and by an *action*, i.e., its influence on the current state (parameters, concentrations). If one wants to enforce the continuity of concentration variables, they can simply exclude them from the variables that can be modified by the *action* part of the events.

Following Biocham syntax, we will write an event as follows: $event(condition, [s_1, \ldots, s_n], [f_1, \ldots, f_n])$, where the s_i indicate the state variables that are modified by the event, the f_i are functions of the state that give the new value to s_i.

There are many possible semantics for events but the basic idea is that an event fires when its condition changes from *false* to *true*. This induces however several issues:

- what happens at the start of the simulation?
- how to find the precise time when a condition becomes true?
- what happens if some events are enabled simultaneously?

The first point is easy to settle, it is an arbitrary decision but does not have a big impact. The simplest choice is to avoid the firing of events at the initial point of the simulation and to reflect initial events by modifying accordingly the initial state.

The second point has been solved in practical tools for a long time: since numerical integration goes by steps, one detects changes in conditions only in the interval of a simulation step. One can simply go back in time until one finds—with a given precision—the first time point where a condition becomes true. Note however that if arbitrarily complex conditions appear in the events, a numerical integrator unaware of the events can hide inside a single step that a condition went from false to true and back to false again. Therefore, a cautious implementation is necessary, and often, fixed step size integration methods are recommended to use, instead of more efficient adaptive step size methods in presence of events.

The final point is again a question with multiple possible answers. Generally, the set of events that are enabled simultaneously at a given time will all be fired, whatever the actions of the events are, but what if several events modify the same variable? It is possible to assume a *synchronous* semantics, where the simultaneous events execute their actions in parallel, but then one must forbid events with conflicting actions, i.e., events that would modify in different ways the same variable at the same time point. The more common choice is an *asynchronous* semantics, that will fire all the events enabled at a given time one after the other, even if some actions invalidate the condition of other enabled events. Conflicts in actions are then solved by the ordering of events, which can be either random, i.e. non-deterministic, or given by the modeller, e.g. by the order of writing or by priorities.

The SBML choice is to keep a very flexible semantics, with asynchronous events, that may be ordered by *priorities*, and that can use either the values at the time they were enabled, or the *current* values at the time they are actually executed, after the execution of the simultaneous events with higher priority.

In Biocham, there are no priorities, the events that are enabled simultaneously are executed in the order of their writing using current values. An event with n assignments of f_i to s_i is therefore equivalent to the sequence of n events with the same condition for each assignment f_i to s_i. The semantics of events implemented in Biocham can thus be defined in SBML using the current value option and priorities corresponding to the order of writing.

2.3   Representation of Hybrid Automata by Reactions and Events

A *hybrid automaton* (HA) is a dynamical system containing both continuous and discrete components [20]. They are therefore commonly used to formalize real-life safety-critical systems and have led to various works on the verification of their different semantics and on their composition (e.g. with Hytech [21]).

Formally, a hybrid automaton is defined by a set of continuous variables, a *control graph* where edges are labelled by *jump conditions and events*, defining the discrete state changes with some labels, and vertices are labelled by *initial, invariant and flow conditions* defining the continuous change in each state. Figure 1 (left) shows the traditional thermostat example.

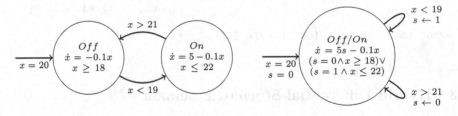

Fig. 1. The classical thermostat example encoded in a single state hybrid automaton

Since the *jumps* describe the possible transitions with a complete description of the resulting state, there are no issues similar to what was described in the previous section to handle conflicting updates.

Note that, it is enough to restrict oneself to hybrid automata with a single state (vertex) with a big parametric system of ODEs corresponding to all the ODEs of the initial states, multiplied by a parameter that is non null only when the corresponding state is *active*. Then the jumps and event labels can remain the same, except that they go from the single state to itself, and change the state variable according to the initial state change. The invariants have as additional condition that the corresponding state must be active to be enforced. One obtains Figure 1 (right) where the ODE system has been factorized for readability.

Now to represent an hybrid automaton in the framework of reaction and events described above, one can first note that the initial and flow conditions simply define an ODE system. Such a system can be represented with reactions, for instance as a synthesis for each variable with rate corresponding to the variable's derivative in the corresponding state.

The jumps can easily be represented as events, however since they do not represent events that should fire, but, unless it violates an invariant, events that may fire, they should be accompanied by another event allowing the state not to change. This event will have as condition the fact that the current invariant is true and that a condition to leave is true. This second part is not necessary but avoids useless firings of events that do not change anything. This event will also need to be able to fire repeatedly, it will thus have a supplementary condition *can_fire* that it will itself make false, another event will always make it true again when it is false.

Note that this encoding relies on a non-deterministic asynchronous semantics for events, as discussed in Section 2.2. Here is the thermostat example as reactions and events:

```
5*s for _ => x.                          0.1*[x] for x => _.

present(x, 20).                          parameter(enabled1, 0).
                                         parameter(enabled2, 0).
parameter(s, 0).                         parameter(can_fire, 1).

event(s = 1 and [x] > 21 and can_fire = 1, [s, can_fire], [0, 0]).
event(s = 1 and [x] > 21 and [x] =< 22 and can_fire = 1,
                                            [s, can_fire], [1, 0]).

event(s = 0 and [x] < 19 and can_fire = 1, [s, can_fire], [1, 0]).
event(s = 0 and [x] < 19 and [x] >= 18 and can_fire = 1,
                                            [s, can_fire], [0, 0]).

event(can_fire = 0, [can_fire], [1]).
```

3 Hybrid Differential-Stochastic Semantics

Chemical reactions, originated from random collisions of particles, are discrete and stochastic in nature. Although there is no way to predict the exact state of a chemical system at a specific time point, its *statistical* behavior can be effectively calculated from known probabilistic properties. A well-mixed, non-linear chemical system can be described by a set of master equations, which in turn can be completely solved by Gillespie's stochastic simulation algorithm (SSA) [15], to be detailed in Section 3.1. Essentially the computation cost of an SSA grows proportional to the number of reaction occurrences. Simulating a system of chemical reactions can be especially slow if one or more of the reactions have fast reaction rates (or high event occurrences) because the next reaction time will be very short due to the high probability of selecting (one of the) fast reactions for firing.

A chemical system may consist of reactions proceeding with significantly different rates. Despite the fact that all reactions are innately stochastic, those with large reactant counts and high reaction rates can be accurately approximated in terms of deterministic behavior expressed by ODEs. By incorporating both differential and stochastic semantics into one simulator, an optimal balance between simulation runtime and accuracy can be achieved. This potentially lifts the scalability of simulating large biological systems. In Section 3.2, we provide an event-based view on the SSA, that serves as basis to a hybrid differential-stochastic simulator built upon an ODE simulator with events.

3.1 Gillespie's Direct Method

A reaction model with kinetic expressions can be interpreted under the stochastic semantics as a continuous-time Markov chain (CTMC). A CTMC can be simulated with a stochastic simulation algorithm (SSA), for example, *Gillespie's direct method* [15]. Rather than solving all possible trajectories' probabilities as in the case of Master equations, the algorithm generates statistically correct trajectories.

Gillespie's direct method first calculates *when* the next reaction will occur, then decides *which* reaction should occur with the help of a random number generator. The probability that a certain reaction μ will take place in the next instant of time is given by its propensity: $\alpha_\mu = (\#\text{combinations of reactants}) \cdot k_\mu$ where k_μ is μ's rate coefficient. The algorithm repeats the following steps.

1. Calculate *how long from now* (Δt) the next reaction will occur.

$$\Delta t = \frac{-1}{\sum_j \alpha_j} \cdot \log(r_1),$$

   where r_1 is a random number within range $(0,1)$ and the α_j are propensities at the current state.

2. Choose which reaction will occur according to the probability distribution of reactions. This is done by generating a random number r_2 within range $(0,1)$, and letting the reaction μ_i be chosen for

$$\frac{\sum_{k=1}^{i-1} \alpha_k}{\sum_j \alpha_j} < r_2 \leqslant \frac{\sum_{k=1}^{i} \alpha_k}{\sum_j \alpha_j}.$$

3. Update the numbers of molecules to reflect the execution of reaction μ_i, and set current time to $t = t + \Delta t$.

3.2 Event Model of Stochastic Simulation

By considering every firing of a chemical reaction as one firing of an event, the *event* semantics of Section 2 enables a direct embedding of stochastic reactions into an intrinsically differential framework without additional implementation of a separate stochastic simulation algorithm. Under this framework, **time** is the only unifying variable to keep track of current state at each instant. This event-based approach permits the simple integration of ODE and stochastic simulation as will be elaborated in Section 3.3.

Notice that, in the SSA of Section 3.1, *when* the next reaction will occur is independent of *which* reaction will occur, and also that only one reaction is chosen each time. These facts make the complete set of stochastic reaction rules be simulated correctly with a single event. Essentially the simulation can be accomplished by compiling the *when* and *which* questions Gillespie's direct method asks into an event. Specifically the event is triggered by the calculated next reaction time (tau); the event obtains a new random variable (ran) and then conditionally updates the molecular counts depending on which reaction is chosen to occur next. To accommodate all stochastic rules in one event, each update entry is composed of conditional expressions over the propensities and the random number that decides which reaction occurs.

Example 1 ([15]). Given the stochastic reaction rules $A + 2B \xrightarrow{k_1} C$ and $C \xrightarrow{k_2} 2A$ we derive their propensities by $\mathtt{alpha1} = k_1 \times (\#A) \times \frac{(\#B) \times (\#B-1)}{2}$, $\mathtt{alpha2} = k_2 \times (\#C)$, where "#" denotes the particle count of a species. Then the next reaction time from the current time point can be decided by $\mathtt{e} = \frac{-1}{\mathtt{alpha\_sum}} \cdot \log(random_1)$ for $random_1$ within $(0, 1)$ and where $\mathtt{alpha\_sum} = \mathtt{alpha1} + \mathtt{alpha2}$. The first reaction is chosen for the next occurring reaction if $0 < (\mathtt{alpha\_sum} \times random_2) \leqslant \mathtt{alpha1}$, which leads to the consumption of one A and two B's and producing one C:

```
event(Time>tau, [tau, ran, #A, #B, #C],
             [Time + e, random,
               if alpha_sum*ran =< alpha1 then #A-1 else #A+2,
               if alpha_sum*ran =< alpha1 then #B-2 else #B,
               if alpha_sum*ran =< alpha1 then #C+1 else #C-1]).
```

Note that the update of the particle counts of the first reaction is reflected in the three then entries, and that of the second reaction is reflected in the three else entries.

This encoding relies on the left to right ordering of the different events associated to a single trigger (see Section 2.2). This ordering is imposed to three kinds of parameters, including the random number for choosing reaction, the lower bound for particle number, and a reaction's propensity function, such that possible errors are avoided. Because these three kinds of parameters all depend on the *current* number of molecules, they are listed in front of molecular species. So their values are not changed before the *completion* of reaction firing, that is, all species' counts have been updated according to the chosen reaction.

3.3 Preprocessor for Composing Differential and Stochastic Models

The purpose of our preprocessor for composing heterogeneous biochemical models is to provide a user-friendly interface to allow users of various backgrounds to conduct hybrid simulation without knowing algorithmic details. The only work required is to decide the semantic model for each of the reactions under simulation.

In classical work on hybrid simulation [3,26], chemical reactions are divided according to their propensities and reactants' concentrations into two groups: one consisting of reactions to be simulated stochastically using SSAs, and the other consisting of reactions to be simulated deterministically using ODEs. The former is referred to as the *stochastic reactions* and the latter *differential reactions*. While differential reactions simply advance with the pass of time, stochastic reactions fire discretely in time with frequency in accordance with their propensities. When the reactant concentrations and the propensity of a reaction are sufficiently large, ODE simulation can be faithfully applied. It avoids frequent simulation updates within a small time interval, thus accelerating simulation speed.

Hybrid species are referred to as those involved in both stochastic and differential rules. This kind of species requires special attention because they are influenced by two different mechanisms: *ODEs* that govern differential behavior by continuously changing related *concentrations*, and *events* that regulate stochastic behavior by modifying *molecule counts* discretely whenever triggered. So a hybrid species is under two kinds of modification: one targets at the evolution of macroscopic concentrations and the other targets at the changes in microscopic particle counts.

In our implementation, a fresh new variable is introduced for each hybrid species to represent its quantity (the summation of the numbers of particles from both differential and stochastic models). In all kinetic expressions, the hybrid species are expressed by the corresponding new variables. It is then a simple matter to put together the ODEs for the continuous part and the events corresponding to the encoding of the stochastic part as described in the previous section.

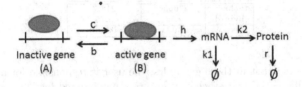

Fig. 2. Gene Regulation Network

Example 2. Let us consider the single gene regulatory model shown in Figure 2. Let the reactions for protein generation and degradation, namely, $mRNA \xrightarrow{k_2} mRNA + protein$ and $protein \xrightarrow{r} \emptyset$ be under the differential interpretation, and all other reactions, namely, $A \underset{b}{\overset{c}{\rightleftharpoons}} B$, $B \xrightarrow{h} mRNA + B$ and $mRNA \xrightarrow{k_1} \emptyset$ be under the stochastic interpretation.

```
% Differential rules              % Stochastic rules
MA(k2) for mRNA => mRNA + protein.   (MA(c), MA(b)) for A <=> B.
MA(r) for protein => _.              MA(h) for B => mRNA + B.
                                     MA(k1) for mRNA => _.
```

Our preprocessor generates a hybrid model composed of reactions and events. Due to the stochastic nature of the reactions, there is no way to check the results point by point. Nevertheless, comparison of mean values and standard deviations shows very good agreement with purely stochastic simulations. The following table shows the CPU time improvement in this example. The number of fired events is about six times smaller and the runtime on a Macbook Pro is about four times faster.

method	step size = 0.01		step size = 0.02	
	#fired_event	CPU time (sec)	#fired_event	CPU time (sec)
stochastic	89066	63.2	83856	51.5
hybrid	14258	15.1	14183	12.9
ratio	0.16	0.24	0.17	0.25

Example 3. The reaction model of bacteriophage T7 infection described in [3] is an interesting example that can be similarly hybridized by partitioning the reactions with differential semantics for protein synthesis and with stochastic semantics for gene activation, as follows:

```
% Differential reaction rules       % Stochastic reaction rules
MA(c5) for tem => tem + struc.      MA(c1) for gen => tem.
MA(c6) for struc => _.              MA(c2) for tem => _.
                                    MA(c3) for tem => tem + gen.
                                    MA(c4) for gen + struc => virus.
```

In this example, tem and struc are hybrid species, while gen and virus are purely stochastic. The following table shows that the hybrid simulation improves by three orders of magnitude the simulation time over a time horizon of 100 hours with a step size of 0.01:

method	#fired_event	CPU time (sec)
stochastic	276556	218.7
hybrid	832	0.75
ratio	0.003	0.003

It is worth noting that in these examples, the user is responsible for a partition of reactions into differential and stochastic groups, that is fixed for the rest of simulation. This restriction may lead to inaccurate or inefficient simulation if the propensity and/or reactants' counts of a reaction change substantially over time and violate the underlying assumptions of differential and stochastic semantics. It is therefore desirable to dynamically adjust the reaction partition along the progress of simulation.

Interestingly, the described framework allows us to easily explore various dynamic partitioning strategies considering the crucial factors of *particle count* and *propensity value* [3]. All species become potentially hybrid and criteria are imposed such that, during the simulation run, the reactions interpreted under the differential semantics are maximized while their current particle counts and propensity values must satisfy some accuracy requirement with respect to the simulation step size.

4   Hybrid Differential-Boolean Semantics

4.1   Preprocessor for Composing Differential and Boolean Models

In this section, we consider the hybrid composition of differential reaction models with Boolean transition models. One typical use of this form of composition is for modeling the interactions between gene expression and metabolism on different time scales. Gene networks can be modeled by simple Boolean regulatory models representing the on/off states of the genes and the possible transitions from one state to another, while metabolic networks are naturally modeled by ODE systems. Hybrid models of gene expression and metabolism can thus be naturally built as hybrid Boolean-differential models, and analyzed and simulated as hybrid automata.

A Differential-Boolean composition necessitates specifying:

- the link between the continuous variables and the Boolean variables, e.g. by fixing concentration threshold values,

- the relationship between the discrete logical time of the Boolean model and the continuous real time of the ODE model, e.g. by adding delays on Boolean transitions,
- the integrity constraints between both dynamics.

There is currently no general method for these tasks. As shown in Section 2.3 however, a set of reactions and events can be interpreted as a hybrid automaton in which there is a state with a particular ODE for each combination of the trigger values, and there is a transition from one state to another state when at least one trigger changes value from false to true in the source state.

This low level mechanism provides all what is necessary to compose differential models with Boolean models, compile them in reaction rules plus events and execute them using hybrid simulations. In the following section we illustrate our composition preprocessor on a hybrid model of the cell cycle.

4.2   Hybrid Composition of Differential-Boolean Cell Cycle Models

In [33], Singhania et al. have proposed a simple hybrid model of the mammalian cell cycle regulation. This cell cycle model of low dimension has been evaluated in terms of flow cytometry measurements of cyclin proteins in asynchronous populations of human cell lines. The few kinetic constants in the model are easier to estimate from the experimental data that the numerous kinetic constants of a single large ODE model. Using this hybrid approach, modelers could thus quickly create quantitatively accurate, computational models of protein regulatory networks in cells.

In this model, Cyclin abundances are tracked by piecewise linear differential equations for cyclin synthesis and degradation. Cyclin synthesis is regulated by transcription factors whose activities are represented by discrete variables (0 or 1) and likewise for the activities of the ubiquitin-ligating enzyme complexes that govern cyclin degradation. The discrete variables change according to a predetermined sequence, with the times between transitions determined by the amount of cyclin presented as well as exponentially distributed random variables.

This model can be reconstructed in our framework as the hybrid composition of a differential reaction model of cyclin activation and degradation, with a Boolean model of cell cycle phase transitions. In our high level interface, this composition is specified by providing as input

1. the differential reaction model of cyclin activation:

   ```
   k_sa for _ => CycA.        MA(k_da) for CycA => _.
   k_sb for _ => CycB.        MA(k_db) for CycB => _.
   k_se for _ => CycE.        MA(k_de) for CycE => _.
   ```

   with initial concentrations and symbolic kinetic expressions

   ```
   k_sa=5+6*B_tfe+20*B_tfb.   k_da=0.2+1.2*B_cdc20a+1.2*B_cdh1.
   k_sb=2.5+6*B_tfb.          k_db=0.2+1.2*B_cdc20b+0.3*B_cdh1.
   k_se=0.02+2*B_tfe.         k_de=0.02+0.5*B_scf.
   ```

2. the Boolean transition system of the cell cycle [33]:

   ```
   states (B_tfe,B_scf,B_tfb,B_cdc20a,B_cdc20b,B_cdh1).
   (0,0,0,0,0,1) ->2 (1,0,0,0,0,1) ->3 (1,0,0,0,0,0)
   ```

```
->4 (1,1,0,0,0,0) ->5 (1,1,1,0,0,0) ->6 (0,1,1,0,0,0)
->7 (0,1,1,1,0,0) ->8 (0,1,1,1,1,0) ->9 (0,1,0,1,1,1)
->1 (0,0,0,0,0,1)
```

3. the specification of the interface between both models as a set conditions and actions associated to the Boolean transitions and macros:

```
delta_t=lambda*exp(random).      tau=Time-delta_t.
masst=mass*exp(0.029*(Time-start_time)).
->2 condition [Time>tau] action [lambda=2,mass=masst/2]
->3 condition [Time>tau and [CycE]*masst>=80] action [lambda=0].
->4 condition [Time>tau and [CycA]>12.5] action [lambda=0.01].
->5 condition [Time>(tau+7)] action[lambda=1].
->6 condition [Time>tau and [CycB]>21.25] action [lambda=0.5].
->7 condition [Time>tau] action [lambda=0.75].
->8 condition [Time>tau] action [lambda=1.5].
->9 condition [Time>tau] action [lambda=0.5].
->1 condition [Time>tau and [CycB]<3] action [lambda=0.025].
```

The compilation process described in Section 2.2 returns the input differential reaction model augmented with a set of events for the Boolean transitions from state 1 to 9 and back to 1, and their synchronization with the differential reaction model. In this form, the simulation over a time horizon of 100 hours takes 60 ms on a MacBook Pro.

4.3   Related Work on Boolean Regulatory Models with Delays

René Thomas's discrete modelling of gene regulatory networks (GRN) is a well known approach to study the logical dynamics of a set of interacting genes. It deals with a graph of positive and negative influences between genes and logical functions that determine the possible trajectories in the state space. Those parameters are a priori unknown, but they may generally be deduced from a large set of biologically observed behaviors in various conditions. Besides, it neglects the time delays for a gene to pass from one level of expression to another one. In [1], it is shown that one can account for time delays depending on the expression levels of genes in a GRN, while preserving powerful enough computer-aided reasoning capabilities. The characteristic of this approach is that, among possible execution trajectories in the model, one can automatically find out both viability cycles and absorption in capture basins. Model-checking techniques developed for hybrid systems are used for this purpose [2]. The authors describe a Hybrid model for the mucus production in the bacterium Pseudomonas aeruginosa and show that they are able to discriminate between various possible dynamics [1,2]. Such a model can be presented and compiled in a set of reaction rules with events as described in the previous section.

Time constraints provide another mean to refine Boolean or discrete models which are often too coarse to be useful. In [28], the authors present a new technique for overapproximating (in the sense of timed trace inclusion) continuous dynamical systems by timed automata for the purpose of efficiently checking timed (as well as untimed) properties. The essence of this technique is the partition of the state space into cubes and the allocation of a clock for each dimension. This is in contrast with other approaches which use only one clock. This idea is close in spirit to rectangular hybrid automata in the sense of separating and bounding the dynamics of each dimension. This makes it possible to get better approximations of the behavior. The timed automata produced by these techniques can be directly composed in our preprocessor for simulation.

5   Conclusion

The combination of kinetic reaction rules with conditional events, as already present in SBML, provides the expressive power of hybrid automata for combining discrete and continuous dynamics. Although introduced in SBML for handling some discrete events, such as for instance the division of the mass by two at each cell division in cell cycle models, SBML events can be used on a large scale as a basic mechanisms allowing for the composition of heterogeneous models and implementing hybrid simulators.

We have presented a high-level interface for composing hybrid models, compiling them in reactions plus events, and running hybrid simulations. In particular we have shown that hybrid differential-stochastic reaction models can be assembled with this interface, compiled in differential reactions plus events for emulating the stochastic reactions, and executed with a *de facto* hybrid simulator with either static or dynamic strategies. This has been illustrated with the hybrid model of bacteriophage T7 infection [3].

We have also shown that hybrid Boolean-differential models can similarly be composed, compiled in reactions plus events, and simulated, through a high-level interface for specifying the input models, the conditions on the continuous variables, and the time delays of the Boolean transitions. This has been illustrated by a reconstruction of the hybrid mammalian cell cycle model of Singhania et al. [33].

This shows the expressive power of SBML events and their possible use as a low-level implementation language for representing and simulating hybrid models. This also shows the need for generating such hybrid models with a preprocessor using a high-level interface as the one prototyped here in Biocham. We are currently improving this interface to use it on more examples and on hybrid models obtained by model reduction using tropicalization methods [30].

Acknowledgements. This work has been supported by the French OSEO Biointelligence and ANR BioTempo projects, and by the European Eranet Sysbio C5Sys project. We acknowledge fruitful discussions with our partners in these projects and with Robin Philip for his early work on this topic during an internship with us.

References

1. Ahmad, J., Bernot, G., Comet, J.-P., Lime, D., Roux, O.: Hybrid modelling and dynamical analysis of gene regulatory networks with delays. ComplexUs 3, 231–251 (2006)
2. Ahmad, J., Roux, O., Bernot, G., Comet, J.-P., Richard, A.: Analysing formal models of genetic regulatory networks with delays. International Journal of Bioinformatics Research and Applications 4(3), 240–262 (2008)
3. Alfonsi, A., Cancès, E., Turinici, G., di Ventura, B., Huisinga, W.: Adaptive simulation of hybrid stochastic and deterministic models for biochemical systems. ESAIM: Proc. 14, 1–13 (2005)
4. Alur, R., Belta, C., Ivančić, F., Kumar, V., Mintz, M., Pappas, G.J., Rubin, H., Schug, J.: Hybrid Modeling and Simulation of Biomolecular Networks. In: Di Benedetto, M.D., Sangiovanni-Vincentelli, A.L. (eds.) HSCC 2001. LNCS, vol. 2034, pp. 19–32. Springer, Heidelberg (2001)

5. Ashburner, M., Ball, C.A., Blake, J.A., Botstein, D., Butler, H., Cherry, J.M., Davis, A.P., Dolinski, K., Dwight, S.S., Eppig, J.T., Harris, M.A., Hill, D.P., Issel-Tarver, L., Kasarskis, A., Lewis, S., Matese, J.C., Richardson, J.E., Ringwald, M., Rubin, G.M., Sherlock, G.: Gene ontology: tool for the unification of biology. Nature Genetics 25, 25–29 (2000)

6. Bockmayr, A., Courtois, A.: Using hybrid concurrent constraint programming to model dynamic biological systems. In: Stuckey, P.J. (ed.) ICLP 2002. LNCS, vol. 2401, pp. 85–99. Springer, Heidelberg (2002)

7. Calzone, L., Fages, F., Soliman, S.: BIOCHAM: An environment for modeling biological systems and formalizing experimental knowledge. Bioinformatics 22(14), 1805–1807 (2006)

8. Chaouiya, C., Remy, E., Thieffry, D.: Petri net modelling of biological regulatory networks. Journal of Discrete Algorithms 6(2), 165–177 (2008)

9. Egerstedt, M., Mishra, B. (eds.): HSCC 2008. LNCS, vol. 4981. Springer, Heidelberg (2008)

10. Fages, F., Gay, S., Jovanovska, D., Rizk, A., Soliman, S.: BIOCHAM v3.4 Reference Manual. INRIA (2012)

11. Fages, F., Soliman, S.: Abstract interpretation and types for systems biology. Theoretical Computer Science 403(1), 52–70 (2008)

12. Feinberg, M.: Mathematical aspects of mass action kinetics. In: Lapidus, L., Amundson, N.R. (eds.) Chemical Reactor Theory: A Review, ch. 1, pp. 1–78. Prentice-Hall (1977)

13. Ghosh, R., Tomlin, C.J.: Lateral inhibition through delta-notch signaling: A piecewise affine hybrid model. In: Di Benedetto, M.D., Sangiovanni-Vincentelli, A.L. (eds.) HSCC 2001. LNCS, vol. 2034, pp. 232–246. Springer, Heidelberg (2001)

14. Gilbert, D., Heiner, M., Lehrack, S.: A unifying framework for modelling and analysing biochemical pathways using petri nets. In: Calder, M., Gilmore, S. (eds.) CMSB 2007. LNCS (LNBI), vol. 4695, pp. 200–216. Springer, Heidelberg (2007)

15. Gillespie, D.T.: General method for numerically simulating stochastic time evolution of coupled chemical-reactions. Journal of Computational Physics 22, 403–434 (1976)

16. Gillespie, D.T.: Exact stochastic simulation of coupled chemical reactions. Journal of Physical Chemistry 81(25), 2340–2361 (1977)

17. Gillespie, D.T.: Approximate accelerated stochastic simulation of chemically reacting systems. Journal of Chemical Physics 115(4), 1716–1733 (2001)

18. Gillespie, D.T.: Deterministic limit of stochastic chemical kinetics. The Journal of Physical Chemistry B 113(6), 1640–1644 (2009)

19. Hellander, A., Lotstedt, P.: Hybrid method for the chemical master equation. Journal of Computational Physics 227(1), 100–122 (2007)

20. Henzinger, T.A.: The theory of hybrid automata. In: Proceedings of the 11th Annual Symposium on Logic in Computer Science (LICS), pp. 278–292. IEEE Computer Society Press (1996), An extended version appeared in Verification of Digital and Hybrid Systems

21. Henzinger, T.A., Ho, P.-H., Wong-Toi, H.: HYTECH: A model checker for hybrid systems. In: Grumberg, O. (ed.) CAV 1997. LNCS, vol. 1254, pp. 460–463. Springer, Heidelberg (1997)

22. Henzinger, T.A., Mikeev, L., Mateescu, M., Wolf, V.: Hybrid numerical solution of the chemical master equation. In: Proceedings of the 8th International Conference on Computational Methods in Systems Biology, CMSB 2010, pp. 55–65. ACM, New York (2010)

23. Hucka, M., et al.: The systems biology markup language (SBML): A medium for representation and exchange of biochemical network models. Bioinformatics 19(4), 524–531 (2003)
24. Ideker, T., Galitski, T., Hood, L.: A new approach to decoding life: Systems biology. Annual Review of Genomics and Human Genetics 2, 343–372 (2001)
25. Kanehisa, M., Goto, S.: KEGG: Kyoto encyclopedia of genes and genomes. Nucleic Acids Research 28(1), 27–30 (2000)
26. Kiehl, T.R., Mattheyses, R.M., Simmons, M.K.: Hybrid simulation of cellular behavior. Bioinformatics 20(3), 316–322 (2004)
27. Kwiatkowska, M., Norman, G., Parker, D.: Using probabilistic model checking in systems biology. SIGMETRICS Performance Evaluation Review 35(4), 14–21 (2008)
28. Maler, O., Batt, G.: Approximating continuous systems by timed automata. In: Fisher, J. (ed.) FMSB 2008. LNCS (LNBI), vol. 5054, pp. 77–89. Springer, Heidelberg (2008)
29. Matsuno, H., Doi, A., Nagasaki, M., Miyano, S.: Hybrid petri net representation of gene regulatory network. In: Proceedings of the 5th Pacific Symposium on Biocomputing, pp. 338–349 (2000)
30. Noël, V.: Modèles réduits et hybrides de réseaux de réactions biochimiques – Applications à la modélisation du cycle cellulaire. PhD thesis, Université de Rennes 1 (2012)
31. Salis, H., Kaznessis, Y.N.: Accurate hybrid stochastic simulation of a system of coupled chemical or biochemical reactions. The Journal of Chemical Physics 122(5), 54103 (2005)
32. Salis, H., Sotiropoulos, V., Kaznessis, Y.N.: Multiscale hy3s: Hybrid stochastic simulation for supercomputers. BMC Bioinformatics 7(1), 93 (2006)
33. Singania, R., Sramkoski, R.M., Jacooberger, J.W., Tyson, J.J.: A hybrid model of mammalian cell cycle regulation. PLOS Computational Biology 7(2) (February 2011)

On the Verification and Correction
of Large-Scale Kinetic Models
in Systems Biology

Attila Gábor[1], Katalin M. Hangos[2,3],
Gábor Szederkényi[2,4], and Julio R. Banga[1]

[1] BioProcess Engineering Group, IIM-CSIC, Vigo, Spain
{attila.gabor,julio}@iim.csic.es
[2] Computer and Automation Research Institute of the
Hungarian Academy of Science, Budapest, Hungary
{hangos,szeder}@sztaki.mta.hu
[3] University of Pannonia, Veszprém, Hungary
[4] Pázmány Péter Catholic University, Budapest, Hungary

Abstract. In this paper we consider the problem of verification of large dynamic models of biological systems. We present syntactical criteria based on biochemical kinetics to ensure the plausibility of a model and the positivity of its solution. These criteria include the positivity of the rate functions, their kinetic type dependence on the reactant species concentrations, and the absence of the negative cross-effects that together guarantee the nonnegativity of the dynamics. Further, the stoichiometric matrix of the truncated reaction system is checked against conservation using its algebraic properties. Algorithmic procedures are then proposed for checking these criteria with emphasis on good scaling up properties. In addition to these verification procedures, we also provide, for certain typical errors, model correcting methods. The capabilities and usefulness of these procedures are illustrated on biochemical models taken from the Biomodels database. In particular, a set of 11 kinetic models related with *E. coli* are checked, finding two with deficiencies. Correcting actions for these models are proposed.

Keywords: verification, model checking, kinetic models.

1 Introduction

Dynamics play a key role in the explanation of complex phenomena occurring in living systems. Therefore, the dynamic modeling and model analysis of biochemical networks is of high importance in systems biology, as quantitative mathematical models allow the description, analysis and/or manipulation of a wide range of biochemical processes.

The mathematical form of these models varies depending on the aim of modeling and on the quality of measured data available. Petri-net models –both deterministic and stochastic– are widely used for the analysis of qualitative dynamic properties, such as persistency [1], stability [2], etc. Qualitative dynamic

A. Gupta and T.A. Henzinger (Eds.): CMSB 2013, LNBI 8130, pp. 206–219, 2013.

models in the form of nonlinear ordinary differential equations (ODEs) are also widely used when good quality measured data are available for model parameter estimation, model verification, validation and detailed dynamic analysis. More in particular, the class of kinetic models [3] (with mass action type or other rate functions) is a widely accepted description form.

In practice, however, many of the medium and large-scale kinetic models in systems biology show problems when the space of parameters is explored. For example, dynamic simulations for certain parameter values result in negative concentrations (suggesting that mass-balance may not be correct), or simply blow-up. Therefore, careful checks should be performed before the use of a published model. This is routinely done in large biochemical model bases (see e.g. [4]), but these checks cannot detect every deficiency that may arise from the many different uses (simulation, parameter estimation, experiment design, etc.) of these models.

A biologically valid model should be valid both from physical and chemical point of view. There are some tools which help the user to avoid making modelling mistakes e.g. by offering predefined rate-functions, tracking the variables, or supporting measurement units and their consistency. These tools serve mostly for syntactic checking purposes. Furthermore, some tools can also check fundamental model properties, such as e.g. mass balance, the existence of admissible steady states, or the characteristics of the dynamic behaviour near a steady state, among others.

The Systems Biology Markup Language (SBML) [5] is a kind of accepted "standard", which offers model *syntax* checking, e.g. checking that the measurement units are correct. Systems Biology Toolbox 2 offers in addition (i) moiety conservation, (ii) steady state calculation, (iii) stoichiometry analysis, and (iv) bifurcation analysis.

As another well-known example, COPASI [6] provides a systematic model building framework to reduce the possibility of making modelling errors. Further functions are: steady state analysis, mass conservation, time scale separation, sensitivity calculation etc.

Despite the above efforts to ensure the acceptable quality of a biochemical model, it is easy to find in the literature such models that do not possess very basic properties, like positivity. This is usually the consequence of model simplification based on assumptions [7]. However these assumptions are sometimes forgotten or not known explicitly. Therefore, our aim was to formulate simple syntactical and semantic criteria of biochemical origin that ensure the plausibility of the studied model and the positivity (more precisely, non-negativity) of its solution. Similar ideas of model checking appear in [8] and [9].

The basic properties of dynamic models that describe reaction kinetic systems are used for this purpose. Roughly speaking, the kinetic property of these models means that the individual reaction rates are non-negative and there cannot be negative cross-effects between the dynamics of species [10]. Besides other important features, the kinetic property implies non-negativity which means that the non-negative orthant of the coordinates system remains invariant for the process dynamics (i.e. the differential variables that typically describe concentrations, always remain non-negative). However, as it is illustrated in this paper, some models published in journals and/or in open access biological databases do not fulfill fundamental kinetic properties, and this can be a serious obstacle in tasks such as parameter estimation, or in the later use or extension of these models.

In addition to the model verification procedures, we aim at localizing the reaction, or set of reactions, that cause a particular problem (for example, possible negative solutions), and at giving advice on how to correct them. Note that verification in this paper is used in the sense of checking for the presence of incorrect dynamic behaviour of the model, and not in the sense of its experimental (in)validation.

2 Plausible Biochemical Models

2.1 Mathematical Models of Biochemical Reactions

Biochemical reactions form an important sub-class in chemical reaction kinetics, that are characterized by the generally large number of reaction steps, and by the potentially complex, e.g. non-monotonous nature of the reaction rate functions. The reaction scheme together with the appropriate reaction constants of the most important biochemical reaction systems is collected in large biochemical databases (such as the Biomodels database [4]), and special description languages (such as SBML [5]) are developed for their standardized representation.

In order to develop a model representation of biochemical reaction systems that is suitable for model verification, the model representation of chemical reaction networks [3] can be used with some adjustments.

2.2 Basic Notions for Describing Biochemical Reactions

Complex biochemical reaction schemes are composed of elementary reaction steps that are irreversible. This means, that a reversible reaction step is represented as two irreversible elementary reaction steps. An elementary reaction step R_ℓ can be formally described using n *species* $X_1, \ldots, X_n$ and associated *stoichiometric coefficients*. The species are classified as reactants (with stoichiometric coefficients $\nu_1, \ldots, \nu_n$) and products (with $\mu_1, \ldots, \mu_n$).

$$R_\ell : \quad \sum_{k=1}^{n} \nu_{ki} X_k \xrightarrow{r_{ij}} \sum_{k=1}^{n} \mu_{kj} X_k. \tag{1}$$

The non-negative linear combinations of the species $\sum_{k=1}^{n} \nu_{ki} X_k$ and $\sum_{k=1}^{n} \mu_{kj} X_k$ are called the *complexes* and are denoted by $C_1, \ldots C_m$, e.g. $C_1 = 2X_1 + X_3$.

It is important to note that some species may appear on both sides of a given reaction with the same stoichiometric coefficients ($\nu_{ki} = \mu_{ki}$).

A *reaction* (elementary reaction step) R_{ij} is an ordered pair of complexes $C_i, C_j \in \mathcal{C}$, which means that the *reactant complex* C_i is transformed to the *product complex* C_j in the chemical reaction network i.e. $R_{ij} = (C_i, C_j)$. Reactant complexes are also called *source complexes*. To each of the reactions, a *reaction rate function* r_{ij} is associated that may depend on the concentration $[X_i] = x_i$ of any species X_i in the biochemical reaction system. The reaction rate is usually measured in units $[\frac{mol}{s}]$ and shows how many moles of a reactant X_k with $\nu_{ki} = 1$ is used, or how many moles of a product X_ℓ with $\mu_{\ell j} = 1$ is produced by the reaction in one second.

2.3 Plausible Reaction Rate Functions

Because of the above chemical meaning of the reaction rate, the reaction rate function should posses the following properties.

1. *Rate positivity.* As the elementary reaction steps are irreversible and the reaction rate is defined as the rate of the consumption (decrease) of the reactant concentrations, the inequality $r_{ij} \geq 0$ should be fulfilled over the entire domain of the reaction rate function, i.e. for all non-negative concentration values in its argument.

2. *Kinetic dependence.* Reaction rate functions in biochemical reactions include the concentrations of the *reactants* such as substrates, which are consumed in the reaction. Some species concentration may not change in the reaction because the same amount is consumed as produced, i.e. $\nu_{ki} = \mu_{kj}$. Further, the reaction function may include other concentrations that modify the reaction rate either in a catalytic way, or in the form of inhibitors, e.g. the concentration of a product of that reaction. However, one only considers the real *reactants* in the source complex which influence the reaction rate in a dominant way that is described by the notion of kinetic dependence. r_{ij} is said to be *kinetic with respect to the species* in the source complex ($X_k \in C_i$) if

$$r_{ij}(x_k = 0) = 0 \quad \text{for all} \quad k = \{1, \ldots, n | X_k \in C_i\} \ , \tag{2}$$

i.e., if the concentration of any species in the source complex is zero, then the reaction rate becomes zero.

2.4 Plausibility of Some Common Biochemical Reaction Rate Functions

Only a limited number of rate function types are usually present in biochemical reaction systems, that are characterized by a functional form and the values of its parameters [11]. A few of the most important ones are analysed for plausibility below.

(i) *Mass action kinetics*
   This is the simplest reaction rate function form $r_{\mathrm{MA},i} = k_i \cdot \prod_{l=1}^{n} x_l^{\nu_{li}}$ where $k_i > 0$ is the reaction rate constant, and the reactant complex is

$C_i = \sum_{l=1}^{n} \nu_{li} X_l$. It is easy to see that $r_{\mathrm{MA},i}$ is kinetic in each of the species in complex C_i.

(ii) *Michaelis-Menten kinetics*

Recall, that elementary reaction steps are irreversible, then the rate function is in the form

$$r_{\mathrm{MM},i} = k_i \cdot \frac{x_i}{(K_i + x_i)} \tag{3}$$

where where $k_i > 0$ and $K_i > 0$ are constant parameters, and the reactant complex $C_i = X_i$. This reaction rate function is kinetic in X_i.

(iii) *Constant level reactions*

Here the rate function is simply a constant, i.e. $r_{\mathrm{C},i} = k_i^M$, where $k_i^M > 0$ is a constant. This rate function does not have kinetic dependence on any specie, thus no reactant species can be associated to this reaction. Consequently *it is not a plausible reaction rate function*, whenever it is a consuming reaction. Note that, when this reaction stands for model input it always occurs with positive sign in the balance equations.

Correcting Non-plausible Reaction Rates. There is unfortunately no general way of correcting non-plausible reaction rates. However, in some cases, such rates can be made plausible. An example of this case is, when a constant level type reaction rate function is present in the kinetic equation of the species X_i with negative sign. Then we can multiply the rate function with x_i that will make this rate function kinetic in X_i.

2.5 Positive (Non-negative) Kinetic Models

The dynamic variables x_k of any biochemical model are species concentrations, that are non-negative. Therefore, any plausible biochemical model should have this property, that is based mathematically on the notion of essentially non-negative functions [12]. A function $f = [f_1 \ldots f_n]^T : [0, \infty)^n \mapsto \mathbb{R}^n$ is called essentially non-negative if, for all $i = 1, \ldots, n$, $f_i(x) \geq 0$ for all $x \in [0, \infty)^n$, whenever $x_i = 0$. In the context of biochemical models, where the components f_i correspond to the right-hand sides of the kinetic differential equations, the non-negativity of individual rate functions and the lack of negative cross-effects between species together guarantee essential non-negativity of the model [10].

2.6 Component Mass Conservation

Kinetic models are constructed based on the conservation of the masses of species assuming *closed systems* and isothermal conditions.

The conservation equations are constructed for species that are either reactants or products of the chemical reactions in the form

$$\frac{dx_k}{dt} = -\sum_{i=1}^{m} \nu_{ki} r_i + \sum_{i=1}^{m} \mu_{ki} r_i = \sum_{i=1}^{m} s_{ki} r_i, \tag{4}$$

where s is the element of the $S \in \mathbb{R}^{n \times m}$ stoichiometric matrix. No dynamic conservation equations are written to species with only catalytic or inhibitory role.

In open systems one has in addition (i) *input terms*, that have positive sign and may depend on externally set concentrations and/or mass flow of certain non-conserved specie, and (ii) *output terms*, that are linear in one conserved specie, have negative sign and appear only in the dynamic equation of that particular specie. Therefore, *all of the input and output terms should be set to zero when checking the conservation property*: this form of the dynamic model will be called the *truncated model*.

A *truncated stoichiometric matrix* $\tilde{S} \in \mathbb{R}^{n \times m}$ is constructed from the truncated model by associating a column S_i to each complex C_i with $[\tilde{S}]_{ki} = \mu_{ki} - \nu_{ki}$ for only the reactant species (but not to the catalytic or inhibitory ones). The truncated biochemical model has the conservation property taking into consideration all species, if there exists a strictly positive vector m such that $m^T \tilde{S} = 0$ (see [13], [14] and for efficient computation methods [15]).

Some biological models do not obey mass conservation on purpose, otherwise the above property enables us *to check the truncated stoichiometric matrix $\tilde{S}$ against conservation*, that is only a partial verification of the values of the stoichiometric coefficients μ_{ki} and ν_{ki} in the model.

Plausible Model Structure. The model structure is said to be plausible, when the stoichiometric constants in the conservation equations (4) are consistent with the reactants and products of the reactions, i.e. ν_{ki} is strictly positive if reaction r_i consumes the species X_k and μ_{ki} is strictly positive if X_k is a product of the reaction r_i. The stoichiometric coefficient of a reaction which neither consumes nor produces a species should be zero in the corresponding balance equation.

3   Model Checking and Correction in Practice

3.1   Steps of Model Verification

Given a biochemical reaction network in terms of the reaction rate functions and the system of ordinary differential equations. The reaction rate functions are assumed to be smooth functions of the time, some concentrations and parameters: $r_i = r_i(t, x, k)$. The explicit time dependency of the reactions permits to incorporate boundary conditions or model inputs for the dynamic system. The ordinary differential equation form of the model is given by Eq. (4).

Inputs of the Algorithm. We can either start with the list of differential equations and the algebraic equations of the reactions, or the model defined in SBML. Since the SBML model does not contain explicitly the differential equations, in this case the SBML import function of the System Biology Toolbox 2 [16] is used to translate the SBML into MATLAB structure and generate the

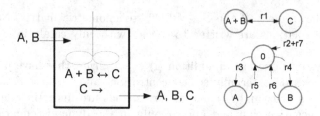

Fig. 1. Continuous flow stirred tank reactor and its reaction graph representation

differential equations. It is important to note that the parameter values of the rate functions are not needed for the verification.

A homogeneous, continuous flow stirred tank bioreactor serves as an tutorial example depicted in Fig. 1. The reaction network consists of three species (A, B, C and their concentrations x_A, x_B and x_C, respectively) and two elementary reactions: a two substrate, one product reversible Michaelis-Menten kinetics (5) and a non-plausible (see subsection 2.4 (iii)) constant reaction (6). The zero complex denoted by a "0" in the reaction graph corresponds to the environment of the system (for clarification see e.g. [17]). The reactor feed contains the substrates A and B with x_A^f and x_B^f constant concentrations, respectively, the corresponding pseudo-reactions [17] are in Eq. (7). The output stream that contains all species is represented by the pseudo-reactions in Eq. (8). The reaction rate functions and the ODEs of the system are

$$r_1 = V_f \frac{\frac{x_A}{K_{x_A}} \frac{x_B}{K_{x_B}}}{1 + \frac{x_A}{K_{x_A}} + \frac{x_B}{K_{x_B}}} - V_r \frac{\frac{x_C}{K_{x_C}}}{1 + \frac{x_C}{K_{x_C}}} \tag{5}$$

$$r_2 = K_d \tag{6}$$

$$r_3 = \zeta x_A^f; \qquad r_4 = \zeta x_B^f \tag{7}$$

$$r_5 = \zeta x_A; \qquad r_6 = \zeta x_B; \qquad r_7 = \zeta x_C \tag{8}$$

$$\frac{dx_A}{dt} = -r_1 + r_3 - r_5 \tag{9}$$

$$\frac{dx_B}{dt} = -r_1 + r_4 - r_6 \tag{10}$$

$$\frac{dx_C}{dt} = r_1 - r_2 - r_7 . \tag{11}$$

Splitting the Reversible Reactions. The irreversible forward and backward reactions are created from the original reactions using regular expressions and the Symbolic Math Toolbox of MATLAB. In this example the algorithm finds the subtraction with two operands in Eq. (5) and separates to

$$r_1^f = V_f \frac{\frac{x_A}{K_{x_A}} \frac{x_B}{K_{x_B}}}{1 + \frac{x_A}{K_{x_A}} + \frac{x_B}{K_{x_B}}} \text{ and } r_1^b = V_r \frac{\frac{x_C}{K_{x_C}}}{1 + \frac{x_C}{K_{x_C}}} . \tag{12}$$

Simultaneously the differential equations are updated to

$$\frac{dx_A}{dt} = -(r_1^f - r_1^b) + r_3 - r_5 \tag{13}$$

$$\frac{dx_B}{dt} = -(r_1^f - r_1^b) + r_4 - r_6 \tag{14}$$

$$\frac{dx_C}{dt} = (r_1^f - r_1^b) - r_2 - r_7 . \tag{15}$$

Model Positivity by Checking the Kinetic Property. Next, the stoichiometric matrix (S) is constructed by parsing the string of the ODEs and collecting the coefficients of the rate functions. Whenever the v_{ij} element of S is negative, i.e. reaction r_j consumes the species x_i, r_j must be kinetic with respect to x_i. This can be checked by substituting zeros for the species x_i in the rate function and evaluating it; the result must be zero.

In our example the model Eqs. (13)-(15) give rise to the stoichiometric matrix

$$S = \begin{bmatrix} -1 & 1 & 0 & 1 & 0 & -1 & 0 & 0 \\ -1 & 1 & 0 & 0 & 1 & 0 & -1 & 0 \\ 1 & -1 & -1 & 0 & 0 & 0 & 0 & -1 \end{bmatrix} \tag{16}$$

and the irreversible reaction vector $R = [r_1^f, r_1^b, r_2, r_3, r_4, r_5, r_6, r_7]^T$. Considering the location of the negative entries of S reaction r_1^f and r_5 must be kinetic to species A, r_1^f and r_6 with respect to B and r_1^b, r_2 and r_7 with respect to C. By substituting zeros for the reactant species in the rate functions –e.g. $r_1^f(x_A = 0)$, $r_5(x_A = 0)$, $r_1^f(x_B = 0)$ etc. – the plausible ones give zeros. At this point reaction r_2 is found to be non-kinetic to the species C, and thus it is a non-plausible reaction. We may correct the rate function by multiplying with its reactant species concentration: $r_2^* = K_d x_C$. This reaction can be regarded as a model output, too.

Component Mass Conservation. The truncated model without the input reactions (Eqs. (7)) and the output reactions (Eqs. (8) and the corrected r_2^*) is represented by the first two columns of S. This sub–matrix is rank deficient and has the $m = [1\,1\,2]^T$ strictly positive vector in the left nullspace indicating the mass conservation law.

3.2   Verified Models

We have checked 11 *E. coli* curated models of the Biomodels database, and some of them turned out to contain non-plausible reactions. Table 1 shows the unique identifiers of the models in the database. The number of species, the number of reactions and the computation time of the algorithm is also included in the following columns. The 5th column contains the non-plausible reaction, while the last column shows whether the truncated model admits mass conservation. In the next section, the verification of two of these models will be presented in detail.

Table 1. Verified models

BioModel ID	No. of species	No. of reactions	Time [s]	Non-plausible reaction	Mass conservation
BIOMD296	4	10	man.*	plausible	no
BIOMD413	5	9	0.3	plausible	no
BIOMD200	22	46	2.3	plausible	yes
BIOMD217	12	22	23	plausible	yes
BIOMD051	18	62	5	reaction vMURSYNTH is not kinetic w.r.t. species CF6P	no
BIOMD066	11	10	man.*	reaction vATPASE is not kinetic w.r.t. species ATP	yes
BIOMD012	6	12	0.8	plausible	no
BIOMD067	7	16	0.6	plausible	no
BIOMD221	8	22	1.9	reaction vSYN is not kinetic w.r.t. species AKG	no
BIOMD222	8	22	1.9	reaction vSYN is not kinetic w.r.t. species AKG	no
BIOMD065	8	16	0.5	plausible	no

*the separation of some reaction rate function needed manual manipulation

3.3 Case Study 1: Central Carbon Metabolism of *E. coli*

Chassagnole et al. [18] describe the central carbon metabolism of the *Escherichia coli*. Although we could reproduce the results in the paper [18] with the published model (BIOMD0000000051), numerical simulations with CVODES [19] during parameter estimation tasks gave errors, because negative concentrations appeared.

About the Model. The metabolism is described by 48 reactions which are grouped into kinetic types: reversible and irreversible Michealis-Menten kinetics, two-substrate reversible and irreversible Michealis-Menten kinetics, allosteric enzyme reactions, allosteric regulation, allosteric activation, ordered uni-bi mechanism, Hill kinetics, constant level reaction and reversible mass action kinetics. Appendix A contains examples of these reactions. The mass balance equations for the 18 metabolites are in the following form

$$\frac{dC_i}{dt} = \sum_j v_{ij} r_j(C, k) - \mu C_i \ , \tag{17}$$

where C is the concentration of the metabolite, v_{ij} is the (i, j)th element of the stoichiometric matrix, $r_j(C, k)$ denotes the j-th reaction rate function, which depends on the concentrations and the k rate function parameters. Finally, μ is the growth factor. The detailed equations are listed in [18] Tables I. and IV.

Checking the Rate Expressions. The first criteria of a plausible model is the non-negativity of the reaction rate functions, for which the reversible reactions have to be cut into a forward and a backward reaction. The separation of the reactions are straightforward in this case study. It is also easy to see that the reactions are always non-negative since the rate expressions contain only such mathematical operators that preserve the positivity. The model have three constant reactions: the Mureine synthesis, the Tryptophan synthesis and the Methionine synthesis

$$r_{\text{MurSynth}} = r_{\text{MurSynth}}^{\max},$$ (18)

$$r_{\text{TrpSynth}} = r_{\text{TrpSynth}}^{\max} \quad \text{and} \quad r_{\text{MetSynth}} = r_{\text{MetSynth}}^{\max},$$ (19)

but only r_{MurSynth} is not kinetic to its source specie, the others are input terms.

Checking the Model Structure. The positivity condition from Section 2.5 together with the model Eq. (17) give rise to the model specific positivity condition

$$\frac{dC_i}{dt} = \sum_j v_{ij} r_j - \mu C_i \geq 0 \quad \text{whenever} \quad C_i = 0, \quad \text{for all} \quad i = 1, \ldots 18.$$

This condition holds for plausible reaction rate functions which have the source kinetic property according to Eq. (2). Furthermore, whenever a reaction is not kinetic w.r.t. its source species and has negative stoichiometric coefficient, depending on the numerical values of the parameters it can violate the condition and cause negative concentrations during simulations.

This is exactly what happens in some parameter domain of this *E. coli* model. From the following model equation ([18] Table I. Eq.(3)):

$$\frac{dC_{\text{f6p}}}{dt} = r_{\text{PGI}} - r_{\text{PFK}} + r_{\text{TKb}} + r_{\text{TKa}} - 2r_{\text{MurSynth}} - \mu C_{\text{f6p}}$$ (20)

one can see that the stoichiometric coefficient of the Mureine synthesis r_{MurSynth} is negative, but it is not kinetic w.r.t. any metabolite. This may result in the appearance of negative concentrations and thus in a non-plausible model.

Correction of the Non-plausible Reaction. There are several ways to correct the non-plausible reaction. A switching function can be included, which turns off the reaction, whenever the concentration of fp6 reach zero. This procedure does not influence the model dynamics in the plausible concentration domain, but the switching function may result in mathematical or numerical simulation issues. Alternatively, one can make the reaction source kinetic by multiplying it with C_{fp6}: $r_{\text{MurSynth}}^{\text{cured}} = r_{\text{MurSynth}}^{\max} C_{\text{f6p}}$. It changes the dynamics of the system, but results in a smooth, plausible reaction rate function.

Mass Conservation. The truncated model is created by omitting the reactions which are either stand for inflows or outflows. We have found three linearly independent non-negative vectors for which $m_i^T S = 0$, for $i = 1, 2, 3$. This implies three moiety conservation laws, but there is no strictly positive m vector in the left kernel of S, and thus the model does not obey to the total mass conservation.

3.4   Case Study 2: Verification of the Model BIOMD0000000221

Singh et al. [20] present two kinetic models of the tricarboxylic acid cycle and glyoxylate bypass in the *Mycobacterium tuberculosis*. Both models are based on a validated *E. coli* model, which is in the focus of this case study. The kinetic model contains 12 metabolites and 12 reactions. The reactions and the differential equations are listed in Appendix B.

The reaction rate functions can be categorized into three kinetic reaction types: one substrate reversible Michaelis-Menten kinetics; two substrate reversible Michaelis-Menten kinetics and ordered uni-bi mechanism. The separation of them to irreversible forward and backward reactions is straightforward. The irreversible forms fulfil the rate positivity condition.

For the positivity of the model the kinetic property of the rate expressions should be analysed. The reactions must be kinetic to the species, which are consumed in that reactions. However, due to a modelling assumption, the r_{11} reaction

$$r_{11}^f = 0.0341 \cdot r_3^f = V_{\text{cell}} \cdot 0.0341 \frac{V_{11}^f \frac{C_{\text{icit}}}{K_{11,\text{icit}}}}{1 + \frac{C_{\text{icit}}}{K_{11,\text{icit}}} + \frac{C_{\text{akg}}}{K_{11,\text{akg}}}} \cdot \tag{21}$$

is not kinetic with respect to the akg source specie, which is consumed in this reaction according to the balance equation:

$$\frac{dC_{\text{akg}}}{dt} = r_3^f - r_3^b - r_4^f + r_4^b - r_{11}^f + r_{11}^b \tag{22}$$

Note that r_{11}^f has a negative stoichiometric coefficient.

Thus, our algorithm detected the consequence of the modelling assumption which lead the model out from the kinetic model system class.

The model has four boundary species, the concentrations of which are hold constant. The omission of the reactions containing these species leads to the truncated model and the corresponding reduced stoichiometric matrix S. There is no strictly positive vector in the left null-space of S: actually, all species but the glyoxylate participate in the mass conservation.

4   Conclusion

Using the syntax and semantics of biochemical models, simple syntactical criteria were formulated in this paper that ensure the plausibility of the studied model and the positivity of its solution.

First, the plausibility of reaction rate function was defined that include its positivity, and its kinetic dependence on the real reactants of the reaction. The absence of the negative cross-effects in the dynamic equations were used to ensure the positivity of the species concentration functions. The stoichiometric matrix of the truncated reaction system was checked against conservation using its algebraic properties.

Algorithmic procedures were proposed for checking these criteria that scale up well with the size of the biochemical model. For certain typical errors, model correcting methods were also proposed.

The developed notions and tools are illustrated on biochemical kinetic models of *E. coli.*

Acknowledgement. This research was supported in part by the Hungarian Research Fund through grant K83440 and by the funding from EU FP7 ITN "NICHE", project no. 289384.

References

1. Angeli, D., De Leenheerb, P., Sontag, E.: A Petri net approach to the study of persistence in chemical reaction networks. Mathematical Biosciences 210, 598–618 (2007)
2. Sontag, E.: Structure and stability of certain chemical networks and applications to the kinetic proofreading model of T-cell receptor signal transduction. IEEE Transactions on Automatic Control 46(7), 1028–1047 (2001)
3. Feinberg, M.: On chemical kinetics of a certain class. Arch. Rational Mech. Anal. 46, 1–41 (1972)
4. Li, C., Donizelli, M., Rodriguez, N., Dharuri, H., Endler, L., Chelliah, V., Li, L., He, E., Henry, A., Stefan, M.I., Snoep, J.L., Hucka, M., Le Novère, N., Laibe, C.: BioModels Database: An enhanced, curated and annotated resource for published quantitative kinetic models. BMC Systems Biology 4, 92 (2010)
5. Hucka, M., et al.: The Systems Biology Markup Language SBML: A medium for representation and exchange of biochemical network models. Bioinformatics 19, 524–531 (2003)
6. Hoops, S., Sahle, S., Gauges, R., Lee, C., Pahle, J., Simus, N., Singhal, M., Xu, L., Mendes, P., Kummer, U.: COPASI — a COmplex PAthway SImulator. Bioinformatics 22(24), 3067–3074 (2006)
7. Hangos, K.M.: Engineering model reduction and entropy-based lyapunov functions in chemical reaction kinetics. Entropy 12, 772–797 (2010)
8. Fages, F., Gay, S., Soliman, S.: Automatic curation of SBML models based on their ODE semantics. Technical report, INRIA Reseqarch Report 8014, hal00723554 (2012)
9. Kaleta, C., Richter, S., Dittrich, P.: Using chemical organization theory for model checking. Bioinformatics 25(15), 1915–1922 (2009)
10. Érdi, P., Tóth, J.: Mathematical Models of Chemical Reactions. Theory and Applications of Deterministic and Stochastic Models. Manchester University Press, Princeton University Press, Manchester, Princeton (1989)
11. Klipp, E., Herwig, R., Kowald, A., Wierling, C., Lehrach, H.: Systems biology in practice: concepts, implementation and application. Wiley-Blackwell (2008)
12. Haddad, W.M., Chellaboina, V., Hui, Q.: Nonnegative and Compartmental Dynamical Systems. Princeton University Press (2010)
13. Famili, I., Palsson, B.O.: The convex basis of the left null space of the stoichiometric matrix leads to the definition of metabolically meaningful pools. Biophysical Journal 85(1), 16–26 (2003)

14. Hangos, K.M., Szederkényi, G.: The underlying linear dynamics of some positive polynomial systems. Physics Letters A 376, 3129–3134 (2012)
15. Sauro, H.M., Ingalls, B.: Conservation analysis in biochemical networks: computational issues for software writers. Biophysical Chemistry 109(1), 1–15 (2004)
16. Schmidt, H., Jirstrand, M.: Systems Biology Toolbox for MATLAB: a computational platform for research in systems biology. Bioinformatics 22(4), 514–515 (2006), http://www.sbtoolbox2.org
17. Feinberg, M.: Chemical Reaction Network Structure and the Stability of Complex Isothermal Reactors – I. The deficiency Zero and Deficiency One Theorems. Chemical Engineering Science 42(10), 2229–2268 (1987)
18. Chassagnole, C., Noisommit-Rizzi, N., Schmid, J.W., Mauch, K., Reuss, M.: Dynamic modeling of the central carbon metabolism of Escherichia coli. Biotechnology and Bioengineering 79(1), 53–73 (2002)
19. Hindmarsh, A.C., Brown, P.N., Grant, K.E., Lee, S.L., Serban, R., Shumaker, D.E., Woodward, C.S.: SUNDIALS: Suite of nonlinear and differential/algebraic equation solvers. ACM Transactions on Mathematical Software (TOMS) 31(3), 363–396 (2005)
20. Singh, V.K., Ghosh, I.: Kinetic modeling of tricarboxylic acid cycle and glyoxylate bypass in Mycobacterium tuberculosis, and its application to assessment of drug targets. Theoretical Biology and Medical Modelling 3, 27 (2006)

Appendix

A   Reaction Rate Functions of the First Case Study

Some examples of the reaction rate functions and their irreversible form can be found in the following list.

1. Reversible mass action kinetics, e.g. Ribose phosphate isomerase reaction

$$r_{\text{R5P1}} = r_{\text{R5P1}}^{max}\left(C_{\text{ribu5p}} - \frac{C_{\text{rib5p}}}{K_{\text{R5P1,eq}}}\right) ,$$

the forward and backward reactions of which are:

$$r_{\text{R5P1}}^{f} = r_{\text{R5P1}}^{max}C_{\text{ribu5p}} \quad \text{and} \quad r_{\text{R5P1}}^{b} = r_{\text{R5P1}}^{max}\frac{C_{\text{rib5p}}}{K_{\text{R5P1,eq}}} .$$

2. Irreversible Michaelis-Menten kinetics, for example Serine synthesis

$$r_{\text{SerSynth}} = \frac{r_{\text{SerSynth}}^{max}C_{3pg}}{K_{\text{SerSynth,3pg}} + C_{3pg}} .$$

3. Allosteric enzyme activation, for example Glucose 1-phosphate adenyltransferase reaction

$$r_{\text{GIPAT}} = \frac{r_{\text{GIPAT}}^{max}C_{\text{glp}}C_{\text{atp}}\left(1 + \left(\frac{C_{\text{fdp}}}{K_{\text{GIPAT,fdp}}}\right)^{n_{\text{GIPAT,fdp}}}\right)}{(K_{\text{GIPAT,glp}} + C_{\text{glp}})(K_{\text{GIPAT,atp}} + C_{\text{atp}})}$$

4. Constant level reaction, such as Mureine synthesis

$$r_{\text{MurSynth}} = r_{\text{MurSynth}}^{max}$$

B   Model Equations of the Second Case Study

This section contains the reaction rate functions and the dynamic model equations of the tricarboxylic acid cycle and glyoxylate bypass model of E. coli. The 1st and 10th reactions have two substrate reversible Michaelis-Menten kinetics:

$$r_i = V_{\text{cell}} \frac{V_i^f \frac{S_1}{K_{i,S1}} \frac{S_2}{K_{i,S2}} - V_i^r \frac{P_1}{K_{i,P1}} \frac{P_2}{K_{i,P2}}}{\left(1 + \frac{S_1}{K_{i,S1}} + \frac{P_1}{K_{i,P1}}\right)\left(1 + \frac{S_2}{K_{i,S2}} + \frac{P_2}{K_{i,P2}}\right)} \quad \text{for } i = \{1, 10\} \ ,$$

where S_- and P_- denote the concentrations of the substrates and products respectively, $K_{-,-}$ are constant parameters and $V^{r/b}$ are the maximal rates of forward and backward reactions.

The 2nd – 8th and 11th reactions belong to the one substrate Michaelis-Menten kinetics

$$r_i = V_{\text{cell}} \frac{V_i^f \frac{S_1}{K_{i,S1}} - V_i^r \frac{P_1}{K_{i,P1}}}{1 + \frac{S_1}{K_{i,S1}} + \frac{P_1}{K_{i,P1}}} \quad \text{for } i = \{2, \ldots 8, 11\} \ .$$

Finally, the 9th reaction has the form:

$$r_i = V_{\text{cell}} \frac{V_i^f \frac{S_1}{K_{i,S1}} - V_i^r \frac{P_1}{K_{i,P1}} \frac{P_2}{K_{i,2}}}{\left(1 + \frac{S_1}{K_{i,S1}} + \frac{P_1}{K_{i,P1}} + \frac{P_2}{K_{i,P2}} + \frac{S_1}{K_{i,S1}} \frac{P_1}{K_{i,P1}} + \frac{P_1}{K_{i,P1}} \frac{P_2}{K_{i,P2}}\right)} \quad \text{for } i = 9 \ .$$

The system of differential equations expressed in terms of the reactions is

$$\frac{dC_{\text{aca}}}{dt} = 0, \qquad \frac{dC_{\text{oaa}}}{dt} = 0, \qquad \frac{dC_{\text{coa}}}{dt} = 0, \qquad \frac{dC_{\text{biosyn}}}{dt} = 0$$

$$\frac{dC_{\text{cit}}}{dt} = r_1^f - r_1^b - r_2^f + r_2^b$$

$$\frac{dC_{\text{icit}}}{dt} = r_2^f - r_2^b - r_3^f + r_3^b - r_9^f + r_9^b$$

$$\frac{dC_{\text{akg}}}{dt} = r_3^f - r_3^b - r_4^f + r_4^b - r_{11}^f + r_{11}^b$$

$$\frac{dC_{\text{sca}}}{dt} = r_4^f - r_4^b - r_5^f + r_5^b$$

$$\frac{dC_{\text{suc}}}{dt} = r_5^f - v_b^5 + r_9^f - r_9^b - r_6^f + r_6^b$$

$$\frac{dC_{\text{fa}}}{dt} = r_6^f - r_6^b - r_7^f + r_7^b$$

$$\frac{dC_{\text{mal}}}{dt} = r_7^f - r_7^b + r_{10}^f - r_{10}^b - r_8^f + r_8^b$$

$$\frac{dC_{\text{gly}}}{dt} = r_9^f - r_9^b - r_{10}^f + r_{10}^b$$

Context-Sensitive Flow Analyses: A Hierarchy of Model Reductions

Ferdinanda Camporesi[2,3], Jérôme Feret[2], and Jonathan Hayman[1,2]

[1] Computer Laboratory, University of Cambridge, UK
[2] DIENS (INRIA/ÉNS/CNRS), Paris, France
[3] Dipartimento di Scienze dell'Informazione, Università di Bologna, Italy

Abstract. Rule-based modelling allows very compact descriptions of protein-protein interaction networks. However, combinatorial complexity increases again when one attempts to describe formally the behaviour of the networks, which motivates the use of abstractions to make these models more coarse-grained.

Context-insensitive abstractions of the intrinsic flow of information among the sites of chemical complexes through the rules have been proposed to infer sound coarse-graining, providing an efficient way to find macro-variables and the corresponding reduced models. In this paper, we propose a framework to allow the tuning of the context-sensitivity of the information flow analyses and show how these finer analyses can be used to find fewer macro-variables and smaller reduced differential models.

1 Introduction

Modellers of molecular signalling networks must cope with the combinatorial explosion of protein states generated by post-translational modifications and complex formations. Rule-based models [13,1] provide a powerful alternative to approaches that require an explicit enumeration of all possible molecular species of a system. Such models consist of formal rules stipulating the (partial) contexts for specific protein-protein interactions to occur. The behaviour of the models can be formally described by stochastic or differential semantics. Yet, the naive computation of these semantics does not scale to large systems, because it does not exploit the lower resolution at which rules specify interactions.

Rules explicitly describe the intrinsic flow of information between sites of complexes. Indeed, rules are contextual: they document only the state of which sites has an influence on the kinetic of the interactions. This can be used to detect some correlations that can be safely ignored. Thus, we can cut molecular complexes into molecular patterns, called fragments, and derive a coarse-grained system which describes exactly the concentration (or population) evolution of these fragments. This method never requires the execution of the concrete rule-based model and the approach is proved exact by abstract interpretation [8].

The so-obtained coarse-graining crucially depends on the accuracy of the analysis of the intrinsic flow of information. In this paper, we introduce a framework to tune the context-sensitivity of the analyses of the flow of information and derive the induced coarse-graining. This way, our analysis can zoom-in or zoom-out

A. Gupta and T.A. Henzinger (Eds.): CMSB 2013, LNBI 8130, pp. 220–233, 2013.

to increase or decrease the accuracy of the description of the flow of information
which passes through each site, according to some conditions about the states of
the other sites around this site. It bridges the gap between fully insensitive analyses (where the information about the sites are summarized according to their
type only) [14,12,5] and fully context-sensitive analyses (that are computed in
the concrete on molecular species) [17] that have been proposed so far, and
provide a whole hierarchy of trade-off between accuracy and efficiency.

Related works. Dependencies between sites and reactions have been used in systematic [7,3] and automatic [1] informal methods for designing coarse-grained
models. These informal methods do not provide exact coarse-graining in several cases, like whenever a site is activated through a binding or in the case of
homo/hetero dimerizations. In [14,12,5], we have introduced a formal framework
which ensures a formal relation between the initial differential semantics and the
reduced one, by the means of abstract interpretation [8]. A similar framework
has been proposed for lumping the stochastic semantics [16,15]. Symmetries can
also be used to reduce the combinatorial complexity of models [6].

These methods are context-insensitive: for each kind of agents, all the information about the agents of this kind is summarised into the single node of a
graph, called the contact map. The framework in [17] is fully context-sensitive:
the abstraction of the information flow is done in the concrete, thanks to a
direct iteration on the molecular species. In comparison, the framework that
is proposed in this paper allows the user to select any trade-off between fully
context-insensitive and fully context-sensitive abstractions.

Context-sensitive approximations of graph-structures have been deeply studied in the field of memory analysis, where complex invariants [18] about recursive
data-structures have to be inferred [19,2]. Our framework is a kind of partitioning
[4], a generic method for refining abstractions.

2   Case Study

Before describing the framework formally, we introduce a case study. We consider one kind of protein P with three numbered phosphorylation sites. Each
site can be phosphorylated, or not. We consider that any site can get phosphorylated or lose its phosphorylation. Configurations and reactions are summarised
in Fig. 1(a). The configuration of a protein P is denoted as a triple (s_1, s_2, s_3)
of symbols among u and p. In general, the rate of phosphorylation (resp. de-
phosphorylation) can depend on the state of the other sites. Here we make the
assumption that only the phosphorylation rate of the third site depends on the
state of the two other sites, but we assume that the phosphorylation rate of
the third site is the same in the configurations (u,u,u) and (u,p,u).

So as to model the behaviour of our system, we assume 1) that the system satisfies the well stirred assumption of mass action law, and 2) that the
population of proteins is large. Under these assumptions, the behaviour of the
system can be formalised by the means of the following system of differential

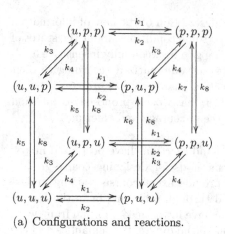

(b) Context-insensitive abstraction of the flow of information.

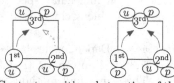

(a) Configurations and reactions.

(c) Context-sensitive abstraction of the flow of information. The dotted arrow represents a dependency that can be dropped if the rate k_1 is zero.

Fig. 1. Case study

equations, which describes the derivatives of the concentrations of each configuration of P as an expression of the concentrations of the configurations of P:

$$[(u,u,u)]' = k_2[(p,u,u)] + k_4[(u,p,u)] + k_8[(u,u,p)] - (k_1+k_3+k_5)[(u,u,u)]$$
$$[(u,u,p)]' = k_2[(p,u,p)] + k_4[(u,p,p)] + k_5[(u,u,u)] - (k_1+k_3+k_8)[(u,u,p)]$$
$$[(u,p,p)]' = k_2[(p,p,p)] + k_3[(u,u,p)] + k_5[(u,p,u)] - (k_1+k_4+k_8)[(u,p,p)]$$
$$[(u,p,u)]' = k_2[(p,p,u)] + k_3[(u,u,u)] + k_8[(u,p,p)] - (k_1+k_4+k_5)[(u,p,u)]$$
$$[(p,p,u)]' = k_1[(u,p,u)] + k_3[(p,u,u)] + k_8[(p,p,p)] - (k_2+k_4+k_7)[(p,p,u)]$$
$$[(p,p,p)]' = k_1[(u,p,p)] + k_3[(p,u,p)] + k_7[(p,p,u)] - (k_2+k_4+k_8)[(p,p,p)]$$
$$[(p,u,p)]' = k_1[(u,u,p)] + k_4[(p,p,p)] + k_6[(p,u,u)] - (k_2+k_3+k_8)[(p,u,p)]$$
$$[(p,u,u)]' = k_1[(u,u,u)] + k_4[(p,p,u)] + k_8[(p,u,p)] - (k_2+k_3+k_6)[(p,u,u)].$$

Providing an initial state mapping each configuration to their initial concentration, this system has a unique smooth solution over the time interval $\mathbb{R}^+$.

Now, we wonder whether or not our model can be coarse-grained: we are looking for a set macro-variables which are defined as a linear combination of the variables of the initial systems (so called micro-variables) that are self-consistent. In previous works [14,12,5], we have introduced frameworks for detecting self-consistent coarse-graining thanks to an over-approximation of the flow of information between the states of the sites of proteins. Indeed the flow of information can be summarised by annotating a contact map (which describes the different kinds of proteins, their sites, their potential phosphorylation states and their potential binding) with an oriented relation over the sites, which summarises how each site may influence the other ones: an arrow from a site s_1 to a site s_2 means that the capability of modifying the state of the site s_2 may change according to the state of the site s_1. The annotated contact map for our case study is given in Fig. 1(b). This is a context-insensitive approximation since all the information about the sites of P is summarised in a single node, regardless the states of its sites. The arrow from the 1st (resp. 2nd) site to the 3rd one comes from the fact that the phosphorilation rate of the 3rd site may depend on the state of the 1st (resp. 2nd). As a result, no coarse-graining can be found in this way.

Indeed, without further assumptions, the model cannot be coarse-grained by any means. But interestingly, if we set the rate k_1 equal to 0, we can abstract away the relation between the state of the 2nd site and the 3rd site in the case when the 1st site is activated, as shown by the following equations:

$$[(?,u,?)]' = k_4[(?,p,?)] - k_3[(?,u,?)]$$
$$[(?,p,?)]' = k_3[(?,u,?)] - k_4[(?,p,?)]$$
$$[(u,?,p)]' = k_2([(p,u,p)] + [(p,p,p)]) + k_5[(u,?,u)] - k_8[(u,?,p)]$$
$$[(u,?,u)]' = k_2([(p,u,u)] + [(p,p,u)]) + k_8[(u,?,p)] - k_5[(u,?,u)]$$
$$[(p,p,u)]' = k_3[(p,u,u)] + k_8[(p,p,p)] - (k_2 + k_4 + k_7)[(p,p,u)]$$
$$[(p,p,p)]' = k_3[(p,u,p)] + k_7[(p,p,u)] - (k_2 + k_4 + k_8)[(p,p,p)]$$
$$[(p,u,p)]' = k_4[(p,p,p)] + k_6[(p,u,u)] - (k_2 + k_3 + k_8)[(p,u,p)]$$
$$[(p,u,u)]' = k_4[(p,p,u)] + k_8[(p,u,p)] - (k_2 + k_3 + k_6)[(p,u,u)],$$

where the macro-variables are intentionally defined as fragments of configurations (question marks denote sites which have been cut away), and extentionally as linear combinations of the configurations which contain these fragments:

$$[(?,u,?)] = [(u,u,u)] + [(u,u,p)] + [(p,u,u)] + [(p,u,p)]$$
$$[(?,p,?)] = [(u,p,u)] + [(u,p,p)] + [(p,p,u)] + [(p,p,p)]$$
$$[(u,?,u)] = [(u,u,u)] + [(u,p,u)]$$
$$[(u,?,p)] = [(u,u,p)] + [(u,p,p)].$$

This coarse-graining can be discovered by tuning the context-sensitivity of the information flow analysis. Indeed, the behaviour of the protein P can be partitioned into two distinct modes. Whenever the 1st site is phosphorylated, the evolution of the state of the 3rd site is controlled by the state of both the 1st and the 2nd sites. But whenever the 1st site is not phosphorylated, the evolution of the state of the 3rd site is not controlled by the state of the 2nd site anymore. This accurate approximation of the flow of information is out of the reach of context-insensitive analysis. Thus we propose to use arbitrary Σ-graphs where different annotations can be written according to the state of well chosen sites, unlike the contact map. An example Σ-graph is given in Fig. 1(c). We notice that two nodes are used to describe the protein P, according to whether or not the 1st site is phosphorylated. The notion of Σ-graph will be formally defined in Sect. 3. Then, we can annotate our Σ-graph with context-sensitive information about the flow of information and obtain the plain arrows in Fig. 1(c). Interestingly, in the left connected component, there is no flow of information from any site into the 2nd site and no flow of information from the 2nd site into the 3rd site. As a consequence, the fragments of proteins that contain the 1st and the 3rd site and the ones that only contain the 2nd site are good candidates as macro-variables. Yet, since in the right connected component there is a potential flow of information from the 1st and the 2nd sites into the 3rd site, any micro-variable where the 1st site is phosphorylated has to be preserved. Thus, we find again the set of macro-variables $\{[(?,u,?)], [(?,p,?)], [(u,?,u)], [(u,?,p)], [(p,p,u)], [(p,p,p)], [(p,u,p)], [(p,u,u)]\}$, which is self-consistent, as we have shown previously.

Then we may wonder why this coarse-graining is not self-consistent when the phosphorylation reaction of the 1st site is not knocked out. This is because configurations of the form $(u,?,?)$ can now be transformed into configurations of the form $(p,?,?)$. Then, so as to express the concentration of the configurations of the form $(p,?,?)$ which are produced this way, we need to express the configurations of the form $(u, ?, ?)$ with at least the same fine-grained level of description. This is captured by the right-gluing construction in [17]. In the present framework, it is necessary to duplicate the arrow between the 2nd site and the 3rd one, from the right connected component into the left one. The resulting arrow, drawn in dotted in Fig. 1(c), prevents any coarse-graining.

The rest of the paper is organised as follows. In Sec. 3, we recall the notion of Σ-graphs and use it to abstract relations between sites in chemical complexes. In Sec. 4, we give an abstract syntax and a formal differential semantics for Kappa. In Sec. 5, we define our generic flow analysis and its induced model reduction.

3 Σ-Graphs

Σ-Graphs with a given signature, Σ-graphs, play a central role in the semantics of Kappa. In this section, we recall the definition of Σ-graphs [10] and show how to annotate them with a relation over their sites.

Definition 1. *A signature is a tuple $\Sigma = (\Sigma_{ag}, \Sigma_{st}, \Sigma_{int}, \Sigma^{int}_{ag\text{-}st}, \Sigma^{lnk}_{ag\text{-}st})$ where Σ_{ag} is a finite set of agent types, Σ_{st} is a finite set of site identifiers, Σ_{int} is a finite set of internal state identifiers, $\Sigma^{lnk}_{ag\text{-}st} : \Sigma_{ag} \to \wp(\Sigma_{st})$ and $\Sigma^{int}_{ag\text{-}st} : \Sigma_{ag} \to \wp(\Sigma_{st})$ are site maps.*

Agent types in Σ_{ag} denote agents of interest, as kinds of proteins for instance. A site identifier in Σ_{st} represents an identified locus for capability of interactions. Internal state identifiers in Σ_{int} are special attributes which encode potential state configurations. Each agent type A is associated with a set of sites which can bear an internal state $\Sigma^{int}_{ag\text{-}st}(A)$ and a set of sites which can be linked $\Sigma^{lnk}_{ag\text{-}st}(A)$. We assume without reducing the expressive power of the framework that $\Sigma^{lnk}_{ag\text{-}st}(A) \cap \Sigma^{int}_{ag\text{-}st}(A) = \emptyset$, for any $A \in \Sigma_{ag}$ and we write $\Sigma_{ag\text{-}st}(A)$ for the set $\Sigma^{lnk}_{ag\text{-}st}(A) \uplus \Sigma^{int}_{ag\text{-}st}(A)$. In our case study, $\Sigma_{ag} = \{P\}$, $\Sigma_{st} = \{1st, 2nd, 3rd\}$, $\Sigma_{int} = \{u, p\}$, $\Sigma^{int}_{ag\text{-}st}(P) = \Sigma_{st}$, and $\Sigma^{lnk}_{ag\text{-}st}(P) = \emptyset$.

Σ-graphs describe both patterns and chemical species. Their nodes are typed agents with some sites which can bear internal states and linking states. We introduce the set Ext as $\{\dashv, -\} \cup \{(A, s) \mid A \in \Sigma_{ag}, s \in \Sigma^{lnk}_{ag\text{-}st}(A)\}$ for describing some linking states.

Definition 2. *A Σ-graph is a tuple $G = (\mathcal{A}, type, \mathcal{S}, \mathcal{L}, p\kappa)$ where $\mathcal{A}$ is a set of agents, type $: \mathcal{A} \to \Sigma_{ag}$ is a function mapping each agent to its type, $\mathcal{S}$ is a set of sites such that $\mathcal{S} \subseteq \{(n, i) \mid n \in \mathcal{A}, i \in \Sigma_{ag\text{-}st}(type(n))\}$, $\mathcal{L}$ is a symmetric relation such that $\mathcal{L} \subseteq (\{(n, i) \in \mathcal{S} \mid i \in \Sigma^{lnk}_{ag\text{-}st}(type(n))\} \cup Ext)^2 \setminus Ext^2$, and $p\kappa$ maps each site $(n, i) \in \mathcal{S}$ such that $i \in \Sigma^{int}_{ag\text{-}st}(type(n))$ to a set of internal states $p\kappa(n, i) \in \wp(\Sigma_{int})$.*

A site $(n, i) \in \mathcal{S}$ such that $i \in \Sigma^{int}_{ag\text{-}st}(type(n))$ is called a property site, whereas a site $(n, i) \in \mathcal{S}$ such that $i \in \Sigma^{lnk}_{ag\text{-}st}(type(n))$ is called a binding site. Whenever $((n, i), \dashv) \in \mathcal{L}$, the binding site (n, i) may be free. Various levels of information can be given about the sites that can be bound. Whenever $((n, i), -) \in \mathcal{L}$, then the binding site (n, i) may be bound to any other site. Whenever $((n, i), (A', i')) \in \mathcal{L}$ for a given agent type $A' \in \Sigma_{ag}$ and a given site identifier $i' \in \Sigma^{lnk}_{ag\text{-}st}(A')$, then the binding site (n, i) can be bound to the site i' of an agent of type A'. Whenever $((n, i), s) \in \mathcal{L}$ with $s \in \mathcal{S}$ then the binding site (n, i) may be bound to the binding site s. We introduce a sub-typing relation $\leq_G$ over binding states, that is defined as the least reflexive relation such that $- \leq_G (type(n), i) \leq_G (n, i)$, for any $n \in \mathcal{A}$ and $i \in \Sigma^{lnk}_{ag\text{-}st}(type(n))$.

For a Σ-graph G, we write as $\mathcal{A}_G$ its set of agents, $type_G$ its typing function, $\mathcal{S}_G$ its set of sites, $\mathcal{L}_G$ its set of links, and $p\kappa_G$ for the internal states map.

Two Σ-graphs can be related by structure-preserving functions, which are called homomorphisms.

Definition 3. *A homomorphism $h : G \to H$ between two Σ-graphs G and H is a function of agents $h : \mathcal{A}_G \to \mathcal{A}_H$ satisfying:*
1. *$type_G(n) = type_H(h(n))$ for all $n \in \mathcal{A}_G$;*
2. *$(h(n), i) \in \mathcal{S}_H$ for all $(n, i) \in \mathcal{S}_G$;*
3. *$p\kappa_G(n, i) \subseteq p\kappa_H(h(n), i)$ for all $(n, i) \in \mathcal{S}_G$ such that $i \in \Sigma^{int}_{ag\text{-}st}(type_G(n))$;*
4. *$((h(n), i), (h(n'), i') \in \mathcal{L}_H$ for all $((n, i), (n', i')) \in \mathcal{L}_G \cap \mathcal{S}^2_G$;*
5. *there exists $y \in \mathcal{S}_H \cup \textbf{Ext}$ such that $((h(n), i), y) \in \mathcal{L}_H$ and $x \leq_H y$ for all $((n, i), x) \in \mathcal{L}_G$ such that $x \in \textbf{Ext}$.*

An injective embedding is called an embedding. The number of embeddings between two Σ-graphs G and H is denoted as $[G, H]$. Whenever $G = H$ and h is a bijection, then h is called an automorphism. We notice that the identity function is always an automorphism. Homomorphisms $f : G \to H$ and $g : H \to K$ compose in the usual way. Moreover, whenever two homomorphisms $f : G \to H$ and $g : H \to G$ are such that $g \circ f$ is the identity homomorphism over G and $f \circ g$ is the identity homomorphism over H, then f and g are called isomorphisms and G and H are said to be isomorphic which is written $G \approx H$. All the constructions in this paper are defined up to isomorphisms.

Now we want to annotate a Σ-graph with a binary relation over its sites, so as to abstract the flow of information among its sites. Two sites can be in relation when they belong to the same agent or when they are linked together.

Definition 4. *An annotated Σ-graph $G^a = (G, \leadsto_{G^a})$ is a pair where G is a Σ-graph and $\leadsto_{G^a} \subseteq \{(n, i), (n, i') \mid n \in \mathcal{A}_G, (n, i), (n, i') \in \mathcal{S}_G\} \uplus (\mathcal{L}_G \cap \mathcal{S}^2_G)$.*

Ordered pairs of sites in $\{(n, i), (n, i') \mid n \in \mathcal{A}_G, (n, i), (n, i') \in \mathcal{S}_G\}$ are called internal edges and are denoted as $(n, i) \overset{\vee}{\leadsto}_{G^a} (n, i')$, whereas ordered pairs in $\mathcal{L}_G \cap \mathcal{S}^2_G$ are called external edges and are denoted as $(n, i) \overset{\wedge}{\leadsto}_{G^a} (n', i')$. An ordered pair of sites can be connected by both an internal edge and an external edge. We omit the symbols $\vee$ and $\wedge$ when they are not important.

Given an annotated Σ-graph G^a, we write as G the Σ-graph and $\leadsto_{G^a}$ the binary relation over its sites.

The set of annotations of a Σ-graph G forms a Boolean lattice isomorphic to the set of the parts of $\{((n, i), (n, i')) \mid n \in \mathcal{A}_G, (n, i), (n, i') \in \mathcal{S}_G\} \uplus (\mathcal{L}_G \cap \mathcal{S}^2_G)$. The least element is the empty annotation and is denoted as $G^\emptyset = (G, \emptyset)$, and the top element relates each pair of sites such that they either belong to the same agent or are linked together and is denoted as $G^\top = (G, \leadsto_{G^\top})$.

4 Differential Semantics

Mixtures, representing the states to which rules are applied, and site graphs, representing patterns, are Σ-graphs. In particular, site graphs are finite, they have no links that immediately loop back to the same site and have at most one link from any site, moreover their sites can bear at most one internal state. Mixtures additionally specify the link state and the internal state of all sites and have no external links.

Definition 5. *A site graph G is a Σ-graph such that: 1) the set $\mathcal{A}_G$ is finite; 2) its link relation $\mathcal{L}_G$ is irreflexive; 3) for any binding site $(n,i) \in \mathcal{S}_G$, $((n,i),x) \in \mathcal{L}_G$ and $((n,i),y) \in \mathcal{L}_G$ implies $x = y$; and 4) for any state site $(n,i) \in \mathcal{S}_G$, $p\kappa(n,i)$ contains at most one element.*

In a site graph G, the states of some sites in $\mathcal{S}_G$ are specified, while others are not. The state of a binding site $(n,i) \in \mathcal{S}_G$ is specified if there exists $x \in \mathcal{S}_G \cup \text{Ext}$ such that $((n,i),x) \in \mathcal{L}_G$, whereas the state of a state site $(n,i) \in \mathcal{S}_G$ is specified if $p\kappa_G(n,i)$ is a singleton.

Definition 6. *A Σ-graph G is said to be fully specified if the three following properties hold: 1) $\mathcal{S} = \{(n,s) \mid n \in \mathcal{A}, s \in \Sigma_{ag\text{-}st}(type(n))\}$; 2) $\mathcal{L} \subseteq (\mathcal{S} \cup \{\dashv\})^2$; and 3) the state of each site in $\mathcal{S}_G$ is specified.*

Definition 7. *A site graph G is said to be connected if for any pair of distinct agents $n_1, n_2 \in \mathcal{A}$, there exists two sites $i_1, i_2 \in \Sigma_{st}$ such that $(n_1,i_1), (n_2,i_2) \in \mathcal{S}$ and $(n_1,i_1) \leadsto^*_{G^\top} (n_2,i_2)$.*

A site graph can be decomposed into a set of connected graphs, called its connected components (*ccs*). A mixture is a fully specified site graph. The variables of the differential semantics are the concentrations of the chemical species, which are defined as isomorphism classes of connected mixtures. Thus we introduce $\mathcal{C}$ as a set of connected mixtures such that for any connected mixture c there exists a unique connected mixture $c' \in \mathcal{C}$ such that $c \approx c'$. We assume that $\mathcal{C}$ is finite.

Transformations between site graphs are described by rules. A rule is a transformation between two site graphs, a left hand side (*lhs*) L and a right hand side (*rhs*) R. In a rule, some agents and some sites are preserved. This is specified by a site graph D which is embedded both into L and into R and which describes anything that is preserved. Not all transformations are allowed: one can remove and add agents, create links between free sites, free pairs of sites that are connected and change the internal state of some sites. The agents that are created have to fully define the state of their sites. We also make extra assumptions to simplify the definition of the approximation of the flow of information: we assume that only the bonds that are shared between two sites can be removed, and that the agents that are removed have to fully define the state of their sites has well. The framework can be easily tuned to relax these two assumptions. Our requirements are formalised in the following definition:

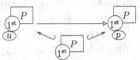 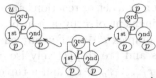

(a) Phosphorylation of the first site (the rate depends on the states neither of the 2nd site, nor the 3rd site).

(b) Phosphorylation of the 3rd site when the two other sites are phosphorylated.

Fig. 2. Examples of rules

Definition 8. *A rule is a span such that:* $L \xleftarrow{f} D \xhookrightarrow{g} R$ *such that :*

1. *for any span* $L \xleftarrow{f'} D' \xhookrightarrow{g'} R$ *and any embedding* $D \xhookrightarrow{h} D'$ *such that* $f = f'h$ *and* $g = g'h$, *then* h *is an isomorphism;*
2. *for any* $x \in \mathbf{Ext} \setminus \{\dashv\}$ *and any site* $(n, i) \in \mathcal{S}_L$, *if* $((n, i), x) \in \mathcal{L}_L$ *then there exists* $m \in \mathcal{A}_D$ *such that* $n = f(m)$, $(m, i) \in \mathcal{S}_D$, *and* $((m, i), x) \in \mathcal{L}_D$;
3. *for any* $x \in \mathbf{Ext} \setminus \{\dashv\}$ *and any site* $(n, i) \in \mathcal{S}_R$, *if* $((n, i), x) \in \mathcal{L}_R$ *then there exists* $m \in \mathcal{A}_D$ *such that* $n = g(m)$, $(m, i) \in \mathcal{S}_D$, *and* $((m, i), x) \in \mathcal{L}_D$;
4. *if* $m \in \mathcal{A}_D$, *then for any* $i \in \Sigma_{ag\text{-}st}(type_D(m))$, $(m, i) \in \mathcal{S}_D$ *iff* $(f(m), i) \in \mathcal{S}_L$ *iff* $((g(m), i)) \in \mathcal{S}_R$ *and, in such a case, the state of the site* $(f(m), i)$ *is specified in the site graph* L *iff the state of the site* $(g(m), i)$ *is specified in the site graph* R;
5. *if* $m \in \mathcal{A}_L$ *and* $m \notin image(f)$, *then, for any* $i \in \Sigma_{ag\text{-}st}(type_L(m))$, $(f(m), i) \in \mathcal{S}_L$ *and the state of the site* $(f(m), i)$ *is specified in the site graph* L;
6. *if* $m \in \mathcal{A}_R$ *and* $m \notin image(g)$, *then, for any* $i \in \Sigma_{ag\text{-}st}(type_R(m))$, $(g(m), i) \in \mathcal{S}_R$ *and the state of the site* $(g(m), i)$ *is specified in the site graph* R.

The first property ensures that D is a local greatest upper bound.

A rule $L \xleftarrow{} D \xhookrightarrow{} R$ is usually denoted as $L \to R$ (leaving the two embeddings and the common region implicit).

Rules can be more or less refined [11] by adding more or less information about the context in which they can be applied. A rule $L' \xleftarrow{} D' \xhookrightarrow{} R'$ is said to be a refinement of the rule $L \xleftarrow{} D \xhookrightarrow{} R$ is and only if there exist three embeddings h_L, h_D, h_R which make the diagram on the right commute. In such a case, the two action maps and the embeddings h_L and h_R form a pushout (e.g. see [10]). Moreover, whenever L' is a mixture, then R' is a mixture as well. Given L' (resp. R') a site graph and an embedding f between L and L' (resp. between R and R') there exists a unique (up to isomorphisms) refinement that is defined by a triple of embeddings $(h_X)_{X \in \{L,D,R\}}$ such that $h_L = f$ (resp. $h_R = f$), this refinement is called the left-refinement (resp. right-refinement) of the rule r by the embedding f. The unicity of the right-refinement does not hold in full Kappa, but follows from our simplifying assumptions. Yet, in general there exists a unique least refined refinement such that $h_L = f$.

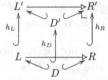

Each rule comes with a kinetic rate k which is denoted as $r : L \to R @k$. We define the corrected rate γ as $k/[L, L]$, where $[L, L]$ is the number of automorphisms (i.e. symmetries) of L. The rule r can be seen as a symbolic representation of a set of reactions among chemical complexes, that is obtained as a left refinement of r by a join of embeddings mapping each cc of L into a chemical complex $c \in \mathcal{C}$. For any such refinement, both the lhs and the rhs are mixtures and are respectively isomorphic to the disjoint union of a tuple of reactants $r_1, \ldots, r_m \in \mathcal{C}^*$ and a tuple of products $p_1, \ldots, p_n \in \mathcal{C}^*$. Each such refinement is associated with the following contribution to the system of differential equations: $x_{r_s}(t)' \stackrel{\pm}{=} -\gamma \cdot \prod_{1 \leq j \leq m} x_{r_j}(t)$ for any integer s such that $1 \leq s \leq m$ and $x_{p_t}(t)' \stackrel{\pm}{=} \gamma \cdot \prod_{1 \leq j \leq m} x_{r_j}(t)$ for any integer t such that $1 \leq t \leq n$.

The differential semantics associated to a set of rules maps each initial state $init \in (\mathbb{R}^+)^{\mathcal{C}}$ to the unique solution $x \in ([0, T) \to \mathbb{R})^{\mathcal{C}}$ of the so obtained system of equations such that $x_c(0) = init_c$ for any $c \in \mathcal{C}$ and T is maximal. By construction, this solution is positive [12].

5   Context-Sensitive Model Reduction

The annotation of a Σ-graph can be viewed as a symbolic representation of a set of patterns, called prefragments. More precisely, given an annotated Σ-graph G^a, a site graph P is a prefragment if we get a directed relation over its sites when we annotate it by the meet of the inverse image of the annotation of G^a by any homomorphism between P and G^a. This is formalised as follows:

Definition 9. *Given an annotated Σ-graph G^a and a Σ-graph H, we define the canonical annotation $\leadsto_{H, G^a}$ of H by the annotated Σ-graph G^a as follows: for any $(a, i), (a', i') \in S_H$ and any $w \in \{\vee, \wedge\}$, $(a, i) \stackrel{w}{\leadsto}_{H, G^a} (a', i')$ if and only if for all homomorphisms $\phi : G \to H$, $(\phi(a), i) \stackrel{w}{\leadsto}_{G^a} (\phi(a'), i')$.*

Definition 10. *Given an annotated Σ-graph G^a, we say that a site graph P is a prefragment for G^a if and only if the set of sites S_P and the transitive and reflexive closure of the relation $\leadsto_{P, G^a}$ form a directed set (i.e. for any $s_1, s_2 \in S_P$, there exists $s \in S_P$ such that $s_1 \leadsto^*_{P, G^a} s$ and $s_2 \leadsto^*_{P, G^a} s$).*

We notice that prefragments are connected and that, since the set S_P is finite, a site graph P is a prefragment if and only if there exists $s^\bullet \in S_P$ such that for any site $s \in S_P$, $s \leadsto^*_{P, G^a} s^\bullet$. In such a case, we call the site $s^\bullet$ (which may not be unique) a root of the prefragment.

A connected site graph P can also be seen extensionally as set of embeddings between itself and any reachable species in $\mathcal{C}$.

Definition 11. *For any connected site graph P, we define y_p as $\sum_{v \in \mathcal{C}, \phi: P \to v} x_v$.*

Thus the set of prefragments define a linear change of variables.

Now we wonder how to annotate, given a set of rules, a Σ-graph such that this change of variables is self-consistent. For this purpose, we will use a special kind of Σ-graphs, the summary graphs. Roughly speaking, summary graphs are used to abstract information about the potential overlaps between the left and right hand sides of rules and connected site graphs such that the common region contains sites that are modified by the rules, so as to express the proper consumption and the proper production of these connected site graphs.

We now formalise the notions of summary graphs:

Definition 12. *A Σ-graph G is summary graph if the three following properties hold: 1) $\mathcal{L} \subseteq (\mathcal{S} \cup \{\dashv\})^2$; 2) for any chemical complex $V \in \mathcal{C}$, there exists a homomorphism $h : V \to G$; 3) for any homomorphism $h : P \to G$ between a connected site graph P and G, there exists a chemical complex $V \in \mathcal{C}$, an embedding $\phi : P \hookrightarrow V$ and a homomorphism $h' : V \to G$ such that $h = h'\phi$.*

The disjoint union of all chemical complexes is a summary graph (the most concrete one), and the contact map as well (the most abstract one).

An overlap between two site graphs is defined by a common region which identifies some nodes in the two site graphs and a merged site graph which ensures that the two site graphs are compatible. The common region can be chosen as a local lower-bound and the merged site graph as a local upper-bound, which ensures that each overlap is defined non-ambiguously:

Definition 13. *An overlap between two site graphs P and C is defined as a pair of a span $P \hookleftarrow X \hookrightarrow C$ and a cospan $P \hookrightarrow Y \hookleftarrow C$ of embeddings where X and Y are non-empty site graphs which make the square commute and such that for any other such pair of a span $P \hookleftarrow X' \hookrightarrow C$ and a cospan $P \hookrightarrow Y' \hookleftarrow C$ where X', Y' are site graphs, there exists a unique pair of embeddings $X' \hookrightarrow X$ and $Y \hookrightarrow Y'$ which makes the diagram on the right commute.*

Flow of information abstracts the relation between the sites that are tested and the sites that are modified. A site is tested in a rule if it occurs in the lhs of this rule. A site $(n, i) \in \mathcal{S}_L$ is modified in the lhs L of a rule $L \xleftarrow{f} D \xhookrightarrow{g} R$ iff either $n \notin image(f)$, or (m, i) is not specified in D (where m is the unique agent $m \in \mathcal{A}_D$ such that $f(m) = n$). We define the same way the sites that are modified in the rhs of a rule. For the sake of simplicity, we assume that each cc in the lhs of a rule has a site that is modified by the rule.

We call a path in a site graph P a sequence $(n_0, i_0) \overset{w_1}{\rightsquigarrow}_{P\top} \ldots \overset{w_k}{\rightsquigarrow}_{P\top} (n_k, i_k)$ of steps in $\mathcal{S}_P$. The path is said alternating if, moreover, for any integer j between 1 and $k - 1$, $w_j = \vee$ if and only if $w_{j+1} = \wedge$.

Some rules induce no direct flow of information. We say that a rule is trivial if it releases a bond between two sites in distinct agents without testing anything (except that the two sites are bound together).

We are now ready to give the constraints for the annotation of a summary graph so as to ensure that its induced change of variables is self-consistent.

Definition 14. *Suppose given a rule $r : L \xleftarrow{f} D \xhookrightarrow{g} R$, an annotated summary graph G^{a}. We say that G^{a} is compatible with the rule r if and only if the three following sets of constraints are satisfied:*
*1. **direct flow.** if r is a non-trivial rule, for any homomorphism $\psi : L \to G$, any alternating path $p = (a_0, i_0) \overset{w_1}{\rightsquigarrow}_{L\top} \ldots \overset{w_k}{\rightsquigarrow}_{L\top} (a_k, i_k)$ in the lhs L of the rule r*

such that the site (a_k, i_k) *is modified by the rule* r, *and any integer* j *such that* $0 \leq j < k$, $(\psi(a_j), i_j) \overset{w_j}{\leadsto}_{G^a} (\psi(a_{j+1}), i_{j+1})$;

2. **backward compatiblity.** *whatever* r *is trivial or not, for any site graph* P, *for any overlap* $(S, \phi_1, \phi_2, \psi_1, \psi_2, X)$ *between the site graph* P *and the site graph* D, *for any ground refinement* $R_L \hookleftarrow R_D \hookrightarrow R_R$ *of* r *defined by a triple of embeddings of the form* $(h_l, h\psi_2, h_r)$, *for any a homomorphism* ϕ *between* R_R *and* G, *for any homomorphism* ψ *between* R_L *and* G, *and for any site* $(n, i) \in \mathcal{S}_S$ *such that the state of the site* $(\phi_2(n), i)$ *is not specified in* D *(ie. the site* $(f(\phi_2(n)), i)$ *(resp.* $(g(\phi_2(n)), i))$ *is modified in* L *(resp.* R), *for any sequence* $w_1, \ldots w_j \in \{\vee, \wedge\}^k$ *and any two integers* j_0, j_1 *such that:*

- $0 \leq j_0 \leq j_1 < k$,
- $(\psi_1(n_{j_0}), i_{j_0}) = (\psi_2\phi_2(n), i)$,
- $([\phi g' h\psi_1](n_j), i_j) \overset{w_{j+1}}{\leadsto}_{G^a} ([\phi g' h\psi_1](n_{j+1}), i_{j+1})$ *for any* j *such that* $0 \leq j < j_1$,
- $([\phi g' h\psi_1](n_{j+1}), i_{j+1}) \overset{w_{j+1}}{\leadsto}_{G^a} ([\phi g' h\psi_1](n_j), i_j)$ *for any* j *such that* $j_1 \leq j \leq k$;

we have that:

- $([\psi f' h\psi_1](n_j), i_j) \overset{w_{j+1}}{\leadsto}_{G^a} ([\psi f' h\psi_1](n_{j+1}), i_{j+1})$ *for any* j *such that* $0 \leq j < j_1$,
- $([\psi f' h\psi_1](n_{j+1}), i_{j+1}) \overset{w_{j+1}}{\leadsto}_{G^a} ([\psi f' h\psi_1](n_j), i_j)$ *for any* j *such that* $j_1 \leq j \leq k$.

3. **cycle.** *Let* $A, B \in \mathcal{A}$ *be two agent types and* $i_A \in \Sigma_{ag\text{-}st}^{lnk}(A)$ *and* $i_B \in \Sigma_{ag\text{-}st}^{lnk}(B)$ *be two site identifiers. For any trivial rule* r *that removes bounds between the sites* i_A *in agents of type* A *and the sites* i_B *in agents of type* B, *if there exists a site* $s \in \mathcal{S}_G$ *and two agents* $n_A, n_B \in \mathcal{A}_G$ *such that :* $type(n_A) = A$, $type(n_B) = B$ *and two distinct paths* $p = (n_A, i_A) \leadsto_{G^a}^* s$ *and* $p' = (n_B, i_B) \leadsto_{G^a}^* s$, *then the rule* r *is considered to be not trivial, and the* direct flow constraints *(see Def. 14(1)) must be applied also with it.*

The set of the annotations of a summary graph that are compatible with a set of rules forms a Moore family. Thus, it has a least element. Seeing each constraint instantiation as an upper closure operator, this least element is also the image of $G_\emptyset$ by the least upper bound in the lattice of the upper closure operators of this set of closure operators [20] and can be computed, whenever the summary graph G is finite, using asynchronous iterations [20,9].

Now we assume that the annotation of the summary graph G is the least solution of the constraints in Def. 14 and that prefragments are defined as in Def. 10. Let us explain the constraints in Def. 14. Direct flow is obtained by taking any homomorphism between the lhs of a rule and the summary graph G and annotating the image of any alternating path between a site that is tested and a site that is modified. For instance, in the Σ-graph that is given in Fig. 1(c) the annotation of the right connected component describes the direct flow that is due to the rule that is given in Fig. 2(b). The existence of the homomorphism ensures the context-sensitivity of the analysis, since only the parts of G that match the lhs are annotated. As a consequence, we report no direct flow for the rule in Fig. 2(b) in the left connected component. The direct flow constraints (see Def. 14(1)) ensure the following property:

Property 1. For any overlap $(X, \psi_1, \psi_2, \phi_1, \phi_2, Y)$ between a cc in the lhs of a non trivial rule and a prefragment such that there exists a site of the form $(\psi_1(n), i)$ that is modified in the rule, the site graph Y is a prefragment as well.

In particular, since each cc in the lhs of a rule contains a site that is modified, each cc in the lhs of a rule is a prefragment.

Backward compatibility ensures that prefragments that overlap with the lhs of rules are always more refined than the ones that overlap with the rhs. Since a prefragment contains at least one root r, any site s of the prefragment is reachable through the annotation by starting from the site that is modified, following a path forward to the root r and following a path backward to the site s. Thus we copy these paths at any place in G which matches with a potential antecedent of the pattern by the rule, which is ensured by the existence of the ground refinement $R_L \hookleftarrow R_D \hookrightarrow R_R$. For instance, the annotation of the right connected component in Fig. 1(c) has to be reported into the left connected component, due to the rule that is given in Fig. 2(a) and the ground refinement when the last two sites are unphosphorylated. Backward compatibility (see Def. 14(2)) ensures the following property:

Property 2. For any overlap $(X, \psi_1, \psi_2, \phi_1, \phi_2, Y)$ between the rhs of a non trivial rule and a prefragment such that there exists a site of the form $(\psi_1(n), i)$ that is modified in the rule, if the cc of the lhs of the right refinement of the rule by the embedding ϕ_1 is such that the number of ccs in the lhs is preserved, then each cc in the lhs of the refined rule is a prefragment.

Backward compatibility is subject to abstraction. Instead of a ground refinement, one may look for a k-depth context around the site graph X.

When a trivial rule that breaks a bond between two sites, it is crucial to express the concentration of prefragments in which the two sites are actually bound together. This is the purpose of constraints in Def. 14(3).

Property 3. Suppose that there exists a trivial rule which breaks a bond between the site i_A of agents of type A and the site i_B of agents of type B. Then for any prefragment $pf = (\mathcal{A}, type, \mathcal{S}, \mathcal{L}, p\kappa)$ that contains two agents n_A and n_B such that $(n_A, i_A) \in \mathcal{S}$, $(n_B, i_B) \in \mathcal{S}$, $((n_A, i_A), (B, i_B)) \in \mathcal{L}$, $((n_B, i_B), (A, i_A)) \in \mathcal{L}$, and $(n_A, i_A) \neq (n_B, i_B)$ and for any agent $n^\bullet$ such that either $n^\bullet \in \mathcal{A}$ and $type(n^\bullet) = B$, or $n^\bullet \notin \mathcal{A}$, the site graph $(\mathcal{A} \cup \{n^\bullet\}, type[n^\bullet \to B], \mathcal{S} \cup \{(n^\bullet, i_B)\}, (\mathcal{L} \cup \mathcal{L}_+) \setminus \mathcal{L}_-, p\kappa)$ with $\mathcal{L}_+ = \{((n_A, i_A), (n_\bullet, i_B)), ((n_\bullet, i_B), (n_A, i_A))\}$, $\mathcal{L}_- = \{((n_\bullet, i_B), (A, i_A)), ((n_A, i_A), (B, i_B)), ((A, i_A), (n_\bullet, i_B)), ((B, i_B), (n_A, i_A))\}$, is a prefragment as well.

We may notice that the *cycle* constraints can be relaxed while still ensuring Prop. 3. But these are technical details that we skip for the sake of simplicity.

Properties 1,2, and 3 are enough to describe the evolution of the concentration of the prefragments by system of differential equations. We consider a set $\hat{\mathcal{F}}$ that contains exactly one prefragment per $\approx$-equivalence class of prefragments.

Definition 15 (Consumption). *For any rule $L \rightarrow R$ @k, L decomposed into ccs $c_1, \ldots, c_n$, and any overlap $(X, \psi_1, \psi_2, \phi_1, \phi_2, Y)$ between a cc c_i and a prefragment pf $\in \hat{\mathcal{F}}$ such that $\psi_1(X)$ contains a site that is modified, then, the proper consumption term for pf due to this overlap can be expressed as follows: $y'_{pf} \stackrel{\pm}{=} -\gamma \cdot y_{pf} \cdot \prod_{1 \leq j \leq n, j \neq i} y_{c_j}$, where $\gamma = k/[L, L]$.*

Definition 16 (Production). *For any rule $r = L \rightarrow R$ @k and any overlap $(X, \psi_1, \psi_2, \phi_1, \phi_2, Y)$ between R of the rule r and a prefragment pf $\in \hat{\mathcal{F}}$ such that $\psi_1(X)$ contains a site that is modified, we consider $L' \rightarrow R'$ @k the right refinement of r by the embedding ϕ_1. Then, the contribution is 0 whenever L and L' have not the same number of ccs, and is given as: $y'_{pf} \stackrel{\pm}{=} \gamma \cdot \prod_{1 \leq j \leq n} y_{c'_j}$ otherwise, where L' is decomposed into ccs $c'_1, \ldots, c'_n$ and $\gamma = k/[L, L]$.*

We have skipped some technical details about the handling of trivial rules.

The following theorem formalises the relation between the initial and the reduced system of differential equations:

Theorem 1. *We consider $x \in ([0, T) \rightarrow \mathbb{R})^{\mathcal{C}}$ the solution of the initial differential system with a given initial state init $\in (\mathbb{R}^+)^{\mathcal{C}}$ and such that T is maximal and $y \in ([0, T') \rightarrow \mathbb{R})^{\hat{\mathcal{F}}}$ the solution of the reduced system with the initial state init$^\sharp$ that is defined as init$^\sharp_{\hat{f}} = \sum_{c \in \mathcal{C}} [\hat{f}, c] \cdot$ init$_{\hat{f}}$ for any $\hat{f} \in \hat{\mathcal{F}}$ and such that T' is maximal. Then, $T = T'$, and for any prefragment $\hat{f} \in \hat{\mathcal{F}}$, at any time $t \in [0, T)$, $y_{\hat{f}}(t) = \sum_{c \in \mathcal{C}} [\hat{f}, c] \cdot x_{\hat{f}}(t)$.*

Thm.1 follows from the proof that can be found in [12] and which only requires the Properties 1, 2 and 3 to hold.

The prefragments which can be refined into a set of prefragments can be eliminated of the system of equations. The others are called fragments.

6   Conclusion

We have introduced a parametric framework for coarse-graining the differential semantics of rule-based models. A summary graph is used to define which contexts are distinguished and allows us to tune the accuracy of our approximation of the flow of information between the sites of chemical complexes. The result of this analysis is used to detect useless correlations between the states of sites, which defines formally our coarse-graining.

As usual with partitioning techniques, the choice of the summary graph can be driven thanks to appropriate strategies. For instance, we can abstract the behaviour of each site by a transition system. Then, we can choose to zoom in the accuracy of the analysis by distinguishing contexts according to the states of the sites the transition system of which is not strongly connected.

Our framework is highly generic and we have focused on the formal foundations so far. In future works, we will address more practical issues: for instance we will define subsets of summary graphs, which make the computation of the set of coarse-grained variables easier.

References

1. Blinov, M.L., Faeder, J.R., Goldstein, B., Hlavacek, W.S.: Bionetgen: software for rule-based modeling of signal transduction based on the interactions of molecular domains. Bioinformatics 20(17), 3289–3291 (2004)
2. Bor-Yuh, E.C., Rival, X.: Relational inductive shape analysis. In: Necula, G.C., Wadler, P. (eds.) POPL, pp. 247–260. ACM (2008)
3. Borisov, N.M., Markevich, N.I., Kholodenko, B.N., Dieter Gilles, E.: Signaling through receptors and scaffolds: Independent interactions reduce combinatorial complexity. Biophysical Journal 89 (2005)
4. Bourdoncle, F.: Abstract interpretation by dynamic partitioning. J. Funct. Program. 2(4), 407–423 (1992)
5. Camporesi, F., Feret, J.: Formal reduction for rule-based models. In: Mislove, M., Ouaknine, J. (eds.) MFPS, Pittsburgh, USA. ENTCS, vol. 276, pp. 29–59. Elsevier (September 2011)
6. Camporesi, F., Feret, J., Koeppl, H., Petrov, T.: Combining Model Reductions. In: Mislove, M., Selinger, P. (eds.) MFPS, Ottawa, Canada. ENTCS, vol. 265, pp. 73–96. Elsevier (September 2010)
7. Conzelmann, H., Saez-Rodriguez, J., Sauter, T., Kholodenko, B.N., Gilles, E.D.: A domain-oriented approach to the reduction of combinatorial complexity in signal transduction networks. BMC Bioinformatics 7 (2006)
8. Cousot, P., Cousot, R.: Abstract interpretation: A unified lattice model for static analysis of programs by construction or approximation of fixpoints. In: Graham, R.M., Harrison, M.A., Sethi, R. (eds.) POPL, pp. 238–252. ACM (1977)
9. Cousot, P., Cousot, R.: Constructive versions of Tarski's fixed point theorems. Pacific Journal of Mathematics 81(1), 43–57 (1979)
10. Danos, V., Feret, J., Fontana, W., Harmer, R., Hayman, J., Krivine, J., Thompson-Walsh, C., Winskel, G.: Graphs, Rewriting and Pathway Reconstruction for Rule-Based Models. In: D'Souza, D., Radhakrishnan, J., Telikepalli, K. (eds.) FSTTCS, Hyderabad, India, vol. 18. IARCS, LIPIcs (2012)
11. Danos, V., Feret, J., Fontana, W., Harmer, R., Krivine, J.: Rule-based modelling, symmetries, refinements. In: Fisher, J. (ed.) FMSB 2008. LNCS (LNBI), vol. 5054, pp. 103–122. Springer, Heidelberg (2008)
12. Danos, V., Feret, J., Fontana, W., Harmer, R., Krivine, J.: Abstracting the differential semantics of rule-based models: exact and automated model reduction. In: LICS, Edinburgh, GB, pp. 362–381. IEEE Computer Society (2010)
13. Danos, V., Laneve, C.: Formal molecular biology. TCS 325(1), 69–110 (2004)
14. Feret, J., Danos, V., Krivine, J., Harmer, R., Fontana, W.: Internal coarse-graining of molecular systems. PNAS 106(16) (April 2009)
15. Feret, J., Henzinger, T., Koeppl, H., Petrov, T.: Lumpability Abstractions of Rule-based Systems. TCS 431, 137–164 (2012)
16. Feret, J., Koeppl, H., Petrov, T.: Stochastic fragments: A framework for the exact reduction of the stochastic semantics of rule-based models. IJSI (to appear)
17. Harmer, R., Danos, V., Feret, J., Krivine, J., Fontana, W.: Intrinsic Information carriers in combinatorial dynamical systems. Chaos 20(3), 037108 (2010)
18. Reynolds, J.C.: Separation logic: A logic for shared mutable data structures. In: LICS, pp. 55–74. IEEE Computer Society (2002)
19. Sagiv, S., Reps, T.W., Wilhelm, R.: Parametric shape analysis via 3-valued logic. In: Appel, A.W., Aiken, A. (eds.) POPL, pp. 105–118. ACM (1999)
20. Ward, M.: The closure operators of a lattice. Annals Math. 42, 191–196 (1942)

ARNI: Abductive Inference of Complex Regulatory Network Structures

Nataly Maimari[1,2], Calin-Rares Turliuc[2], Krysia Broda[2], Antonis Kakas[3], Rob Krams[1], and Alessandra Russo[2]

[1] Dept. of Bioengineering, Imperial College London, UK
[2] Dept. of Computing, Imperial College London, UK
[3] Dept. of Computer Science, University of Cyprus

Abstract. Physical network inference methods use a template of molecular interaction to infer biological networks from high throughput datasets. Current inference methods have limited applicability, relying on cause-effect pairs or systematically perturbed datasets and fail to capture complex network structures. Here we present a novel framework, ARNI, based on abductive inference, that addresses these limitations.

Keywords: abductive inference, logic-based modeling, gene networks.

1 Introduction

Amongst the approaches proposed to tackle the task of network reconstruction are methods based on physical network models. These approaches explain experimental observations on a template of protein-protein and transcription factor-DNA interactions. The links in the inferred networks represent molecular interactions and can capture biological mechanism of action. Despite the contributions of current approaches in inferring causal networks with hidden regulatory elements, so far there has not been an approach able to capture complex regulatory structures which govern fundamental properties of biological systems. We present a general framework, Abductive Regulatory Network Inference (ARNI), for regulatory network inference that addresses these limitations. Logical rules use prior knowledge from online databases, and a signal propagation model, expressed as constraints, to determine how affected genes are organized in causal network. ARNI extents the use of physical network inference methods to datasets where the source of perturbation is unknown (e.g in the case of environment factors). Using a network controlling T-cell differentiation, we illustrate that ARNI can effectively capture complex structures not detected by existing methods.

2 Methods

Abduction is commonly defined as the problem of finding a set of hypotheses of a specified form that, when added to a given (partial) knowledge, allows a given set of observations to be explained, whilst satisfying predefined domain specific

A. Gupta and T.A. Henzinger (Eds.): CMSB 2013, LNBI 8130, pp. 235–237, 2013.
© Springer-Verlag Berlin Heidelberg 2013

integrity constraints [2]. The problem of network reconstruction naturally maps to an abductive framework: i) the gene expression data constitute the observations; these are expressed as binary variables, that if equal to 1 (resp. −1) denote that the expression value of the gene has increased (resp. decreased). ii) the given (partial) knowledge is a logic-based representation of the interactome and gene ontologies. The template of all possible interactions between genes is captured by logical facts of the form *interactive_potential(G1, G2)*. The prior knowledge also includes information on the known function of specific genes. This is modelled by *regulatory_potential(R1, RPV)*, where $R1$ is a specific regulator gene and RPV is a binary variable over the set $\{1, -1\}$. Integrity constraints capture signal propagation principles to ensure consistency with the experimental data and the biological priors stated above. The edges in the network constitute the abducible sentences of our inference process and they can be of two forms: *compatible_regulator(G1, Qx, E)* which infers complementary gene influences and *overpowered_regulator(G2, Qx, E2)* which infers competitive gene influences. The first two arguments are genes, whereas the third argument E is a binary variable over the set $\{1, -1\}$ denoting the causal effect of the interaction between two genes. Hence, the abductive computation task of our approach is the inference of possible signed-directed networks, in terms of complementary and competitive gene regulations, that, together with the prior biological knowledge, fully explain the observations. Compatible regulators are consistent with the data, whereas overpower regulators are allowed to be inconsistent iff there exist sufficient compatible regulators in the inferred network that overpower them. ARNI implements a function to quantify the influence of each regulator.

3   Results

A key dimension of our work is the definition of a rule-based model that caters for the notions of feedback loop detection, correct assignment of overpowered influences and post-translational regulations so enabling the inference of more complete biological networks. Using a known network active in T-cells we assessed ARNI's faithfulness to biological reality in terms of inferring the complete network. We compared our results with those achieved by an existing approach [1]. ARNI was able to infer the entire gold standard network, whereas the approach in [1] was able to infer only a partial network. The missing and mislabeled links can be attributed to specific limitations in the assumptions used in [1].

4   Conclusions

In its current form, ARNI offers a computational method with improved expressiveness and wider applicability. Future direction of the work includes evaluation of ARNI in terms of scalability, thorough comparison with existing methods and a probabilistic extension of the work.

Acknowledgments. This work is funded by BHF Centre of Excellence.

References

1. Yeang, C.-H., Ideker, T., Jaakkola, T.: Physical network models. Journal of Computational Biology 11, 243–262 (2004)
2. Kakas, A.C., Kowalski, R.A., Toni, F.: The Role of Abduction in Logic Programming. Handbook of Logic in Artificial Intelligence and Logic Programming, 235–324 (1998)

A Systems Biology and Ecology Framework for POPs Bioaccumulation in Marine Ecosystems

Marianna Taffi[1,3], Nicola Paoletti[1], Pietro Liò[2], Luca Tesei[1],
Emanuela Merelli[1], and Mauro Marini[3]

[1] School of Science and Technology, University of Camerino, Camerino, Italy
[2] Computer Laboratory, University of Cambridge, Cambridge, United Kingdom
[3] National Research Council, Institute of Marine Science, Ancona, Italy

Abstract. We propose a modelling framework for studying bioaccumulation of Persistent Organic Pollutants (POPs) and microbial bioremediation in the Adriatic food web. The integration of network estimation methods, ODE simulation and sensitivity analysis and tools from synthetic biology allows investigating multiscale effects and biological responses to POPs contamination, from the molecular level (bacteria metabolism) to the ecosystem level (food web) of a marine ecosystem.

Keywords: bioaccumulation modelling, PCBs, bioremediation, FBA.

When a chemical compound is released into an ecosystem, its ecological impact on living organisms and environment is hard to predict. Due to their biochemical and biophysical characteristics, *POPs* (*Persistent Organic Pollutants*) enter protein pathways at the cell surface or inside organisms, in which *bioaccumulation* occurs as the result of the uptake from contaminated environment and food. The marine ecosystem is a sink and a source of POPs that, being resistant to degradation, remain persistently into the environment and bind permanently with the fat tissue of fish. Thus, contaminants follow the same paths as biomass flows, making every species in a polluted ecosystem prone to bioaccumulation, a phenomenon that increases at higher trophic levels. What is important is not just estimating contamination levels, but also identifying which species has the largest effect on the diffusion of a pollutant through a food web (*keystones*).

On the other hand, microbial communities constitute the most prominent marine compartment in terms of abundance and diversity and, more importantly, are able to degrade POPs by using them as growth substrates in their metabolic pathways (*bioremediation*). Despite of that, the role of micro-organisms in bioaccumulation modelling has been poorly considered so far.

In this work, we investigate the systems biology of *Polychronated Biphenyls* (*PCBs*, a class of POPs) bioaccumulation in the Adriatic ecosystem, by integrating the classical food web of macro-organisms with the complex chemical pathways of the many micro-organisms involved in bioremediation. We model the microbial pool as a unique super-organism where a continuous exchange of genetic information occurs among bacteria by means of conjugative plasmids, prophages and DNA uptake [2].

A. Gupta and T.A. Henzinger (Eds.): CMSB 2013, LNBI 8130, pp. 238–239, 2013.

Procedurally, we have estimated the food web structure in terms of trophic and contaminant flows from literature data with the *Linear Inverse Modelling (LIM)* [3] method (Fig. 1 a). Estimated rates have been used to parametrize a ODE dynamic bioaccumulation model (Fig. 1 b). Keystones have been identified with network analysis tools (trophic and topological centrality indices), and with a newly introduced index, *Sensitivity Centrality (SC)*, based on the sensitivity analysis of the ODE model. Using *Flux Balance Analysis (FBA)* we have reconstructed the metabolic pathways of PCB bioremedation, by extending a FBA model of *P. Putida* [1]. In this way, we investigate the multiscale effects of the optimization of bacterial functions and perturbations (e.g. gene knockout) in the metabolic network on the bioaccumulation dynamics in the food web.

The combination of synthetic biology and ecological analysis tools provided insights into both the key species in a contaminated network through a novel network index; and the role that the bioengineering of bacterial metabolism plays in the remediation of polluted environments.

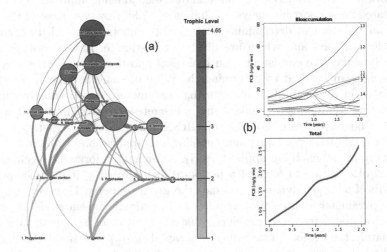

Fig. 1. PCBs bioaccumulation in Adriatic food web (a). Nodes represent species (size proportional to PCBs concentration). Edges represent feeding links. (b) Dynamic bioaccumulation model (x-axis, time; y-axis, PCB concentration).

References

1. Nogales, J., et al.: A genome-scale metabolic reconstruction of pseudomonas putida kt2440: ijn746 as a cell factory. BMC Systems Biology 2(1), 79 (2008)
2. Sobecky, P.A., Hazen, T.H.: Horizontal gene transfer and mobile genetic elements in marine systems. In: Horizontal Gene Transfer, pp. 435–453. Springer (2009)
3. van Oevelen, D., et al.: Quantifying food web flows using linear inverse models. Ecosystems 13(1), 32–45 (2010)

A Symbolic Approach Based on Model Checking and Constraint Solving Techniques for Reverse Engineering of Thomas Networks Parameters

Emmanuelle Gallet[1], Matthieu Manceny[2],
Pascale Le Gall[1], and Paolo Ballarini[1]

[1] MAS Laboratory, Ecole Centrale Paris, 92195 Châtenay-Malabry, France
{emmanuelle.gallet,pascale.legall,paolo.ballarini}@ecp.fr
[2] LISITE Laboratory, ISEP, 28 Rue Notre-Dame-des-Champs 75006 Paris, France
matthieu.manceny@isep.fr

Biological Context. To understand the dynamics of a Genetic Regulatory Network (GRN), various continuous and discrete modeling approaches have been advocated for supporting analysis techniques [9,7,8]. However, most of them suffer from the need of determining biological parameters on which depend the possible dynamics and which are difficult to estimate. Indeed, not all of the GRN dynamics are consistent with biological knowledge or observations. This knowledge can be used to determine the value of some parameters or can be translated in the form of constraints that parameters should comply with. By abstracting continuous dynamics using a discrete-step asynchronous dynamics, the R. Thomas discrete modeling of GRN has the double advantage of highlighting qualitative reasoning and enabling the application of formal methods, especially model checking approaches [1]. After having formally specified a biological observation in form of a temporal logic property, it becomes possible to verify if a target dynamics satisfies the given property. However, the problem of parameter identification requires to investigate the entire set of possible dynamics, that is to consider each possible combination of parameter values. Unfortunately, the number of such dynamics rapidly grows with the size of the GRN and the key question becomes the design of effective techniques for analyzing a family of models parameterized by unknown parameters.

Related Work. Bernot et al.[3] is a pioneering work where model checking techniques are applied to verify whether a given discrete Thomas model fulfills some relevant biological temporal properties given as *Computation Tree Logic* (CTL) formulas [1]. The need to test each dynamics one after the other makes this approach only usable with small networks. [6,2] define an approach to share computations between different models. Algorithms are optimized for the particular case of *time series*, that are sequences of dynamic states observed one after the other. In [4], Constraint Logic Programming techniques are used: once GRN dynamics and biological knowledge are described by means of declarative rules and constraints on parameters, target behaviors are expressed as some kind of finite paths that models have to verify.

A. Gupta and T.A. Henzinger (Eds.): CMSB 2013, LNBI 8130, pp. 240–241, 2013.

Our Contribution. Similarly to [6], our approach considers LTL properties and represents several models of a GRN in a unique representation. Nevertheless, our models are not manipulated using an explicit enumeration, but are implicitly referenced as solutions of constraints defined over parameters. Indeed, we follow the same creed as the one advocated in [4]: model sets have to be handled intentionally through some logical language both to avoid combinatorial explosion and to take benefit of the expressive power of a logical language and of constraint solving techniques. For that, we represent a set of dynamics with one structure called *Parametric GRN* where biological parameters are processed as symbols within constraints. We developed an algorithm, inspired by LTL model checking, combining symbolic execution and constraint solving techniques. Our method combines the advantage of identifying values of biological parameters by symbolically manipulating them as constraints and by expressing biological knowledge in the form of LTL properties or direct constraints over parameters. Moreover, we consider the full LTL language while [5,4] focus on finite paths and [6] focuses on time series (that represent finite paths of arbitrary length). We demonstrated our methodology by analyzing a real case study (*i.e.* cytotoxicity of *P. aeruginosa*). Such analysis has been carried out through the SPuTNIk tool, a prototype software implementation of the proposed method.

References

1. Baier, C., Katoen, J.-P.: Principles of Model Checking. The MIT Press (2008)
2. Barnat, J., Brim, L., Krejci, A., Streck, A., Safránek, D., Vejnar, M., Vejpustek, T.: On parameter synthesis by parallel model checking. IEEE/ACM Trans. Comput. Biology Bioinform. 9(3), 693–705 (2012)
3. Bernot, G., Comet, J.-P., Richard, A., Guespin, J.: Application of formal methods to biological regulatory networks: extending Thomas' asynchronous logical approach with temporal logic. Journal of Theoretical Biology 229(3), 339–347 (2004)
4. Corblin, F., Fanchon, E., Trilling, L.: Applications of a formal approach to decipher discrete genetic networks. BMC Bioinformatics 11, 385 (2010)
5. Corblin, F., Tripodi, S., Fanchon, E., Ropers, D., Trilling, L.: A declarative constraint-based method for analyzing discrete genetic regulatory networks. BioSystems 98, 91–104 (2009)
6. Klarner, H., Streck, A., Šafránek, D., Kolčák, J., Siebert, H.: Parameter identification and model ranking of thomas networks. In: Gilbert, D., Heiner, M. (eds.) CMSB 2012. LNCS(LNBI), vol. 7605, pp. 207–226. Springer, Heidelberg (2012)
7. Thieffry, D., Colet, M., Thomas, R.: Formalisation of regulatory networks: a logical method and its automation. Math. Modelling and Sci. Computing 2, 144–151 (1993)
8. Thieffry, D., Thomas, R.: Dynamical behaviour of biological regulatory networks immunity control in bacteriophage lambda. Bull. Math. Biol. 57(2), 277–297 (1995)
9. Thomas, R.: Logical analysis of systems comprising feedback loops. Journal of Theoretical Biology 73(4), 631–656 (1978)

Compositionality Results
for Cardiac Cell Dynamics

Md. Ariful Islam[1], Abhishek Murthy[1], Ezio Bartocci[2],
Antoine Girard[3], Scott A. Smolka[1], and Radu Grosu[1,2]

[1] Department of Computer Science, Stony Brook University
[2] Department of Computer Engineering, Vienna University of Technology
[3] Université Joseph Fourier, Grenoble, France

Abstract. We show that the 13-state sodium channel component of the
Iyer et al. cardiac cell model can be replaced with a previously identified δ-bisimilar 2-state Hodgkin Huxley-type abstraction by appealing
to a small gain theorem. To prove this feedback compositionality result,
we construct quadratic-polynomial exponentially decaying bisimulation
functions between the two sodium channel models and also for the rest of
a simplified version of the Iyer et al. model using the SOSTOOLS toolbox. Our experimental results validate the analytical ones. To the best
of our knowledge, this is the first application of δ-bisimilar, feedback-
assisting, compositional reasoning in biological systems.

The Iyer et al. model (IMW) [3] is a physiologically detailed cardiac myocyte
(ventricular) model that can be used to to simulate the change in a cell's trans-
membrane potential in response to an external electrical stimulus, also known
as the *Action Potential (AP)*. In this work, we ask *"assuming that the AP is
the only observable, can we replace the sodium current component, M_I, of IMW
with an equivalent model-order reduced Hodgkin Huxley (HH)-type model M_H?"*
The HH model [2], uses two variables m and h to model a squid neuron's trans-
membrane sodium current. In [4], we proposed an algorithm to identify M_H that
is δ-bisimilar (equivalent) to the 13-state voltage-controlled M_I.

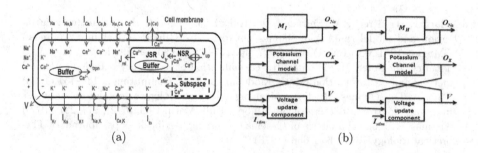

(a)   (b)

Fig. 1. (a)The IMW model, showing various currents. (b)The sodium channel components M_I (detailed) and M_H (abstract) composed with the potassium and voltage
components forming the rest of a simplified version (IMW') of the IMW model.

A. Gupta and T.A. Henzinger (Eds.): CMSB 2013, LNBI 8130, pp. 242–244, 2013.
© Springer-Verlag Berlin Heidelberg 2013

Compositionality of the equivalent sodium channels with the rest of the simplified IMW model (IMW-RT') can be established using Bisimulation Functions (BFs) and a small gain theorem based on them.

Definition 1. *Consider two dynamical systems Σ_i, as per [1], but with g_i : $\mathcal{X}_i \to \mathcal{Y}_i$, being the output functions that map a state to $\mathbf{y}_i \in \mathcal{Y}_i \subseteq \mathbb{R}^p$. Let $R_\delta = \{(\mathbf{x}_1, \mathbf{x}_2) | \parallel g_1(\mathbf{x}_1) - g_2(\mathbf{x}_2) \parallel \leq \delta\}$. A smooth function $S : R_\delta \to \mathbb{R}_0^+$ is a δ-Restricted BF (δ-RBF) over Σ_1 and Σ_2 if:*

$$\parallel g_1(\mathbf{x}_1) - g_2(\mathbf{x}_2) \parallel \; \leq S(\mathbf{x}_1, \mathbf{x}_2) \tag{1}$$

and there exists $\lambda > 0$, $\gamma \geq 0$ such that $\forall \mathbf{u}_1 \in \mathcal{U}_1, \mathbf{u}_2 \in \mathcal{U}_2$,

$$\frac{\partial S}{\partial \mathbf{x}_1} f_1(\mathbf{x}_1, \mathbf{u}_1) + \frac{\partial S}{\partial \mathbf{x}_2} f_2(\mathbf{x}_2, \mathbf{u}_2) \leq -\lambda S(\mathbf{x}_1, \mathbf{x}_2) + \gamma \parallel \mathbf{u}_1 - \mathbf{u}_2 \parallel \tag{2}$$

Theorem 1. *Let Σ_1, Σ_2 and Σ_3 be three dynamical systems. Let Σ_{13} and Σ_{23} be interconnections (as defined in [1]) of Σ_3 with Σ_1 and Σ_2 respectively. Let S_{12} be a δ-RBF between Σ_1 and Σ_2 and S_3 be δ-RBF for Σ_3. We denote by λ_{12} and γ_{12} (λ_3 and γ_3 respectively) the constants such that Eq. (2) holds. If $\frac{\gamma_{12}\gamma_3}{\lambda_{12}\lambda_3} < 1$, then there exists a BF S between Σ_{13} and Σ_{23} of the form $S(\mathbf{x}_{13}, \mathbf{x}_{23}) = \alpha_1 S_{12}(\mathbf{x}_1, \mathbf{x}_2) + \alpha_2 S_3(\mathbf{x}_3, \mathbf{x}'_3)$ where, $\mathbf{x}_{13} = [\mathbf{x}_1, \mathbf{x}_3]$, and $\mathbf{x}_{23} = [\mathbf{x}_2, \mathbf{x}'_3]$ The real constants α_1 and α_2 can be chosen as in Eq.4 of [1] by replacing $\lambda_1 = \lambda_{12}$, $\gamma_1 = \gamma_{12}$, $\lambda_2 = \lambda_3$ and $\gamma_2 = \lambda_3$.*

The two BFs, 1) between M_I and M_H and 2) for IMW-RT' were identified in the SOSTOOLS toolbox [5] by adding the following constraint along with the ones that define a BF: $S(\mathbf{x}, \mathbf{x}') - \parallel g_1(\mathbf{x}) - g_2(\mathbf{x}') \parallel \leq \delta$. The parameter λ was fixed to either 10^{-4} and 10^{-5} for the two BFs and γ was found to be 10^{-6}, which resulted in the small-gain condition being satisfied. Fig. 2 shows experimental evidence of the model equivalence on replacing M_I by M_H.

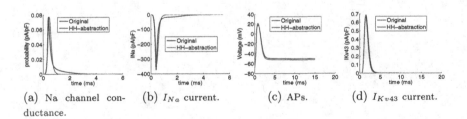

(a) Na channel conductance.   (b) I_{Na} current.   (c) APs.   (d) I_{Kv43} current.

Fig. 2. IMW' was stimulated using -100 pA/pF stimulus with M_I and then M_H. The resulting mean L1 errors were O_{Na} : 9.15×10^{-4}, I_{Na}: $3.8pA/pF$, I_{Kv43}: $0.0078pA/pF$, V: $2.29mV$.

References

1. Girard, A.: A composition theorem for bisimulation functions. Preprint (2007) arXiv:1304.5153
2. Hodgkin, A.L., Huxley, A.F.: A quantitative description of membrane current and its application to conduction and excitation in nerve. Journal of Physiology 117, 500–544 (1952)
3. Iyer, V., Mazhari, R., Winslow, R.L.: A computational model of the human left-ventricular epicardial myocytes. Biophysical Journal 87(3), 1507–1525 (2004)
4. Murthy, A., Islam, M. A., Bartocci, E., Cherry, E.M., Fenton, F.H., Glimm, J., Smolka, S.A., Grosu, R.: Approximate bisimulations for sodium channel dynamics. In: Gilbert, D., Heiner, M. (eds.) CMSB 2012. LNCS, vol. 7605, pp. 267–287. Springer, Heidelberg (2012)
5. Prajna, S., Papachristodoulou, A., Seiler, P., Parrilo, P.A.: SOSTOOLS: Sum of squares optimization toolbox for MATLAB (2004)

Quantification of Biological Network Perturbations: Impact Assessment Using Causal Biological Networks

Florian Martin, Alain Sewer, Marja Talikka, Yang Xiang,
Julia Hoeng, and Manuel C. Peitsch

Biological Systems Research, R&D, Philip Morris International,
Neuchatel, NE, Switzerland

High-throughput profiling of gene expression has opened new avenues for the understanding of biological processes at the molecular level. However, the amount of information collected can be overwhelming, making interpretation of the data difficult and subsequent detailed biological understanding elusive. Reducing the complexity of such data by evaluating them in a relevant biological context is required to gain meaningful insight. We propose that "cause-and-effect" network approaches to pharmacology and toxicology are valuable to quantify network perturbations caused by bio-active substances, and to identify mechanisms and biomarkers modulated in response to exposure (Hoeng & al., Drug Discov Today, 2012). The underlying concept is that transcriptional changes are the consequences of the biological processes described in the network.

We have recently built an ensemble of network models that consist of cause-and-effect relationships (typically activation or inhibition) between molecular entities and activities (e.g. kinase activation or increased protein abundance) (Westra & al, BMC Sys Biol 2011, Schlage & al, ibid, Gebel & al, Bioinformatics and Biology Insights, 2013, Westra & al, ibid). The description of the biological context has been manually built into the network models using prior knowledge extracted from both relevant literature and published datasets after a large-scale knowledge mining effort. Some network nodes are also related to mRNA abundance entities that they positively or negatively regulate. Thus, our biological network models have a two-layer structure, where the functional level is explicitly distinguished from the transcriptional level. Using transcriptional downstream effects to infer the activity of upstream entities has its advantages, because the activity of a node is inferred based on the differential expression of many genes known to be regulated by a given entity, even the ones encoding proteins with unknown functions. This is unlike the networks derived from other pathway databases, which rely upon the "forward assumption" stating that changes in gene expression induce changes in the activity and abundance of the gene product.

We present a novel framework for the quantification of the amplitude of network perturbations to enable comparisons between different exposures and systems. Also, our approach enables quantification of each biological entity (nodes) in the network, among which key contributors, referred to as leading nodes, can be identified to unravel biological mechanisms. It efficiently integrates

A. Gupta and T.A. Henzinger (Eds.): CMSB 2013, LNBI 8130, pp. 245–246, 2013.

transcriptomics data and network models to enable a mathematically coherent framework from quantitative impact assessment to data interpretation and mechanistic hypothesis generation. The gene expression fold-changes are translated into differential values for each node of the network (denoted by f) by fitting the functional layer relationships with respect to the boundary constraint given by the observed fold-changes. The node differential values are in turn summarized into a quantitative measure of network perturbation amplitude (NPA). The NPA is computed as a (semi-) Sobolev norm on the signed directed graph underlying the network, which can be expressed as a quadratic form $f^T Q f$. In addition to the confidence intervals of the NPA scores, which account for experimental error, companion statistics were derived to inform on the specificity of the NPA score with respect to the biology described in the network. The network is considered to be specifically perturbed if all P-values are low (typically < 0.05).

An *in-vivo* dataset was generated to study cessation effect upon smoking in C57/Bl6 mouse emphysema model. Mice were exposed to mainstream cigarette smoke (CS) from the Kentucky reference cigarette, 3R4F, or to fresh air for up to 7 months. Following 2- and 3-month exposure to 3R4F, subgroups of mice were exposed to fresh air for a 3-, 4-, or 5-month cessation period.

Cell proliferation mechanism was investigated based on lung transcriptomics data using a mechanistic network model (Westra & al, BMC Sys Biol 2011). The growth factor subnetwork is clearly perturbed and released from perturbation, following CS exposure and cessation, respectively. Of the growth factors that could be measured in BronchoAlveolar Lavage Fluid (BALF), bFGF and VEGF levels are similarly altered in CS exposed mice. The Vegfa node in the network follows the same trend as the analyte level in the BALF, being most affected at 7 months of exposure. On the other hand, Egf signaling shows no consistent behavior at the network node level and similarly, the protein measured in BALF fails to serve as a marker for disease progression/reversal.

Next we investigated the activation of cell cycle using the submodel within the cell proliferation network. While there are no BALF analytes that could serve as a surrogate for cell cycle progression, NPA can provide meaningful insights into the disease progression. CS is known to affect cell cycle, but its role is not clear in emphysema development. Similar to other biological processes, the perturbation of the cell cycle model is decreased upon smoking cessation. The comparison of the leading nodes of the network perturbation, for the 3 months CS exposed group and two cessation groups, reveals nodes that persist highly ranked even after smoking cessation. Interestingly, while the overall perturbation of cell cycle after cessation is low as compared to 7 month exposure, the leading nodes that persist are essentially the same.

In summary, the NPA of the network models offers mechanistic understanding on the biological impact of the CS exposure, revealing multiple network models, subnetworks and nodes, whose scores are consistent with measured experimental endpoints. Moreover, even when there is no phenotypic information available, network scoring provides valuable mechanistic insight and testable hypotheses.

Deciphering the Transcriptional Landscape of *Caulobacter crescentus* at Base Pair Resolution

Bo Zhou, Jared Schrader, Beat Christen, Harley McAdams, and Lucy Shapiro

Department of Developmental Biology, Stanford University School of Medicine,
Stanford, CA, USA

Abstract. A key to a systems-level understanding of the regulatory circuitry and the complex flow of genetic information that guides the development of a single cell is to understand when, where, and how is genomic information extracted through the process of transcription as well as all essential components of its genome. Using *Caulobacter crescentus* as a model organism and taking advantage of high-throughput sequencing, high-density microarrays, and hyper-saturated transposon mutagenesis, we have mapped, at the resolution of single base pairs, all sites of transcriptional initiation as a function of the *Caulobacter* cell cycle as well as all essential elements of its genome including 480 essential genes, 8 small RNAs, 402 regulatory regions, and 90 intergenic regions. Our study has shown that the transcriptional landscape of *Caulobacter* is much more complex than previously thought. We have discovered many novel transcriptional elements including open reading frames (ORFs) with internal transcriptional initiation, small non-coding RNAs, ORFs with multiple promoters, as well as ORFs with antisense transcription, one of these is the cell cycle master regulator *dnaA*. We have also made the surprising discovery that operon structures are dynamic and cell cycle regulated. Furthermore, our study has also enhanced the resolution of the cell-cycle expression of genes down to the level of individual promoters, a drastic improvement over standard microarrays, which shows gene expression as the result of multiple regulatory phenomena.

Keywords: *Caulobacter*, transcriptome, essential genome, non-coding sRNA, antisense, transcription, promoter.

A. Gupta and T.A. Henzinger (Eds.): CMSB 2013, LNBI 8130, p. 247, 2013.
© Springer-Verlag Berlin Heidelberg 2013

Identifiablity Analysis and Improved Parameter Estimation of a Human Blood Glucose Control System Model

Eszter Lakatos, Domokos Meszéna, and Gábor Szederkényi

Pázmány Péter Catholic University, Faculty of Information Technology
Práter u. 50/a, H-1083 Budapest, Hungary

Introduction. Quantitative dynamical mathematical models are very useful in the thorough understanding and possible targeted manipulation of biological processes. However, determining the model parameters from available data is often a challenging task for such models, typically given in the form of nonlinear ordinary differential equations. Global structural identifiability of parameterized ODE models means that there is (at least a theoretical) possibility to uniquely determine system parameters from appropriate measurement data [2]. The aim of this paper is to study structural identifiability for a published molecular level model of human blood glucose control, to achieve improvement in model fit compared to published results, and thus to obtain a model that will be suitable to examine the effect of natural and artificial feedbacks.

Model and Goals. We selected the mathematical model published in [1] that has the following form:

$$\frac{dp_i}{dt} = -(a_{1,i} + a_{2,i})p_i + u_i(g_2), \quad \frac{dh_i}{dt} = -a_{4,i}h_i(R_i^0 - r_i) - a_{3,i}h_i + a_{1,i}p_i\frac{V_p}{V}$$

$$\frac{dr_i}{dt} = a_{4,i}h_i(R_i^0 - r_i) - a_{5,i}r_i, \quad \frac{dg_1}{dt} = \frac{k_1 r_2}{1 + k_2 r_1}\frac{V_{max}^{gs}g_2}{K_m^{gs} + g_2} - k_3 r_1\frac{V_{max}^{gp}g_1}{K_m^{gp} + g_1} \quad (1)$$

$$\frac{dg_2}{dt} = -\frac{k_1 r_2}{1 + k_2 r_1}\frac{V_{max}^{gs}g_2}{K_m^{gs} + g_2} + k_3 r_1\frac{V_{max}^{gp}g_1}{K_m^{gp} + g_1} - f_u(g_2, h_2) + G_{in},$$

where $f_u(g_2, h_2) = U_b(1 - \exp\left(\frac{-g_2}{C_2}\right)) + \frac{g_2}{C_3} \cdot \left(U_0 + \frac{(U_m - U_0)\left(\frac{h_2}{C_4}\right)^\beta}{1 + \left(\frac{h_2}{C_4}\right)^\beta}\right).$

The state variables of the model are the following: p_i, the plasma hormone, h_i the cellular hormone and r_i the hormone-bound receptor concentration, where $i = 1$ and 2 stand for glucagon and insulin, respectively. g_1 represents blood glycogen and g_2 blood glucose levels, the latter being the measured output. The model in a simplified form contains the hormone dynamics, glycogen-glucose transition in the liver, insulin-independent and dependent utilization ($f_u(g_2, h_2)$) of glucose. Feedback was incorporated in the glucose-dependent hormone infusion rates via u_i (see the detailed explanation of the model and its parameters in [1]). Based on literature data, our current study focuses on the most uncertain parameters, namely: the plasma hormone transitional rates $a_{1,i}$; the feedback gains for glycogen-glucose transition k_i ($i = 1, 2, 3$); and the two most crucial parameters of glucose utilization, C_2 and β.

A. Gupta and T.A. Henzinger (Eds.): CMSB 2013, LNBI 8130, pp. 248–249, 2013.
© Springer-Verlag Berlin Heidelberg 2013

Methods and Tools. First of all, the set of parameters was divided into two subsets: the first four parameters in which the system is linear ($\theta_1 = \{k_1, k_3, a_{1,i}\}$) and the three remaining parameters in which the dependence is non-linear ($\theta_2 = \{k_2, C_2, \beta\}$). Structural identifiability was studied using the GenSSI toolbox available in Matlab [2].

The parameter estimation cost function was the standard normed quadratic error between the experimental data taken from literature and the simulated output. The estimation procedure was an iterative process, where θ_1 was estimated using a least squares procedure, while θ_2 was estimated by the pattern search minimization method.

Results and Discussion. Identifiability analysis showed that considering only θ_1, the model is structurally globally identifiable. On the other hand, including any of the remaining parameters of θ_2 into the unknown parameter vector, global identifiability could not be proved with the applied tools. This result further justifies the separation of parameters into θ_1 and θ_2. Moreover, the value of the estimation objective function was 2% lower than in [1]. The model was validated using a new input based on the widely used oral glucose tolerance test (OGTT) [3]. The main results are shown in Fig. 1. The model output shows the well-known features of healthy OGTT test results, such as a downstroke in glucose level due to the temporary increase of insulin (which can lead to a hypoglycemic state in patients with reactive hypoglycemia).

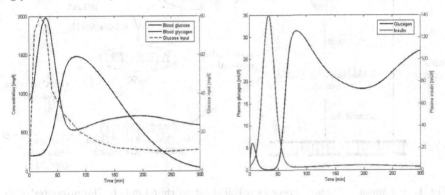

Fig. 1. (a) glucose input, glucose and glycogen levels, (b) glucagon and insulin levels

References

1. Liu, W., Tang, F.: Modeling a simplified regulatory system of blood glucose at molecular levels. Journal of Theoretical Biology 252, 608–620 (2008)
2. Chis, O., Banga, J.R., Balsa-Canto, E.: GenSSI: a software toolbox for structural identifiability analysis of biological models. Bioinformatics 27(18), 2610–2611 (2011)
3. Definition, diagnosis and classification of diabetes mellitus and it's complications. World Health Organisation, Department of Noncommunicable Disease Surveillance (1999)

RNA Interference in Cancer and Cell Cycle Networks: A Case Study of E2F Proteins*

Jesús Miró-Bueno

Research Institute of the IT4Innovations Centre of Excellence,
Faculty of Philosophy and Science, Silesian University in Opava,
74601 Opava, Czech Republic
jesus.mirobueno@gmail.com

RNA interference (RNAi) is a cellular process for silencing gene expression. This process is driven by the RNA-induced silencing complex (RISC). Here, we present a mathematical model that shows that RNA silencing of a single gene with positive feedback can produce bistability and oscillatory behaviour. We focus our study on a specific gene network: RISC acting on the so-called E2F proteins. These E2F proteins are an important family of transcription factors related with cancer and cell cycle [1]. Members of the E2F family positive

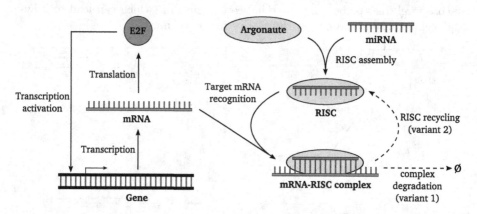

Fig. 1. The model. A gene expresses a E2F protein that binds to the promoter of its own gene increasing the transcription rate. RISC molecules are assembled at constant rate (from miRNAs and argonaute proteins) and bind to target mRNA molecules. Then, RISC can be degraded together with mRNA (variant 1) or recycled (variant 2).

regulate their own transcription creating a positive feedback. Recently, microRNAs (miRNAs) that downregulate E2F gene expression have been discovered. The model presented here takes into account these two basic parts of E2F

* This work was supported by the European Regional Development Fund in the IT4Innovations Centre of Excellence project (CZ.1.05/1.1.00/02.0070) and EU project Development of Research Capacities of the Silesian University in Opava (CZ.1.07/2.3.00/30.0007).

A. Gupta and T.A. Henzinger (Eds.): CMSB 2013, LNBI 8130, pp. 250–252, 2013.

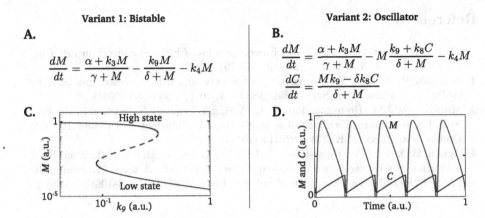

Fig. 2. Variant 1 (**A** and **C**) and variant 2 (**B** and **D**) of the model[1]. (**A**) Differential equation of variant 1. (**B**) Differential equations of variant 2. (**C**) Hysteresis diagram obtained from **A**. The two stable states *high* and *low* (solid lines) depends on RISC assembly rate (k_9). (**D**) Oscillations obtained from numerical solution of **B**. The oscillations are mainly driven by M and C.

proteins regulated by miRNAs: 1) a gene with a positive transcriptional feedback loop and 2) a negative interaction carried out by RISC (Fig. 1). We study two variants of the model depending on whether RISC is recycled or not. If RISC is not recycled (variant 1) the model produces bistability and the dynamics is described by one ODE (Fig. 2A,C). If RISC is recycled (variant 2) the gene network produces oscillations and two ODEs describe the dynamics (Fig. 2B,D). The variants 1 and 2 of this model are modifications of refs. [2] and [3], respectively. Here the molecule repressed is the mRNA, whereas in refs. [2] and [3] the molecule repressed is the activator protein. This means that here the dynamics is mainly driven by mRNAs instead of proteins. In this model, transcription activation of miRNAs or other genes are not necessary to obtain these two behaviours in contrast to other more complex models [4]. The predictions of our simple model can be useful to clarify the role of E2F gene networks in cancer and cell cycle, and as well in other positive autoregulatory gene networks silenced by RNAi. The model can be also interesting in synthetic biology for engineering new genetic circuits based on RNAi.

---

[1] The model can be reduced by QSSA if standard values for the rates are assumed [3]. The slow variable in the variant 1 is the mRNA (M). In variant 2, the slow variables are the mRNA (M) and mRNA-RISC complex (C). The parameters are as follows: $\alpha = k_{-1}k_2k_6/k_1k_5$, $\gamma = k_{-1}k_6/k_1k_5$ and $\delta = k_{10}/k_7$. (k_1: binding rate of E2F to its gene promoter, k_{-1}: unbinding rate of $E2F$ from its gene promoter, k_2: basal transcription rate, k_3: activated transcription rate, k_4: mRNA degradation rate, k_5: translation rate, k_6: E2F degradation rate, k_7: binding rate of RISC to mRNA, k_8: RISC cleavage rate, k_9: RISC assembly rate, k_{10}: RISC degradation rate).

References

1. Chen, H.Z., Tsai, S.Y., Leone, G.: Emerging roles of E2Fs in cancer: an exit from cell cycle control. Nat. Rev. Cancer 9, 785–797 (2009)
2. François, P., Hakim, V.: Design of genetic networks with specified functions by evolution *in silico*. Proc. Natl. Acad. Sci. USA. 101(2), 580–585 (2004)
3. Miró-Bueno, J.M., Rodríguez-Patón, A.: A Simple Negative Interaction in the Positive Transcriptional Feedback of a Single Gene Is Sufficient to Produce Reliable Oscillations. PLoS ONE 6(11), e27414 (2011)
4. Aguda, B.D., Kim, Y., Piper-Hunter, M.G., Friedman, A., Marsh, C.B.: MicroRNA regulation of a cancer network: Consequences of the feedback loops involving miR-17-92, E2F, and Myc. Proc. Natl. Acad. Sci. USA. 105(50), 19678–19683 (2008)

A Balancing Act: Parameter Estimation for Biological Models with Steady-State Measurements

Andrzej Mizera[1], Jun Pang[1], Thomas Sauter[2], and Panuwat Trairatphisan[2]

[1] Computer Science and Communications, University of Luxembourg, Luxembourg
[2] Life Sciences Research Unit, University of Luxembourg, Luxembourg

Problem Statement. Constructing a computational model for a biological system consists of two main steps: (1) specifying the model structure and (2) determining the values for the parameters of the model. Usually, the model structure is represented in the form of a biochemical reaction network and the parameters are the reaction rate constants. The values of the reaction rates can be determined by fitting the model to experimental data by performing parameter estimation. However, the question remains whether the experimental data allow for unique identification of the parameters. To address this problem, one could perform a number of independent parameter estimations and investigate the range of obtained values among those parameter sets that result in a good fit. From the correlation of the obtained parameter sets one could, e.g., study whether only certain parameters are identifiable. This approach requires an *effective, efficient* and *automatic* way of performing estimation. In this study we concentrate on the case of fitting a deterministic mathematical model of a biological process, i.e., expressed in terms of a system of ordinary differential equations (ODEs), and a number of its variants to *multi-experiment steady-state data*. We propose a computational pipeline involving available software packages for achieving this goal while keeping a balance between the optimisation time and accuracy.

Our Approach. The number of steady-state measurements may be insufficient to identify all the parameters of a model, especially when the number of parameters is larger than the number of measured species. To address this difficulty, we take into account data of the so called *knockout mutant models*, i.e., variants of the original model obtained by eliminating one or more interactions between species. Since the knockout mutants of real biological systems can be obtained and investigated in experimental practice as well as the physical/chemical properties are common for all variants, the steady-state measurements of the mutants can enrich the set of data available for parameter estimation and make possible the identification of model parameters. In our approach we assume that such multi-experiment, steady-state data are available. The aim is to gather in an automatic way (to the possibly largest extent) a number of parameter sets that result in a good simultaneous fit of all the mutants. The collected parameter sets could further be used to investigate the parameter identifiability question.

We propose MATLAB as a control environment for the task of executing many independent parameter estimation runs. In the proposed computational pipeline the ODE-based knockout model variants are compiled with the SBTOOLBOX2

A. Gupta and T.A. Henzinger (Eds.): CMSB 2013, LNBI 8130, pp. 253–254, 2013.

for MATLAB to C MEX files for efficiency. Parameter estimation is performed using a pattern search optimisation algorithm provided in the PSwarm global optimisation solver [1], mainly because (1) the solver provides a pattern search algorithm, which assures a local minimum convergence and does not require any information on the gradient of the score function and (2) the search step of the algorithm is implemented as the particle swarm algorithm [2], which makes it to a global optimisation algorithm.

Since the SBTOOLBOX2 does not provide any methods for an efficient and direct computation of the steady-state, we estimate the steady-state values by independently integrating each of the C MEX models to a point where a necessary steady-state condition is satisfied, i.e., the norm of the difference between points on the trajectory is less than a threshold. Since the accuracy depends on this threshold, the main challenge is to find a balance between the computational time and the accuracy of steady state estimation. To reduce the computational time, we first perform model fitting with a threshold that results in a steady state or a state close to a steady state but which can be reached with a relatively small number of integration steps. Next, the obtained parameter values are taken to COPASI for further optimisation with another direct method algorithm, i.e., PRAXIS, and a direct, efficient computation of a steady state.

Preliminary Results. We apply the proposed approach to fit an ODE-based model of the PDGF signalling pathway [3] to steady-state experimental data. The model consists of 31 species and 40 reactions. Nine different variants of the model are considered and 31 unknown parameters common for all the variants need to be estimated. The experimental data are on the concentration of two species at steady state. In total the data set consists of 18 measurements. The models are implemented both in MATLAB and COPASI. The simultaneous implementation of all nine mutants in COPASI results in 279 ODEs.

One parameter estimation run in MATLAB lasts for approximately 30 mins and the cost function is evaluated for 40000 times. Then, the resulting parameter values are given to COPASI as a starting point for further optimisation with PRAXIS and 2000 cost function evaluations. This requires up to 2 hours of computational time of four cores. The improvement in the fit score is up to 30% of the score obtained in MATLAB. Our experiments show that this level of fit quality is unreachable in comparable amount of time if parameter estimation is performed from scratch in COPASI alone, even if a combination of COPASI optimisation algorithms is applied.

References

1. Vaz, A.I.F., Vicente, L.N.: A particle swarm pattern search method for bound constrained global optimization. J. Global Optimization 39(2), 197–219 (2007)
2. Kennedy, J., Eberhart, R.: Particle swarm optimization. In: Proc. IEEE International Conference on Neural Networks, pp. 1942–1948. IEEE (1995)
3. Yuan, Q., Trairatphisan, P., Pang, J., Mauw, S., Wiesinger, M., Sauter, T.: Probabilistic model checking of the PDGF signaling pathway. T. Comp. Sys. Biology 14, 151–180 (2012)

A Fusion Approach Linking Signaling Logic and Metabolic Mass-Flow Kinetics in Hepatocytes

Anke Ryll[1], Joachim Bucher[2], Jens Niklas[2], and Steffen Klamt[1]

[1] Max Planck Institute for Dynamics of Complex Technical Systems,
Sandtorstraße 1, D-39106 Magdeburg, Germany
{ryll,klamt}@mpi-magdeburg.mpg.de
[2] Insilico Biotechnology AG, Meitnerstraße 8, D-70563 Stuttgart, Germany
{joachim.bucher,jens.niklas}@insilico-biotechnology.com

Abstract. Starting from (i) an ODE-based kinetic model, covering metabolic processes and mass-flows in hepatocytes and (ii) a *Boolean* network, encompassing associated intracellular hormonal signaling and gene-regulatory events, we introduce a new formalism to integrate signal transduction processes, gene regulation, and metabolism. The integrated model was eventually tested for physiologic representativity by *in silico* simulations, the latter qualitatively addressing and demonstrating the hepatocyte's switch-like behaviour upon nutrient-dependent changes in extracellular insulin and glucagon levels.

Keywords: model integration, kinetic modeling, logical modeling, hormonal signaling, gene regulation, liver metabolism.

1   Introduction

Given the variety of different modeling approaches and implementation strategies in systems biology (reviewed in [1]) with each being increasingly forced to cope with physiological questions spanning a multitude of intra- and intercellular organization scales, integration concepts gain vital importance. Model integration is, however, a delicate task as computational complexity needs to be minimized while individual model characteristics (such as complex regulatory network structures) might, on the other hand, be essential with respect to certain aspects or questions and should therefore be preserved, for instance, when computing potential intervention points of pharmacological concern.

2   Method

We therefore introduce a new fusion approach exemplarily addressing the linking of intracellular signaling and metabolic processes whereas technically realizing the integration of qualitative logical transitions and quantitative mass-flow kinetics within the following steps (manuscript in preparation): An (i) ODE-based kinetic model

A. Gupta and T.A. Henzinger (Eds.): CMSB 2013, LNBI 8130, pp. 255–256, 2013.
© Springer-Verlag Berlin Heidelberg 2013

representing glycolytic *vs.* gluconeogenic reactions and mass-flows and an (ii) apriori *Boolean* network, which covers the associated signaling and gene-regulatory events controlling the metabolic fluxes constitute the starting points. The logical signaling model is (iii) subsequently transformed into a set of qualitative ODEs using a method by Wittmann *et al.* [2]. Model interfaces, *i.e.* continuous metabolic enzyme activities depicting signaling outputs to metabolism and normalized metabolic compound concentrations denoting metabolic outputs to signaling, are (iv) eventually coupled, linking signaling to metabolism and *vice versa*.

3   Results and Perspective

The integrated model was finally parameterized to qualitatively reproduce a hepatocyte's response to extracellular nutritional changes in insulin *vs.* glucagon levels, the latter characterizing hormonal regulation of glucose homeostasis. Subsequent *in silico* simulations demonstrated the coupled system to successfully respond in a physiologically representative manner, as hormonal stimuli adaptation and switch-like behavior with respect to glycolytic *vs.* gluconeogenic processes could be qualitatively resembled. However concerning the practical model consistency, further efforts have to aim at model validation and verification based on data from targeted experiments.

References

1. Goncalves, E., Bucher, J., Ryll, A., Niklas, J., Mauch, K., Klamt, S., Rocha, M., Saez-Rodriguez, J.: Bridging the layers: towards integration of signal transduction, regulation and metabolism into mathematical models. Molecular bioSystems (2013) doi: 10.1039/c3mb25489e
2. Wittmann, D.M., Krumsiek, J., Saez-Rodriguez, J., Lauffenburger, D.A., Klamt, S., Theis, F.J.: Transforming Boolean models to continuous models: methodology and application to T-cell receptor signaling. BMC Systems Biology 3, 98 (2009)

Esther: Introducing an Online Platform for Parameter Identification of Boolean Networks

Adam Streck[1], Juraj Kolčák[2], Heike Siebert[1], and David Šafránek[2]

[1] Freie Universität, Berlin, Germany
[2] Masaryk University, Brno, Czech Republic

1 Introduction

The framework of boolean networks and its derivatives have proven to be a useful tool for getting insights into biological processes of gene regulation and signal transduction [1]. Using such a framework one describes a system as a *regulatory graph* where a node stands for a component of the system, each of which can adopt one of finitely many discrete levels. Directed edges represent component interactions. The knowledge captured in this graph is generally insufficient to infer the system dynamics. In particular, to be able to simulate a behavior of the system, one must also add *logical parameters* that describe how a level of a component changes based on the levels of its regulators. The set of possibilities is usually quite sizable and the process of *parameter identification*—finding parametrizations that provide high correspondence to the modeled system—is laborious both methodologically and computationally.

In [2] we have introduced a parameter identification tool-chain consisting of a model-checker and data management and visualization tools. The volume of the data obtained—possible parameters, simulation traces etc.—is usually extensive. The complexity and heterogeneity of the results are increasing even more as we develop new tools, making the usage of the tool-chain nontrivial even for an experienced user.

To tackle this problem we have developed Esther—an on-line service (available at http://esther.fi.muni.cz/) with a visual interface and server-side data management, providing an instant access to our existing tools.

2 Esther

Results of the model checking and other analytical processes are all stored within an SQLite database, generally dividing the tools into three groups based on whether they:

1. Enumerate all the parametrizations that satisfy imposed constraints and thus create a database.
2. Change the database, either by computing additional data or by filtering the current content based on the demands of a user.
3. Read the database to create human-readable plots of the data.

A. Gupta and T.A. Henzinger (Eds.): CMSB 2013, LNBI 8130, pp. 257–258, 2013.

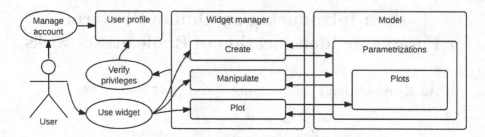

Fig. 1. A scheme of the Esther architecture

Currently only a single tool is available through Esther for each of the steps. The design of the service however expects other tools to be employed as well—each tool can be connected to the server using a so-called *Esther Widget* (EW). Each EW spawns an individual tab within a user session and provides a meta-layer between a user and a tool or a datafile present on the server. The functionality of an EW then depends on its purpose and may include execution of a tool, data editing, viewing results, etc. A scheme of the architecture is given in Fig. 1. The platform comprises file management utilities allowing for simple manipulation, storing and sharing files between users.

3   Conclusion

Currently available tools mostly do not provide any means of manipulating multiple parametrizations. In this sense, the closest relative to Esther is the SMBioNet tool [3], which is however restricted only to the process of parameter filtering and moreover is not currently available for download. Considering existing web platforms, the closest one would be BMA [4]. This user-friendly service utilizes a similar framework and relies also on model-checking, however, BMA requires complete specification of parameters *a priori*.

Therefore we believe that our web service makes a significant contribution. Here we present its first instance, which we plan to further refine and equip with new tools, some of which are currently in late stages of their development.

References

1. Wang, R.-S., Saadatpour, A.: Boolean modeling in systems biology: an overview of methodology and applications. Physical Biology 9(5) (2012)
2. Klarner, H., Streck, A., Šafránek, D., Kolčák, J., Siebert, H.: Parameter identification and model ranking of Thomas networks. In: Gilbert, D., Heiner, M. (eds.) CMSB 2012. LNCS, vol. 7605, pp. 207–226. Springer, Heidelberg (2012)
3. Khalis, Z., Comet, J.-P., Richard, A., Bernot, G.: The SMBioNet method for discovering models of gene regulatory networks. Genes, Genomes and Genomics 3(1), 15–22 (2009)
4. Benque, D., Bourton, S., Cockerton, C., Cook, B., Fisher, J., Ishtiaq, S., Piterman, N., Taylor, A., Vardi, M.Y.: BMA: visual tool for modeling and analyzing biological networks. In: Madhusudan, P., Seshia, S.A. (eds.) CAV 2012. LNCS, vol. 7358, pp. 686–692. Springer, Heidelberg (2012)

Identifying Latent Dynamic Components in Biological Systems

Ivan Kondofersky, Christiane Fuchs, and Fabian J. Theis

Institute of Computational Biology, Helmholtz Zentrum München, Germany
Institute for Mathematical Sciences, Technische Universität München, Germany
ivan.kondofersky@helmholtz-muenchen.de
hmgu.de/icb

Abstract. In systems biology, a general aim is to derive regulatory models from multivariate readouts, thereby generating predictions for novel experiments. However any model only approximates reality, leaving out details or regulations. These may be completely new entities such as microRNAs or metabolic fluxes which have a substantial contribution to the network structure and can be used to improve the model describing the regulatory system and thus produce meaningful results. In this poster, we consider the case where a given model fails to predict a set of observations with acceptable accuracy. In order to refine the model, we propose an algorithm for inferring additional upstream species that improve the prediction as well as the model fit and at the same time are subject to the model dynamics. In the studied context of ODE-based models, this means systematically extending the network by an additional latent dynamic variable. This variable is modeled by splines in order to easily access derivatives; the influence vector of the variable onto the species is then estimated from the data via model selection.

Keywords: Dynamical modeling, Differential Equations, Splines, Model Selection, Maximum Likelihood Estimation.

A central objective in systems biology is to identify components of biological system networks and their relation to one another. For the prediction of network behavior, mathematical models are employed, typically involving several unknown parameters in addition to the network components. A popular modeling approach for time-resolved measurements are ordinary differential equations (ODEs), representing the dynamics of and dependencies between the components of the network. The parameters describing the dynamics in the ODE have to be inferred statistically, and in case of several competing network models, the most appropriate one can be chosen by model selection methods.

In such an analysis, the ODEs directly arise from the network topology, i.e. the modeller specifies the components of the network and possible interactions. In many applications, key elements of the dynamics of interest have been determined in various studies and are well-known in the literature. It is possible, however, that some interaction partners or possible connections are still missing. For example, in addition to transcription factors modulating gene regulation,

A. Gupta and T.A. Henzinger (Eds.): CMSB 2013, LNBI 8130, pp. 259–260, 2013.
© Springer-Verlag Berlin Heidelberg 2013

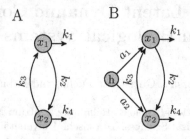

Fig. 1. Example networks (circles: observed and hidden components, triangle: unobserved or indirectly observed components). **A**: network without hidden components and all components observed. **B**: network with one hidden component (h) and all other components observed.

there is strong evidence that microRNAs play an important role in transcription and translation processes . Translation could also be influenced by external influences like drugs. Consequently, a considered mathematical model might be insufficient to explain the dynamics of interest, i. e. even the best model fit can show discrepancies to the measured data which are not simply due to measurement error.

In our work, we address this problem by considering the effect of hidden influences to the network. We do not assume a functional form of the putative time courses of such hidden processes but flexibly estimate their dynamics and interaction strengths. If we find a hidden influence that substantially improves the model explaining the data, we will also provide its biological meaning with the help of experimental collaborators. Thus, we can guide design of additional experiments in a detailed manner by providing exact quantification of the hidden time courses as well as relative reaction rates between the hidden components and the existing network.

Figure 1 demonstrates the central idea of our work. Existing biological networks (Figure 1A) are systematically extended with the addition of a latent component h (Figure 1B). We develop a two-step method in which we use splines and maximum likelihood estimation in order to estimate and identify the time course of the hidden component as well as its reaction rate constants a_i.

Application of our method on several artificially created examples suggest a very good performance in terms of prediction of the unknown time course of the hidden component as well as the produced estimates of the parameters of interest. Additionally our method can be used as a tool for recovering previously misspecified networks.

Next-Newtomics: The Next Generation Repository for Bioinformatical Interpreted Omics Datasets from the Newt *Notophthalmus viridescens*

Marc Bruckskotten, Mario Looso, and Thomas Braun

Max-Planck-Institute for Heart and Lung Research, Bad Nauheim, Germany

Abstract. Comprehensive bioinformatics repositories for the standard model organisms like mouse [9], rat [8], zebrafish [12], *Drosophila* [10] and *Xenopus* [4] provide access to all levels of sequence data sets, including genome, transcriptome and proteome data. For non-standard model organisms, very little information from publically accessible data have been collected and organized. This situation prevents dissemination of useful research information to a broader research community and keeps such model organisms in isolation.

One of these organisms is the red spotted newt *Notophthalmus viridescens*, known for its exceptional regenerative capabilities for more than 200 years. The newt possesses the ability to entirely replace lost appendages [5, 13] and regenerate the lens [7], parts of the central nervous system [2], and the heart [1, 11].

These unique features make the newt an excellent model to study fundamental processes of tissue regeneration. Challenging is the fact, that the estimated genome size of the newt is up to 10 times larger than that of humans. These circumstances have severely impeded genome projects despite the increasing speed and capacities of modern sequencing machines and assembly algorithms. As a result of these drawbacks, approximately only 100 non-redundant protein sequences for the newt are available in the NCBI-NR database, although a set of almost 11 000 sequenced Expressed Sequence Tags (ESTs) from regenerating hearts of the newt *Notophthalmus viridescens* exists [3].

In this context the 'Newtomics Resource' (http://newtomics.mpi-bn.mpg.de/) [6] is developed as a bioinformatics tool with an integrated database, which enables researchers to analyze, retrieve and store data sets dedicated to the molecular characterization of this organism in a data-warehouse like manner. Newtomics has a unique transcript-centered database design, which refers to the biological reality and allows analyzing, storing, managing and data-mining of complex high-throughput datasets, as well as meta-information. The integrated design combines high-throughput data from NGS and traditional sequencing, annotation and functional characterization as well as quantitative expression data from time-series microarray-experiments and RNA-seq approaches. Furthermore, Newt-omics is also capable to work with large sets of identified peptides derived from a mass spectrometry approach.

The design is open to additional datasets from different sources, without the need to change the database-structure or the data within. The integrated information is analyzed and data-mined by bioinformatics tools and pipelines, which processes the external data from operational sources. A web based graphical

A. Gupta and T.A. Henzinger (Eds.): CMSB 2013, LNBI 8130, pp. 261–262, 2013.
© Springer-Verlag Berlin Heidelberg 2013

user interface allows access to the sets of molecular data. The implemented tools and the transcript-centered view combines and visualizes all kinds of data and allows a live view of the data on the transcriptomic and proteomic level.

The open design and the bioinformatics tools allow a transfer of these achievements and use this important bio-computational tool in similar projects, which also focus on the characterization of niche model organisms.

Keywords: Next-Generation Sequencing, *N. viridescens*, heart regeneration, database, data-warehouse, data-mining, high-throughput techniques.

References

1. Bader, D., Oberpriller, J.O.: Repair and reorganization of minced cardiac muscle in the adult newt (Notophthalmus viridescens). Journal of Morphology 155, 349–357 (1978)
2. Berg, D.A., Kirkham, M., Beljajeva, A., et al.: Efficient regeneration by activation of neurogenesis in homeostatically quiescent regions of the adult vertebrate brain. Development 137, 4127–4134 (2010)
3. Borchardt, T., Looso, M., Bruckskotten, M., et al.: Analysis of newly established EST databases reveals similarities between heart regeneration in newt and fish. BMC Genomics 11, 4 (2010)
4. Bowes, J.B., Snyder, K.A., Segerdell, E., et al.: Xenbase: a Xenopus biology and genomics resource. Nucleic Acids Research 36, D761–D767 (2008)
5. Brockes, J.P.: Amphibian limb regeneration: rebuilding a complex structure. Science 276, 81–87 (1997)
6. Bruckskotten, M., Looso, M., Reinhardt, R., et al.: Newt-omics: a comprehensive repository for omics data from the newt Notophthalmus viridescens. Nucleic Acids Research 40, D895–D900 (2012)
7. Call, M.K., Grogg, M.W., Tsonis, P.A.: Eye on regeneration. Anatomical Record. Part B, New Anatomist 287, 42–48 (2005)
8. Dwinell, M.R., Worthey, E.A., Shimoyama, M., et al.: The Rat Genome Database 2009: variation, ontologies and pathways. Nucleic Acids Research 37, D744–D749 (2009)
9. Eppig, J.T., Blake, J.A., Bult, C.J., et al.: The Mouse Genome Database (MGD): comprehensive resource for genetics and genomics of the laboratory mouse. Nucleic Acids Research 40, D881–D886 (2012)
10. Marygold, S.J., Leyland, P.C., Seal, R.L., et al.: FlyBase: improvements to the bibliography. Nucleic Acids Research 41, D751–D757 (2013)
11. Piatkowski, T., Muhlfeld, C., Borchardt, T., et al.: Reconstitution of the Myocardium in Regenerating Newt Hearts is Preceded by Transient Deposition of Extracellular Matrix Components. Stem Cells and Development (2013)
12. Sprague, J., Bayraktaroglu, L., Clements, D., et al.: The Zebrafish Information Network: the zebrafish model organism database. Nucleic Acids Research 34, D581–D585 (2006)
13. Tsonis, P.A.: Amphibian limb regeneration. In Vivo 5, 541–550 (1991)

miRNA Expression Analysis during Heart Regeneration of *N. viridescens*

Mario Herzog, Stefan Günther, Thomas Braun, and Mario Looso

Max-Planck-Institute for Heart and Lung Research, Bad Nauheim, Germany

Abstract. *Notophthalmus viridescens*, a member of the salamander family, is an excellent model organism to study regenerative processes. Recently we gained first molecular insights into its ability to regenerate parts of its heart after injury by generating and analyzing the transcriptome.[1] This study now reveals additional information about miRNAs, a small class of non-coding RNA which play an important role in post-transcriptional gene regulation in many biological processes, including development and regeneration.[2]

Small RNA libraries including several time points during heart regeneration were prepared and analysed by next-generation sequencing. Due to the lack of a genomic sequence of the newt MiRNAs have been identified by searching homologue sequences contained in miRBase[3], leading to set of 588 annotated sequences. Several potential new mi-RNAs could be detected but were not used for further analysis. Expression profiles of candidates included in each library were generated, indicating several miRNAs with distinct expressions. Among those mir-451 was the most abundant one and manifold higher expressed in response to heart damage. All miRNAs were clustered regarding to their profiles in order to find functional similarities.

Keywords: MiRNA, Next-Generation Sequencing, *N. viridescens*, heart regeneration.

References

1. Bruckskotten, M., Looso, M., Reinhardt, R., Braun, T., Borchardt, T.: Newt-omics: a comprehensive repository for omics data from the newt Notophthalmus viridescens. Nucleic Acids Res. 40, D895–D900 (2011)
2. Porrello, E.: microRNAs in cardiac development and regeneration. Clinical Science 125, 151–166 (2013)
3. Kozomara, A., et al.: miRBase: integrating microRNA annotation and deep-sequencing data. Nucleic Acids Res. 39(Database issue), D152–D157 (2011)

A. Gupta and T.A. Henzinger (Eds.): CMSB 2013, LNBI 8130, p. 263, 2013.

Algorithm to Predict G-Quadruplex Folding through Score Computing

Bedrat Amina[1,2], Guédin Aurore[1,2], Amrane Samir[1,2],
Renaud de la Faverie Amandine[1,2], and Mergny Jean-Louis[1,2]

[1] Université Bordeaux Segalen , 33076 Bordeaux cedex, France
a.bedrat@iecb.u-bordeaux.fr
[2] Institut Européen de Chimie et de Biologie, Laboratoire ARNA, 2 rue Robert
Escarpit , 33608 Pessac Cedex, France

The Watson-Crick model for the B-DNA type double-helix has always been regarded as the biologically relevant structure of DNA. In addition to this canonical duplex structure, single-stranded DNAs can fold into a wide variety of non-canonical base pairs such as: hairpin, triplex, G-quadruplex, and i-motif structures. Polypurine and polypyrimidine tracts and other repetitive sequences can form non-duplex structure, which are related to a wide variety of biological activities. It has been previously found that some G-rich DNA (and RNA) sequences are able to forme stable four-stranded structures known as G-quadruplexes. The topology of the G-quadruplexes (or G4) consists of stacks of a square arrangement of four guanines (named a tetrad or a G-quartet) in a planar Hoogsteen hydrogen bonded form. This structure is stabilized by monovalent cation e.g. K^+ and Na^+. The core guanines are linked by three nucleic acid sequences (loops) of varying composition and topology [1]. Our goal is to accurately predict which genomic sequences are able to form G4.

The high thermodynamic stability of G4 under near-physiological conditions suggests that these structures may form in genomic DNA *in vivo*. In addition a structure-specific antibody has been employed to quantitatively visualize DNA G-quadruplex in human cell [2]. Altogether, these findings suggest that G4 structures are physiologically relevant.

Bioinformatic approaches have played an important role by identifying genomes candidate sequences with G-quadruplex forming potential. A number of algorithms have been used to date such us pattern-based sequences algorithms. They are used to predict G4 based on data from DNA experiments, with limited structural information available and comparative analyses. The most important limitation is that the prediction is generally limited to the standard description $G_{3+}N_{(1-7)}G_{3+}N_{(1-7)}G_{3+}N_{(1-7)}G_{3+}$ needed for G4 formation [3],[4]. Approximately 370000 sequences with putative G-quadruplex-forming motifs are dispersed in the human genome [5], [4]. They are concentrated in promoter regions [6], 5' and 3' UTR' s [4]. Both the β subunit of the *Oxytricha* telomere end-bending protein (βTBP)[7] and repressor activator protein 1 (RAP1) in *Saccharomyces cerevisieae* [8] promotes G-quadruplex formation. G4 structures are also associated with a number of important aspects of genome function, which include transcription, recombination and replication [4].

A. Gupta and T.A. Henzinger (Eds.): CMSB 2013, LNBI 8130, pp. 264–265, 2013.
© Springer-Verlag Berlin Heidelberg 2013

Based on these findings [3], the pattern used in most current bioinformatics searches for identifying putative G-quadruplex-forming sequences could be reformulated, and the number of putative G4 in the genome is expected to be larger than previously reported.

We propose an objective score function that would predict G4 folding propensity from a linear nucleic acid sequence. The new method focus on Guanines clusters and GC asymmetry, taking into account the whole genomic region rather than individual quadruplexes sequences. In parallel with this computational technique, a large scale *in vitro* experimental work has also been developed to test the prediction of our algorithm *in silico* on some genes of interest and small prokaryotes and eukaryotes genomes. It is relatively straightforward to experimentally determine the thermodynamic stability using Ultra Violet melting[9], the structural insight of the nucleic acid sequence by Circular Dichroism, Thermal Difference Spectra and NMR [10],[11]. Finally, the accuracy of our prediction method has already been proved and compared to previously predicted sequences. In addition this methodology has found new quadruplex putative sequences in HIV and *Dictyostelium*, which could not be identified by previous computational methods.

References

1. Luedtke, N.W.: Targeting G-quadruplex DNA with Smal Molecules. Chimia 63, 134–139 (2009)
2. Biffi, G., Tannahill, D., McCafferty, J., Balasubramanian, S.: Quantitative visualization of DNA G-quadruplex structures in human cells. Nat. Chem. 5, 182–186 (2013)
3. Mukundan, V.T., Tuân, P.A.: Bulges in G-Quadruplexes: Broadening the Definition of G-Quadruplex-Forming Sequences. J. Am. Chem. Soc. (2013)
4. Huppert, J.L., Balasubramanian, S.: Prevalence of quadruplexes in the human genome. Nucleic Acids Res. 33, 2908–2916 (2005)
5. Todd, A.K., Johnston, M., Neidle, S.: Highly prevalent putative quadruplex sequence motifs in human DNA. Nucleic Acids Res. 33, 2901–2907 (2005)
6. Hershman, S.G., Lee, C.Q., Kozak, M.L., Johnson, F.B.: Genomic distribution and functional analyses of potential G-quadruplex-forming sequences in Saccharomyces cerevisiae. Nucleic Acids Res. 36, 144–156 (2008)
7. Fang, G., Cech, T.R.: The β subunit of Oxytricha telomere-binding protein promotes G-quartet formation by telomeric DNA. Cell 74, 875–885 (1993)
8. Giraldo, R., Rhodes, D.: The yeast telomere-binding protein RAP1 binds to and promotes the formation of DNA quadruplexes in telomeric DNA. EMBO J. 13, 2411–2420 (1994)
9. Mergny, J.L., Lacroix, L.: Analysis of thermal melting curves. Oligonucleotides 13, 515–537 (2003)
10. Mergny, J.L., Lacroix, L., Amrane, S., Chaires, J.B.: Thermal difference spectra: a specific signature for nucleic acid structures. Nucleic Acids Res. 33, e138 (2005)
11. Masiero, S., Spada, G.P.: A non-empirical chromophoric interpretation of CD spectra of DNA G-quadruplex structures. Org. Biomol. Chem. 8, 2683–2692 (2010)

Optimization Based Inference of Metabolic Networks from Metabolome Data

Melik Öksüz[1,2], Hasan Sadikoğlu[2], and Tunahan Çakır[1,*]

[1] Department of Bioengineering, Gebze Institute of Tehnology, 41400, Gebze-Kocaeli, Turkey
{tcakir,moksuz}@gyte.edu.tr
[2] Department of Bioengineering, Gebze Institute of Tehnology, 41400, Gebze-Kocaeli, Turkey
sadikoglu@gyte.edu.tr

Abstract. An optimization-based network inference approach was developed and applied to in silico metabolome data of *Escherichia coli* and *Saccharomyces cerevisiae*. The steady-state metabolome data used were generated in silico by simulating kinetic models belonging to the investigated microorganisms. Lyapunov equation, which puts a link between Jacobian matrix of the system and the covariance matrix is the basis for the optimization based approach. Data-derived covariance matrix is the input to the underdetermined Lyapunov equation, which is used for the prediction of Jacobian matrix based on an objective function. Taking into account the sparsity of biological networks as cellular objective, a consistent mathematical objective function was chosen. Inference of the underlying metabolic network was performed based on a genetic-algorithm formulation. The approach results in promising inference of the metabolic networks in question. Sensitivity of the results to the approach is also investigated.

Keywords: Network Inference, Lyapunov equation, Metabolome Data, Reverse Engineering.

1 Introduction

Network Inference based on high-throughput data is a common approach with many examples on transcriptome data-based inference of gene-regulatory networks [1, 2]. On the other hand, application to metabolome data to reverse-engineer metabolic networks is scarce [3, 4, 5]. These examples have used synthetic metabolome data belonging to different organisms to infer mostly undirected metabolic networks. Metabolic network inference has the potential to identify both enzymatic and regulatory interactions in metabolism.

A proper experimental design is important for the inference of cellular networks. The main focus is on dynamic data in literature since it is considered to be more informative in terms of reverse engineering. The mathemathical approaches are also abundant, from statistical analyses to Bayesian based models and optimization approaches [6]. This study focuses on an optimization based network inference of metabolic networks based on steady-state type of metabolome data.

---

[*] Corresponding author.

A. Gupta and T.A. Henzinger (Eds.): CMSB 2013, LNBI 8130, pp. 266–267, 2013.
© Springer-Verlag Berlin Heidelberg 2013

2   Results

In silico data based on available kinetic models of *Escherichia coli* (central carbon metabolism) [7] and *Saccharomyces cerevisiae* (glycolysis) [7] was used as an input to the optimization based approach. Simulations were performed in MATLAB 2013a, with the help of Global Optimization and Parallel Computing Toolboxes. The genetic-algorithm based optimization approach takes the data as an input to Lyapunov equation, and uses maximal sparsity as the cellular objective function. The calculated Jacobian matrix holds the information on directed metabolic network. The demonstration of the approach on in silico metabolome data is promising, with close to one true positive and false positive rates. The effect of noise on data was also investigated to see the sensitivity of the approach. Compared to an undirected statistics-based approach (Graphical Gaussian Model), the approach was shown to have acceptable prediction rates, with directionality information. The approach was shown to be applicable to small metabolic systems successfully. One other promising aspect of the approach is the inference of a directed network from steady-state data.

Acknowledgements. The financial support by TUBITAK (Project Code: 110M464) is gratefully acknowledged.

References

1. Bansal, M., Belcastro, V., Ambesi-Impiombato, A., di Bernardo, D.: How to infer gene networks from expression profiles. Mol. Sys. Biol. 3, 78 (2007)
2. Soranzo, N., Bianconi, G., Altafini, C.: Comparing association network algorithms for reverse engineering of large-scale gene regulatory networks: Synthetic versus real data. Bioinformatics 23, 1640–1647 (2007)
3. Nemenman, I., Escola, G.S., Hlavacek, W.S., Unkefer, P.J., Unkefer, C.J., Wall, M.E.: Reconstruction of metabolic networks from high-throughput metabolite profiling data: In silico analysis of red blood cell metabolism. Ann. NY Acad. Sci. 1115, 102–115 (2007)
4. Çakır, T., Hendriks, M.M.W.B., Westerhuis, J.A., Smilde, A.K.: Metabolic network discovery through reverse engineering of metabolome data. Metabolomics 5, 318–329 (2009)
5. Hendrickx, D.M., Hendriks, M.M.W.B., Eilers, P.H.C., Smilde, A.K., Hoefsloot, H.C.J.: Reverse engineering of metabolic networks, a critical assessment. Mol. Biosys. 7, 511–520 (2011)
6. Markowetz, F., Spang, R.: Inferring cellular Networks-A review. BMC Bioinformatics 8(suppl. 6), S5 (2007)
7. Chassagnole, C., Noisommit-Rizzi, N., Schmid, J.W., Mauch, K., Reuss, M.: Dynamic modeling of the central carbon metabolism of Escherichia coli. Biotechnol. Bioeng. 79, 53–73 (2002)
8. Teusink, B., Passarge, J., Reijenga, C.A., Esgalhado, E., van der Weijden, C.C., Schepper, M., et al.: Can yeast glycolysis be understood in terms of in vitro kinetics of the constituent enzymes? Testing biochemistry. Eur. J. Biochem. 267, 5313–5329 (2000)

PHISTO: A New Web Platform for Pathogen-Human Interactions

Saliha Durmuş Tekir[1], Tunahan Çakır[2,*], Emre Ardıç[3], İlknur Karadeniz[4],
Arzucan Özgür[4], Fatih Erdoğan Sevilgen[3], and Kutlu Ö. Ülgen[1]

[1] Department of Chemical Engineering, Boğaziçi University, 34342, Bebek-İstanbul, Turkey
{saliha.durmus,ulgenk,arzucan.ozgur,
ilknur.karadeniz}@boun.edu.tr
[2] Department of Bioengineering, Gebze Institute of Tehnology, 41400, Gebze-Kocaeli, Turkey
{tcakir,sevilgen,eardic}@gyte.edu.tr
[3] Department of Computer Engineering, Gebze Institute of Tehnology, 41400, Gebze-Kocaeli,
Turkey
[4] Department of Computer Engineering, Boğaziçi University, 34342, Bebek-İstanbul, Turkey

Abstract. The interactions between the proteins of infectious microorganisms, pathogens, and their human hosts are the cause behind the manipulation of human cellular mechanisms by the microorganisms to their own advantage, resulting in infection in the host organism. Improved understanding of pathogen-host interactions (PHIs) will significantly contribute to our knowledge of the mechanisms involved in infection, and allow novel therapeutic solutions to be devised. In the post-genomic era, following the advances in genomics, proteomics, and then interactomics, interspecies protein interaction data of pathogen-human systems could be produced in large-scale within very recent years. PHISTO (Pathogen-Host Interaction Search Tool, www.phisto.org) is a new Web platform that provides relevant information about pathogen-host protein-protein interactions. It enables access to the most up-to-date PHI data for all pathogen types for which experimentally-verified protein interactions with human are available. The platform also offers integrated tools for visualization of PHI networks, graph theoretical analysis of human proteins targeted by pathogens and BLAST search. PHISTO aims to facilitate PHI studies that provide potential therapeutic targets for infectious diseases by offering up-to-date data through its database functionality as well as computational analysis tools.

Keywords: pathogen-human interaction, web-accessible platform, bioinformatics, infection mechanism, therapeutic target.

1 Introduction

The recent advances in high-throughput protein interaction detection methods have led to the production of large-scale interspecies protein-protein interaction (PPI) data of pathogen-human systems [1]. Currently, there are a number of pathogen–host interaction (PHI) resources that are specific to some pathogens. The only available

A. Gupta and T.A. Henzinger (Eds.): CMSB 2013, LNBI 8130, pp. 268–269, 2013.

resource to access all PHI data in a single database [2] does not offer any additional functionality to analyze PHI networks.

PHISTO has been developed to serve as an up-to-date and functionally enhanced source of PHI data through a user-friendly interface [3]. PHIs in PHISTO are imported from several PPI databases using the PSICQUIC tool [4]. Text mining is used to label PHIs extracted without any information on interaction detection method. Tools for visualization of small PHI networks and graph-theoretical analysis of targeted human proteins may enable users to gain crucial insights on roles of pathogen/human proteins within infection mechanisms [5; 6]. The BLAST interface offers to search for orthologous PHIs for pathogens lacking experimental data.

2   Results

PHISTO is designed as a Web-accessible platform with two-tier architecture. The back tier is a MySQL-based database. The front tier is a PHP- and Javascript-based user interface that runs on an Apache Web server.

With regular data updates and complete coverage of all data available for each pathogen type, PHISTO will always provide unified access to up-to-date PHI data. PHISTO is aimed to provide a centralized and up-to-date platform for studying pathogen–host protein interaction systems with future curation of PHIs from literature by text mining and additional advanced analysis tools for PHI networks.

Acknowledgements. The Research Funds of Boğaziçi University, Project 5554D, Marie Curie FP7-Reintegration-Grants within the 7th European Community Framework Programme, and Ekin Kimya Tic. Ltd. Şti gratefully are acknowledged for the financial support.

References

1. Durmuş Tekir, S., Ülgen, K.Ö.: System Biology of Pathogen-Host Interaction: Networks of Protein-Protein Interaction within Pathogens and Pathogen-Human Interactions in the Post-Genomic Era. Biotechnol. J. 8, 85–96 (2013)
2. Kumar, R., Nanduri, B.: HPIDB – a unified resource for host-pathogen interactions. BMC Bioinformatics 11, S16 (2010)
3. Durmuş Tekir, S., Çakır, T., Ardıç, E., Sayılırbaş, A.S., Konuk, G., Konuk, M., Sarıyer, H., Uğurlu, A., Karadeniz, İ., Özgür, A., Sevilgen, F.E., Ülgen, K.Ö.: PHISTO: Pathogen-Host Interaction Search Tool. Bioinformatics 29, 1357–1358 (2013)
4. Aranda, B., Blankenburg, H., Kerrien, S., Brinkman, F.S., Ceol, A., Chautard, E., Hermjakob, H.: PSICQUIC and PSISCORE: accessing and scoring molecular interactions. Nat. Methods. 8, 528–529 (2011)
5. Durmuş Tekir, S., Çakır, T., Ulgen, K.: Infection strategies of bacterial and viral pathogens through pathogen-human protein-protein interactions. Front. Microbiol. 3, 46 (2012)
6. Dyer, M.D., Murali, T.M., Sobral, B.W.: Landscape of human proteins interacting with viruses and other pathogens. PLoS Pathog. 4, 32 (2008)

Frameshift Correction in De Novo Assembled Transcriptome Data Using Peptide Data, Blast Sequence Alignments and Hidden Markov Models

Stephan Neese and Mario Looso

Max Planck Institute for Heart and Lung Research, Department of Bioinformatics,
Bad Nauheim, Germany

Abstract. Frameshift errors in de novo sequenced transcriptome data are difficult to find since not all transcripts are covered with publicly available and curated reference data. One approach for finding frameshift errors are hidden markov models. HMMs are a widely used approach in bioinformatics to identify patterns in sequences such as coding or conserved regions [1]. However there haven't been made many efforts using HMMs for detecting frameshift errors in nucleotide sequences.

Here we introduce an approach for frameshift correction using hidden markov models, blast alignment data and peptide data derived from mass spectrometry [2]. This algorithm is implemented as pipeline with three correction steps for transcriptomic sequences. First, it employs peptide data for a preliminary correction, followed by aligning the investigated sequences against publicly available protein databases such as the SwissProt database utilizing the BlastX alignment tool. Finally, the resulting alignment files for correction are used for creating training data sets for the HMM using known coding and shifted areas on the sequence. The trained HMM is then used to perform the final identification.

Keywords: Hidden markov models, Frameshift errors, Transcriptome data, de novo.

References

1. Gouzy, J., Carrere, S., Schiex, T.: FrameDP: sensitive peptide detection on noisy matured sequences. Bioinformatics 25(5), 670–671 (2009)
2. Bruckskotten, M., Looso, M., Reinhardt, R., Braun, T., Borchardt, T.: Newt-omics: a comprehensive repository for omics data from the newt Notophthalmus viridescens. Nucleic Acids Res. 40, 895–900 (2011)

A. Gupta and T.A. Henzinger (Eds.): CMSB 2013, LNBI 8130, p. 270, 2013.
© Springer-Verlag Berlin Heidelberg 2013

From Prokaryote Genome Sequencing to Pan-Genomic Modeling

Carsten Künne[*]

Max-Planck-Institute for Heart and Lung Research, Bad Nauheim, Germany
carsten.kuenne@mpi-bn.mpg.de

Abstract. Exciting technological developments including rapid and cost-effective DNA sequencing have led to an explosion of information considering prokaryotic genome sequences, annotation, classification, genomic epidemiology, taxonomy, and pathogen detection. While production of these extensive data sets summarizing a pan-genome has become a standard procedure, interpretation frequently necessitates massive manual intervention by a trained bioinformatician in order to correlate and extract relevant data [1].

We have devised an integrated software suite serving as a centralized hub for comparative genomic assessment called GECO, which is able to streamline further analyses using third-party tools and can be operated using a web browser [2]. GECO allows fully automatic classification of genes by pan-genomic conservation into core, accessory and specific clusters based on flexible sequence homology criteria. Resulting correlations can be exported graphically or in the form of tab-delimited lists. Among the latter are matrices sorted for conservation in selected replicons or for synteny according to a reference strain. These can be employed to create concise and congruent batch annotations of new genome sequences.

The size and distribution of the pan-genome of any prokaryotic set of genomes can be surveyed in order to identify insertions or deletions supporting taxonomic or phenotypic divisions, evolutionary patterns, pathogenicity determinants, or genomic loci valuable for typing purposes [3]. Putatively horizontally transferred genes can be identified by deviation of GC content from the average of the genome, as well as a deviating codon composition [4]. A simple procedure for extraction of gene and protein sequences of single genes and complete homology clusters supports further analyses like multiple sequence alignments and phylogenetic reconstructions [5, 6]. Finally, diverging regions inside related replicons can be identified and visualized in publication quality images to recognize hyperdynamic hotspots, mobile elements and prophages.

All of these methods to analyze, annotate and model prokaryotic pan-genomes are offered by GECO to an audience of researchers without detailed knowledge of computational sciences, and can be installed on local hardware. Streamlined integration and correlation of pan-genomic data will be paramount to the effective synthesis of available knowledge.

---

[*] Corresponding author.

A. Gupta and T.A. Henzinger (Eds.): CMSB 2013, LNBI 8130, pp. 271–272, 2013.
© Springer-Verlag Berlin Heidelberg 2013

Keywords: prokaryote, pan-genome modeling, annotation, mobile elements, phylogeny, correlation, homology, evolution, adaptation, software suite.

References

1. Tettelin, H., Riley, D., Cattuto, C., Medini, D.: Comparative genomics: the bacterial pan-genome. Current Opinion in Microbiology 11, 472–477 (2008)
2. Kuenne, C.T., Ghai, R., Chakraborty, T., Hain, T.: GECO–linear visualization for comparative genomics. Bioinformatics 23, 125–126 (2007)
3. Kuenne, C., Billion, A., Mraheil, M.A., Strittmatter, A., Daniel, R., Goesmann, A., Barbuddhe, S., Hain, T., Chakraborty, T.: Reassessment of the Listeria monocytogenes pan-genome reveals dynamic integration hotspots and mobile genetic elements as major components of the accessory genome. BMC Genomics 14, 47 (2013)
4. Hasan, M.S., Liu, Q., Wang, H., Fazekas, J., Chen, B., Che, D.: GIST: Genomic island suite of tools for predicting genomic islands in genomic sequences. Bioinformation 8, 203–205 (2012)
5. Angiuoli, S.V., Salzberg, S.L.: Mugsy: fast multiple alignment of closely related whole genomes. Bioinformatics 27, 334–342 (2011)
6. Tamura, K., Peterson, D., Peterson, N., Stecher, G., Nei, M., Kumar, S.: MEGA5: molecular evolutionary genetics analysis using maximum likelihood, evolutionary distance, and maximum parsimony methods. Molecular Biology and Evolution 28, 2731–2739 (2011)

A Simulation Approach to Detect Oscillating Behaviour in Stochastic Population Models

Jorge Júlvez*

Universidad de Zaragoza, Spain
julvez@unizar.es

System Parameters and Deterministic Limit. This work focuses on biological systems modelled as density dependent Markov processes[2]. The dynamics of such systems is often studied by considering the deterministic limit, which is obtained as the solution of a set of Ordinary Differential Equations (ODEs)[1]. The deterministic limit might not capture important system behaviours such as oscillations[2]. The method presented here averages the distances and angles of a number of stochastic simulations to easily detect oscillating behaviours.

System parameters: a) $s \in \mathbb{N}$ and $n \in \mathbb{N}$ are the number of species and events; c) $\mathbf{X}(t) \in \mathbb{N}_{\geq 0}^q$ is the state of the system at time t ($X_i(t)$ denotes the number of elements of species i at time t); d) $\nu \in \mathbb{N}_{\geq 0}^{q \times n}$ is the stoichiometry matrix, i.e., ν_i^j is the change produced in species i by event j; e) $V \in \mathbb{R}_{>0}$ is the system size; f) $W_j : \mathbb{R}_{\geq 0}^q \times \mathbb{R}_{>0} \to \mathbb{R}_{\geq 0}$ is the transition rate function, i.e, $W_j(\mathbf{X}(t), V)$ is the rate associated to event j for population $\mathbf{X}(t)$ and system size V (for conciseness, we will use $\mathbf{X}$ rather than $\mathbf{X}(t)$, and $W_j(\mathbf{X})$ rather than $W_j(\mathbf{X}(t), V)$).

The system is modelled as a jump Markov process in which events are exponentially distributed with rates $W_j(\mathbf{X})$. The occurrence of an event j changes the system state from $\mathbf{X}$ to $\mathbf{X} + \nu^j$. Functions $W_j(\mathbf{X})$ are assumed to be differentiable, nonnegative, time independent and to satisfy the mass-action law[2].

Deterministic limit: Under some conditions[1] on $W_j(\mathbf{X})$, the deterministic limit behaviour is given by the following set of ODEs: $\frac{dX_i}{dt} = \sum_{j=1}^n \nu_i^j W_j(\mathbf{X})$.

Method. Consider the trajectories obtained for two stochatic simulations. When computing the mean populations, one averages the cartesian coordinates of the populations in the phase space. Nevertheless, other coordinate systems, e.g., polar coordinates if $s = 2$, can be considered. Figure 1(a) shows the result of averaging the cartesian and polar coordinates of two states.

Let us describe how to average the polar coordinates of a number of stochastic simulations (for systems with $s > 2$, hyperspherical coordinates can be used). Assume that M stochastic simulations have been performed, and the trajectories have been resampled at same sampling times. Let (X_q^0, Y_q^0), (X_q^1, Y_q^1), ..., be the cartesian coordinates of simulation $q \in \{1 \ldots M\}$ at the sampling times. Let the origin of the polar coordinate system be the reference point a with cartesian coordinates (a_x, a_y). Each (X_q^k, Y_q^k) can be transformed to polar

* This work has been partially supported by CICYT - FEDER project DPI2010-20413. The Group of Discrete Event Systems Engineering (GISED) is partially co-financed by the Aragonese Government (Ref. T27) and the European Social Fund.

A. Gupta and T.A. Henzinger (Eds.): CMSB 2013, LNBI 8130, pp. 273–275, 2013.

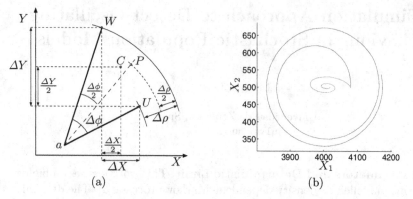

Fig. 1. (a) Average cartesian (C) and polar (P) coordinates of U and W with respect to a; (b) Average cartesian and polar trajectories

coordinates (ρ_q^k, θ_q^k) with origin at a by using: $\rho_q^k = \sqrt{(X_q^k - a_x)^2 + (Y_q^k - a_y)^2}$, $\theta_q^k = \text{atan}(Y_q^k - a_y, X_q^k - a_x)$ where $\text{atan}(y, x) : \mathbb{R} \times \mathbb{R} \to \mathbb{R}$ is the arctangent of a point with cartesian coordinates (x, y) that takes into account the quadrant. We will assume that the range of $\text{atan}(y, x)$ is $(-\pi, \pi]$ and that $\text{atan}(0, 0) = 0$. This straightforwad transformation to polar coordinates poses a problem when averaging θ: if at step k, θ_i^k is positive and close to π while θ_j^k is negative and close to $-\pi$, the mean of will be close to 0 what is not a useful average. To overcome this problem, we define a new value ϕ_q^k to account for the overall angular distance run by the trajectory. Let us define $\phi_q^0 = \theta_q^0$, and for each $k \geq 0$, let us express ϕ_q^k as $\phi_q^k = z_q^k 2\pi + h_q^k$, with $z_q^k \in \mathbb{Z}$ and $-\pi < h_q^k \leq \pi$, i.e, z_q^k is the number of completed loops and h_q^k is the angular distance run on the current loop. The value of z_q^k is positive(negative) if the angular distance was run anticlockwise(clockwise). Then, for $k > 0$, ϕ_q^k can be computed as follows:

$$\phi_q^k = \begin{cases} z_q^{(k-1)} 2\pi + \theta_q^k + 2\pi & \text{if } h_q^{(k-1)} > \frac{\pi}{2} \text{ and } \theta_q^k < -\frac{\pi}{2} \\ z_q^{(k-1)} 2\pi + \theta_q^k - 2\pi & \text{if } h_q^{(k-1)} < -\frac{\pi}{2} \text{ and } \theta_q^k > \frac{\pi}{2} \\ z_q^{(k-1)} 2\pi + \theta_q^k & \text{otherwise} \end{cases}$$

The first(second) case of the expresion account for the discontinuity of the angle returned by atan when the trajectory moves from the second to the third(from the third to the second) quadrant. An average trajectory in polar coordinates is obtained as the mean of ρ_q^k and ϕ_q^k over all simulations.

Results. Consider the following system[2]: $s = 2$; $n = 5$; $\nu = \begin{pmatrix} 1 & -1 & -1 & 1 & 0 \\ 0 & 0 & 1 & -1 & -1 \end{pmatrix}$; $V = 5 \cdot 10^3$; $W_1 = \frac{X_1 + X_2}{1 + (0.4 \cdot (X_1 + X_2))/V}$, $W_2 = 0.2 \cdot X_1$, $W_3 = 10 \cdot X_1 \cdot X_2/V$, $W_4 = 3 \cdot X_2$ and $W_5 = 5 \cdot X_2$ with initial populations $X_1(0) = 4080$ and $X_2(0) = 500$. The system has a unique non extinction fixed point $a = (4000, 502)$ which is taken as origin of the polar coordinate system. Figure 1(b) shows the average trajectories of 5000 simulations. The trajectory tending to a is the average of the cartesian coordinates, while the trajectory tending to a steady oscillation is the average of the polar coordinates. The interpretetation is that simulation trajectories tend to loop around the fixed point at an average distance of 170. Thus, while the cartesian mean informs about the trajectory of the *center of mass* of the simulations, the polar mean informs about the average *circular motion* what uncovers the undamped oscillations reported in[2].

References

1. Ethier, S., Kurtz, T.: Markov Processes: Characterization and Convergence. John Wiley (1986)
2. Natiello, M., Solari, H.: Blowing-up of Deterministic Fixed Points in Stochastic Population Dynamics. Mathematical Biosciences 209(2), 319–335 (2007)

On Bistability Causing Structures in *Escherichia Coli*'s Metabolism

Bernhard Kramer and Carsten Conradi

Max Planck Institute for Dynamics of Complex Technical Systems,
Sandtorstrasse 1, 39106 Magdeburg, Germany
{kramer,conradi}@mpi-magdeburg.mpg.de

Abstract. It is in many cases not known which structures of a metabolic network are responsible for qualitative phenomena like bistability. Here we examine a model of *E. coli*'s metabolism including the methylglyoxal pathway but without regulation. We use a method called subnetwork analysis to look for subnetworks that may cause multistationarity (a prerequisite for bistability). As there are no such subnetworks we conclude that the unregulated metabolic network contains no bistability causing structures.

Keywords: Multistationarity, Bistability, Methylglyoxal.

1 Introduction

Complex dynamical phenomena like bistability and oscillations have been observed experimentally in *Escherichia coli*. In [1], for example, bistability has been observed in the lactose utilisation network, while [2] describes oscillations in the methylglyoxal pathway. The methylglyoxal pathway is a particularly interesting object of study, as it plays an important role in energy production and free radical generation in procaryotes and eucaryotes alike [3]. Methylglyoxal remarkably shows versatile toxicity to the cell because of its high reactivity to DNA, RNA and proteins. Furthermore methylglyoxal has been reported to influence the cell division machinery [4]. The methylglyoxal detoxification pathway is highly conserved in the metabolism of procaryotes and eucaryotes [3]. Hence studying it in the metabolism of the well known organism *E. coli* might have consequences for eucaryotes as well.

Recent experimental studies concerning the role of methylglyoxal in the metabolism of *E. coli* deal with the participation of methylglyoxal in the anaplerotic pathway, the regulatory impact of methylglyoxal on the triosephosphate balancing in carbohydrate metabolism (especially connected to nutrition imbalancing) and the methylglyoxal detoxification pathways. It is, however, unknown which parts of the methylglyoxal pathway (or the metabolism as a whole) are responsible for qualitative phenomena like bistability and oscillations. With respect to bistability we propose to use *E. coli* metabolic models and the results of [5] to identify structures that may cause bistability.

A. Gupta and T.A. Henzinger (Eds.): CMSB 2013, LNBI 8130, pp. 276–277, 2013.
© Springer-Verlag Berlin Heidelberg 2013

2   Method

Numerous dynamical models have been developed describing the metabolism of *E. coli* including the central carbon metabolism around glycolysis, diauxic growth, catabolite repression, and combining metabolism and regulation. For an overview on models covering these topics see [6]. For our investigation we created a model by extending the dynamic model of Kotte et al. [7].

The publication [5] is concerned with multistationarity, a prerequisite for bistability. The reference shows a method where it is easy to partition a complex network into a collection of smaller networks for which it is not difficult to establish multistationarity. It furthermore discusses conditions which allow to infer multistationarity in the overall network from multistationarity arising in a subnetwork. We consider these subnetworks as bistability causing structures.

3   Results

Using the aforementioned method we partitioned the network into 24869 subnetworks (of which 24800 contain at least 5 reactions and subsequently were subjected to the algorithm analysing these subnetworks towards multistationarity). Taken with mass action kinetics none of these subnetworks showed multistationarity and hence bistability. This indicates that the unregulated network contains no bistability causing structures. Hence bistability arises either from the complete metabolic network or from influences of a higher level regulatory mechanism.

References

1. Ozbudak, E.M., Thattai, M., Lim, H.N., Shraiman, B.I., van Oudenaarden, A.: Multistability in the lactose utilization network of *Escherichia coli*. Nature 427, 737–740 (2004)
2. Weber, J., Kayser, A., Rinas, U.: Metabolic flux analysis of *Escherichia coli* in glucose-limited continuous culture. II. Dynamic response to famine and feast, activation of the methylglyoxal pathway and oscillatory behaviour. Microbiology 151(3), 707–716 (2005)
3. Ferguson, G.P., Tötemeyer, S., MacLean, M.J., Booth, I.R.: Methylglyoxal production in bacteria: suicide or survival? Archives of Microbiology 170(4), 209–219 (1998)
4. Kalapos, M.P.: On the promine/retine theory of cell division: now and then. Biochimica et Biophysica Acta 1426, 1–16 (1999)
5. Conradi, C., Flockerzi, D., Raisch, J., Stelling, J.: Subnetwork analysis reveals dynamic features of complex (bio)chemical networks. Proceedings of the National Academy of Sciences of the United States of America 104(49), 19175–19180 (2007)
6. Lee, S.Y. (ed.): Systems Biology and Biotechnology of *Escherichia coli*. Springer (2009)
7. Kotte, O., Zaugg, J.B., Heinemann, M.: Bacterial adaptation through distributed sensing of metabolic fluxes. Molecular Systems Biology 6(355), 1–9 (2010)

Mathematical Modelling of the Function of Ubiquitylation in TNFR1-Mediated NF-κB Signalling

Leonie Amstein[1], Nadine Schöne[1], Simone Fulda[2], and Ina Koch[1]

[1] Molecular Bioinformatics Group, Cluster of Excellence "Macromolecular Complexes", Johann Wolfgang Goethe-University Frankfurt am Main, Robert-Mayer-Straße 11-15, 60325 Frankfurt am Main, Germany
[2] Institute of Experimental Cancer Research in Pediatrics, Cluster of Excellence "Macromolecular Complexes", Johann Wolfgang Goethe-University Hospital Frankfurt am Main, Komturstraße 3a, 60528 Frankfurt am Main, Germany

The tumor necrosis factor receptor 1 (TNFR1) pathway plays a crucial role in immune signalling and development by controlling cell growth and death [7]. The binding of tumor necrosis factor-α (TNFα) to TNFR1 can either trigger two different forms of cell death - apoptosis and necroptosis - or promote cell survival due to the activation of the transcription factor nuclear factor-κB (NF-κB) [5]. A dysregulation of this pathway can result in chronic diseases and cancer-related inflammation and features a strictly controlled regulatory network [6]. Therefore, the logic of the pathway regulation is of interest for cancer research to recognise the mechanisms that determine the outcome of death receptor stimulation.

The TNFR1 pathway displays a complex signalling network with different regulatory features like feedforward, feedback, and crosstalk. The membrane bound receptor signalling complex (RSC) builds a crucial part in the signalling cascade, since the duration and composition of its formation determine the activity of the effector kinases and thereby the capacity of gene expression. In this context, ubiquitylation plays a pivotal role as an important post-translational modification process [2]. Just recently a novel component of the NF-κB pathway was discovered being responsible for linear ubiquitylation events, which enhance the activation of the transcription factor [6]. In order to elucidate the complex dynamics of these interwoven pathways, we established an interaction network in a systems biology approach using the Petri net formalism. Therefore, we constructed a Petri net with focus on TNFR1-mediated NF-κB pathway to examine the assembly of macromolecular complexes orchestrating the decision between survival and cell death. Here, we especially consider ubiquitylation events and their effect on signalling to NF-κB.

Petri nets permit to model systems with concurrent processes at different levels of abstraction within a unique and well defined formalism. Many analysis methods have been developed that allow for static as well as dynamic analysis, so that the network dynamics can be predicted without the knowledge of kinetic data. Moreover, Petri nets provide an intuitive graphical representation allowing for optimal interdisciplinary communication [4]. As a consequence, Petri net theory is of particular interest for the investigation of signalling processes, since

A. Gupta and T.A. Henzinger (Eds.): CMSB 2013, LNBI 8130, pp. 278–279, 2013.
© Springer-Verlag Berlin Heidelberg 2013

the determination of kinetic parameters is still difficult for signal transduction pathways. Due to this, we apply a P/T-Petri net instead of a kinetic or stochastic Petri net, since no quantitative data are available to describe the biochemical processes. We model at qualitative level of detail according to literature while taking into account previously gained insights about signalling to NF-κB. Our Petri net model analyses the pivotal regulatory processes of the NF-κB pathway such as RSC formation-, ubiquitylation-, and feedback mechanisms. The Petri net was modelled and analysed by usage of MonaLisa software [1].

The mathematical analysis of the model detects the basal regulatory processes. Since the Petri net is covered by elementary modes, it fulfils the necessary CTI property confirming its consistency [4]. The analysis of the P-Invariants [3] reveals feedback mechanisms of the NF-κB pathway. Furthermore, the animation of the Petri net allows for investigating the basal behaviour and regulatory dynamics of the signalling cascade. The model describes the sequential assembly of the RSC in detail while having regard to the necessary ubiquitylation events. Hence, it considers the effect of linear ubiquitylation for subsequent gene expression and points out the inhibitory regulation of cell death signalling. Additionally, regulatory feedback modules are incorporated along with the dissociation of the RSC resulting in the termination of signal transduction.

In this study, new insights regarding the role of ubiquitylation for NF-κB and cell death processes were examined in a Petri net approach. The established Petri net of NF-κB signalling delivers a valuable tool to investigate the regulatory mechanisms along with their dynamics on a basic molecular level, even though the processes are described without any knowledge of kinetic data. Thus, the constructed pathway model reflects the current understanding of signalling to NF-κB emphasising the pivotal role of post-translational modifications for the conduction of the cell response following TNFR1 activation.

References

1. Einloft, J., Ackermann, J., Noethen, J., Koch, I.: MonaLisa - visualization and analysis of functional modules in biochemical networks. Bioinformatics 29, 1469–1470 (2013)
2. Fulda, S., Rajalingam, K., Dikic, I.: Ubiquitylation in immune disorders and cancer: from molecular mechanisms to therapeutic implications. EMBO Molecular Medicine 4, 545–556 (2012)
3. Koch, I., Reisig, W., Schreiber, F.: Modeling in Systems Biology: The Petri net approach. Springer (2011)
4. Sackmann, A., Heiner, M., Koch, I.: Application of Petri net based analysis techniques to signal transduction pathways. BMC Bioinformatics 7, 482–499 (2006)
5. Vandenabeele, P., Galluzzi, L., Vanden Berghe, T., Kroemer, G.: Molecular mechanism of necroptosis: an ordered cellular explosion. Nature Reviews Molecular Cell Biology 11, 700–714 (2010)
6. Walczak, H.: TNF and ubiquitin at the crossroads of gene activation, cell death, inflammation, and cancer. Immunological Reviews 244, 9–29 (2011)
7. Wertz, I., Dixit, V.: Signaling to NF-κB: Regulation by Ubiquitination. Cold Spring Harbor Perspectives in Biology (2010)

Organizing SIV Gene Network Modules into Graph Database

Raquel L. Costa[1,2], Jean-Marc Schwartz[2], David L. Robertson[2],
and Fabio Porto[1]

[1] National Laboratory of Scientific Computing, 25651-075 Petropolis, RJ, Brazil
[2] Faculty of Life Sciences, University of Manchester, Manchester M13 9PT, UK
{quelopes,fporto}@lncc.br,
{jean-marc.schwartz,david.robertson}@manchester.ac.uk
http://www.lncc.br,http://www.ls.manchester.ac.uk/

Abstract. Temporal microarray data of Simian Immunodeficiency Virus (SIV) infection make it possible to infer gene interaction networks and help unveil new infection mechanisms. In this work, we show that the inference and analysis of gene interaction networks can be enhanced using graph databases. Gene interaction network modules and the provenance associated to the inference process are modeled into the graph data and high-level queries can help comparing different results and studying individual modules.

Keywords: SIV interactions, Module network, Graph Database.

1 Introduction

The SIV infects many African Nonhuman Pimates (NHP) in the wild. While some monkey species such as African Green Monkey (AGM) and Sooty Mangabey (SM) do not develop the immunodeficiency syndrome (AIDS), the Rhesus Macaque (RM), a non-natural host, is affected by the disease [1].

In [2], the complexity involved in inferring gene interactions network was discussed. Currently, various tools are used during this process, each with different assumptions leading to different, and possibly complementary, results (ibid). Thus, researchers may find themselves with several output networks that must be interpreted, compared, and analyzed. It is clear that there is an urgent need to support the storage and high level access to inferred gene networks.

In order to support the analysis of interactions inferred from gene expression data, we investigated the adoption of graph model databases [3]. We have conceived a graph representation for inferred gene interaction networks and implemented it using the Neo4j database system. Queries may compare the results on different networks and explore relationships in individual modules.

2 Methods

In this work, AGM, RM and SM transcript expression data were obtained from Jacquelin *et al.*, (2009) and Bosinger *et al.*, (2009). The dataset consists in

A. Gupta and T.A. Henzinger (Eds.): CMSB 2013, LNBI 8130, pp. 280–281, 2013.

biological replicates for each species providing time series of gene transcripts spanning three phases: before infection, acute infection and chronic phase. First, we selected probes showing significant differential expression and fitted a linear model to each probe. Modular structures emerged from the application of correlation analysis, hierarchical clustering and gene set enrichment analysis. We devised a graph data representation (Fig. 1), in which provenance information describes the process of generating the network. We extended the basic model proposed by the PROV initiative (http://www.w3.org/TR/prov-primer/). The gene interaction network was translated into the graph representation.

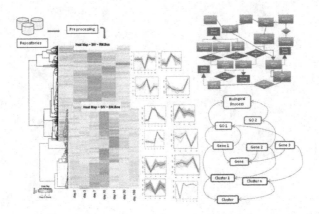

Fig. 1. From expression data to graph database

3   Conclusion

We discussed the adoption of graph databases technology to help unveiling new mechanisms of genetic regulation in SIV infection. The graph data model natively supports the representation of gene networks, as well as the provenance information describing the process used to derive each network module. Our approach enables comparisons between different inferred networks and the exploration of gene interaction paths, using queries expressed in the high-level language offered by the database system.

References

1. Bosinger, S.E., et al.: Global genomic analysis reveals rapid control of a robust innate response in SIV-infected sooty mangabeys. JCI 12, 3556–3572 (2009)
2. De Smet, R., Marchal, K.: Advantages and limitations of current network inference methods. Nature Reviews. Microbiology 8, 717–729 (2010)
3. Angles, R., Gutierrez, C.: Survey of graph database models. ACM Computing Surveys 40, 1–39 (2008)

Author Index